Volume 32

# Advances in Genetics

## Incorporating Molecular Genetic Medicine

# Volume 32

# Advances in Genetics

## Incorporating Molecular Genetic Medicine

Edited by

**Jeffrey C. Hall**
Department of Biology
Brandeis University
Waltham, Massachusetts

**Jay C. Dunlap**
Department of Biochemistry
Dartmouth Medical School
Hanover, New Hampshire

Associate Editors

**Theodore Friedmann**
Department of Pediatrics
Center for Molecular Genetics
School of Medicine
University of California, San Diego
La Jolla, California

**Francesco Giannelli**
Division of Medical and
  Molecular Genetics
United Medical and Dental
  Schools of Guy's and
  St. Thomas' Hospital
London Bridge, London SE1 9RT
United Kingdom

**Academic Press**
San Diego  New York  Boston  London  Sydney  Tokyo  Toronto

This book is printed on acid-free paper. ∞

Academic Press, Inc.
A Division of Harcourt Brace & Company
525 B Street, Suite 1900, San Diego, California 92101-4495

*United Kingdom Edition published by*
Academic Press Limited
24-28 Oval Road, London NW1 7DX

International Standard Serial Number: 0065-2660

International Standard Book Number: 0-12-017632-7

PRINTED IN THE UNITED STATES OF AMERICA
95  96  97  98  99  00  QW  9  8  7  6  5  4  3  2  1

# Contents

Contributors     ix
Foreword     xi
Preface     xiii

## 1 After Gene Therapy: Issues in Long-Term Clinical Follow-Up and Care     1
Fred D. Ledley

    I. Introduction     1
    II. Assessing the Long-Term Consequences of Somatic
        Gene Therapy     3
    III. Ongoing Therapy for Adverse Experiences     8
    IV. Ongoing Therapy for Clinical Benefit     12
    V. Informed Consent for Follow-Up     13
    VI. Conclusion     14
        References     15

## 2 Gaucher Disease     17
Ernest Beutler

    I. Introduction     17
    II. Clinical Manifestations of Disease     18
    III. Molecular Biology     20
    IV. Population Genetics     24
    V. Treatment     27
    VI. Unresolved Issues     36
    VII. Controversies     39
        References     41

3  **The Genetics of Non-insulin-Dependent**
   **Diabetes Mellitus**    51
   T. S. Pillay, W. J. Langlois, and J. M. Olefsky

    I. Introduction    51
    II. Natural History of the Development of NIDDM    52
    III. The Role of Genetic Factors in the Etiology of NIDDM    54
    IV. Strategies for Identifying Diabetes-Susceptibility Genes in
        NIDDM    55
    V. Candidate Genes    63
    VI. The "Thrifty" Gene and Silent Polymorphism
        Hypotheses    88
    VII. Conclusion    88
        References    89

4  **The Hemophilias**    99
   P. M. Green, J. A. Naylor, and F. Giannelli

    I. Two Distinct Forms of Hemophilia    99
    II. Genetics of the Hemophilias    101
    III. Factor IX: The Gene and the Protein    102
    IV. The Hemophilia B Mutations    104
    V. Functional Interpretation of Observed Sequence Changes and
        Genotype/Phenotype Correlations in Hemophilia B    108
    VI. Progress in Carrier and Prenatal Diagnosis
        of Hemophilia B    110
    VII. Factor VIII: The Gene and the Protein    113
    VIII. Hemophilia A Mutations    116
    IX. Functional Interpretation of Observed Sequence Changes and
        Genotype/Phenotype Correlations in Hemophilia A    126
    X. Progress in Carrier and Prenatal Diagnosis
        in Hemophilia A    129
    XI. Contribution of Molecular Biology to Treatment    130
    XII. Conclusion    130
        References    131

5  **The Influence of Molecular Biology on our Understanding**
   **of Lipoprotein Metabolism and the Pathobiology**
   **of Atherosclerosis**    141
   Thomas P. Knecht and Christopher K. Glass

    I. Introduction    141
    II. Background and Historical Perspective    142

III. Atherosclerosis Research in the Molecular Biology Era 147
IV. Summary and Future Directions 190
References 191

## 6 Molecular Genetics of Phenylketonuria: From Molecular Anthropology to Gene Therapy 199

Randy C. Eisensmith and Savio L. C. Woo

I. General Background 199
II. Characterization of the Human Phenylalanine
Hydroxylase Gene 204
III. Molecular Genetics of PAH-Deficient PKU
and Related HPAs 208
IV. Population Genetics of Phenylketonuria 228
V. Gene Therapy for Phenylketonuria 244
VI. Conclusions 253
References 254

## 7 The Proterminal Regions and Telomeres of Human Chromosomes 273

Nicola J. Royle

I. Introduction 273
II. The Functions of Telomeres and Proterminal Regions 273
III. The Isolation and Structure of Human Telomeres 278
IV. The Structure of Proterminal Regions and Telomeres 281
V. Chromosome Ends from Other Primates 294
VI. The Role of Proterminal Regions and Telomeres
in Senescence, Tumorigenesis, and Genetically
Determined Diseases 296
VII. Conclusions 305
References 306

Index 317

# Contributors

Numbers in parentheses indicate the pages on which the authors' contributions begin.

**Ernest Beutler** Department of Molecular and Experimental Medicine, The Scripps Research Institute, La Jolla, California 92037 (17)

**Randy C. Eisensmith** Department of Cell Biology, Baylor College of Medicine, Houston, Texas 77030 (199)

**F. Giannelli** Division of Medical and Molecular Genetics, United Medical and Dental Schools of Guy's and St. Thomas' Hospitals, London Bridge, London SE1 9RT, United Kingdom (99)

**Christopher K. Glass** Division of Cellular and Molecular Medicine, and Division of Endocrinology/Metabolism, University of California at San Diego, La Jolla, California 92093 (141)

**P. M. Green** Division of Medical and Molecular Genetics, United Medical and Dental Schools of Guy's and St. Thomas' Hospitals, London Bridge, London SE1 9RT, United Kingdom (99)

**Thomas P. Knecht** Division of Cellular and Molecular Medicine, and Division of Endocrinology/Metabolism, University of California at San Diego, La Jolla, California 92093 (141)

**W. J. Langlois** Division of Endocrinology and Metabolism, Department of Medicine, University of California at San Diego, La Jolla, California 92093 (51)

**Fred D. Ledley** GeneMedicine, Inc., Houston, Texas 77054 (1)

**J. A. Naylor** Division of Medical and Molecular Genetics, United Medical and Dental Schools of Guy's and St. Thomas' Hospitals, London Bridge, London SE1 9RT, United Kingdom (99)

**J. M. Olefsky** Division of Endocrinology and Metabolism, Department of Medicine, University of California at San Diego, La Jolla, California 92093 (51)

**T. S. Pillay** Division of Endocrinology and Metabolism, Department of Medicine, University of California at San Diego, La Jolla, California 92093 (51)

**Nicole J. Royle** Department of Genetics, University of Leicester, Leicester, LE1 7RH, United Kingdom (273)

**Savio L. C. Woo** Department of Cell Biology, Baylor College of Medicine, Houston, Texas 77030 (199)

# Foreword

The practice of genetics is based on the observation of visually discernible phenotypes. The better the observer and the observer's sense of the organism, the better and finer the distinctions in phenotype can be, and the better the underlying genetic principles governing the phenotypes can be described, hence McClintock's emphasis on acquiring "a feeling for the organism." Among the organisms that scientists observe most closely, particularly in the population sense, there are few that are scrutinized as closely as other humans. Thus, the genetic analysis of humans may add not only to our general health in the medical sense, but also to promote the understanding of the basic genetic principles governing complex phenotypes, especially those related to mental processes. As the molecular era of genetics continues to expand, it is not surprising that the area of human genetics has expanded to such a degree.

We are thus pleased that Volume 32 of *Advances in Genetics* will initiate increased coverage of the field of human genetics. As ever, our job as editors is to identify and promote both breadth and quality of coverage within the general area of genetics, and, as regards human genetics, we will be greatly aided both by the formal incorporation of *Molecular Genetic Medicine* into *Advances in Genetics* and by the addition of Theodore Friedmann and Francesco Giannelli as Associate Editors.

As before, our goals for the *Advances in Genetics* series remain the identification of emerging problems in genetics as they coalesce and the recruitment and promotion of contributions that are at the same time comprehensive and comprehensible, informed and informative, critical, insightful, and readable. In this volume of *Advances in Genetics* we present a mixture of synoptic and topical reviews describing advances in research areas that are largely focused on the genetics and molecular genetics of human disease. These are more fully described and put into context in the Preface to this volume, written by our Associate Editors.

*Advances in Genetics* was first published in 1946 under the editorship of Milislav Demerec who was then acting as the director of the Genetics Department of the Carnegie Institute of Washington in Cold Spring Harbor. His goal was to produce a series such that "critical summaries of outstanding genetic problems, written by prominent geneticists in such form that they will be useful as reference material for geneticists and also as a source of information to nongeneticists, may appear in a single publication." Although times have changed and the face of genetic research is remarkably different from how it looked 49 years ago, Demerec's goals are still a good touchstone. We trust the present volume continues to fulfill this model.

Jeffrey C. Hall
Jay C. Dunlap

# Preface

In many ways, the hemophilias represent ideal paradigms for testing the impact of molecular genetics on our understanding of the pathogenesis and treatment of human disease. They are genetically simple and well characterized and have therefore been targeted by many investigators for extensive study at both the basic and clinical levels. It has been known for many centuries that in some cases hereditary factors are responsible for serious bleeding disorders. The effectiveness of Queen Victoria and her family in popularizing the disorder through the royal houses of Europe and the interventions of the monk Rasputin in the care of Tsar Nicholas's son crown prince Alexei have helped to make the hemophilias some of the most widely recognized genetic disorders.

The biochemical and physiological functions of the clotting factors that regulate the entire complicated clotting cascade have been understood for decades, and the genes responsible for the two most common forms of hemophilia, those caused by deficiencies of clotting factors VIII (classical hemophilia, or hemophilia A) and IX (Christmas disease or hemophilia B), have now been identified and characterized. Based on an understanding of the underlying physiological defects of these and other bleeding disorders, effective forms of therapy involving replacement of the missing function through transfusion of blood and blood products have long been available. Nevertheless, the treatments have been imperfect and extraordinarily expensive.

These facts combined with recent treatment debacles caused by contamination of the blood supply with the HIV virus have made the search for truly curative forms of therapy more urgent than ever. For that reason, hemophilia has become one of the principal targets for gene therapy. In this volume, P. M. Green, J. A. Naylor, and Francesco Giannelli of Guy's Hospital, London, review the molecular genetics of hemophilias A and B and describe the ways in which knowledge of the structure of the gene and gene product promise improved understanding of the disease and its treatment.

Like the hemophilias, Gaucher disease has been recognized for more than a century and it represents what is now recognized to be one of the more common human genetic disorders. In some patients, the manifestations of the disease are severe and life threatening, in some they are quite mild, and in some the disease is virtually undetectable. Gaucher disease, like the hemophilias, is genetically simple, but unlike the hemophilias the mechanisms by which the biochemical defects (i.e., lysozomal accumulation of the glycolipid glucocerebroside) came about were largely unknown until the discovery of the causative genetic defect. The one feature that has made Gaucher disease a particularly interesting model disorder has been that it is one of the very few human genetic diseases that is truly amenable to treatment by enzyme replacement therapy. Of course, lysozomal storage diseases lend themselves well, at least in principle, to enzyme replacement, but replacement of the missing enzyme has not been as convincingly demonstrated in any other disease as it has been in Gaucher disease. That is not to say that this approach is the ideal form of therapy; the current financial burden on patients and their families for effective enzyme replacement therapy is enormous. As with the hemophilias, this has led to growing interest in this disorder as a target for genetic intervention. Ernest Beutler from the Scripps Clinic in La Jolla summarizes the nature of the biochemical and genetic defects in Gaucher disease and evaluates the potential for application of gene therapy.

When gene therapy becomes clinically useful, as it surely will for the hemophilias, for Gaucher disease, or for any other human disorder, these new scientific and medical procedures will bring with them difficult new problems of evaluation and long-term follow-up. The urgency of treating otherwise intractable and devastating human diseases with the newly emerging tools of gene therapy has led to a rapidly growing number of clinical studies in which little thought has been given to the long-term consequences of genetic manipulation in human patients with inborn errors of metabolism, somatic genetic disorders such as cancer, degenerative diseases, and other gene therapy targets. How do we keep track of the patients; how do we determine the effect of a genetic manipulation on a disease phenotype and assure ourselves that we have done more good than harm; how do we compare patients treated by replacement of gene X in one study with those treated by reconstitution of gene X in another study? These are the normal questions posed and answered every day in clinical investigations, but they have only recently been applied to the language of gene therapy. Fred Ledley of GeneMedicine in Houston paints for us a picture of the problem and identifies the need for careful, long-term, and comparative evaluation of gene therapy patients.

A class of human disorders that is just now being elucidated is the collection of diseases that result from defects in lipoprotein metabolism and their relationships to atherosclerosis. As with the hemophilias, we have determined a great deal regarding the biochemical aberrations associated with the lipoprotein disorders, but the mechanisms of pathobiology were not clear until the important work of Brown and Goldstein and their colleagues in elucidating the role of the low-density lipoprotein receptor in cholesterol metabolism and in the pathogenesis of familial hypercholesterolemia. The clarification of molecular genetic components of human disease and the illumination of mechanisms of disease pathogenesis have been as impressive and important in the field of lipoprotein disorders as in any other area of medicine. The nature and evolutionary implications of underlying mutations, their effects on basic processes of cell biology, the resulting physiological aberrations, and the design of new approaches to treatment have all contributed to significant progress in understanding this class of human diseases. The history of this adventure and its likely future directions are exhaustively presented by Thomas Knecht and Christopher Glass from the University of California, San Diego, in La Jolla.

One set of human diseases that screams for greater understanding is collectively known as diabetes mellitus. The underlying concept of insulin insufficiency has been with us since the days of Banting and Best, and has been recognized for a longer period of time than the biochemical aberrations of any other human disease. Furthermore, the advent of specific gene product replacement with exogenous insulin has provided an effective, if not a definitive, therapy for some forms of the disease. Even to this day, the treatment of insulin-deficient diabetes is fairly straightforward and effective, and while the complicating microangiopathies associated with diabetes still occur at an alarmingly unaltered rate, new approaches to more effectively controlled insulin delivery promise to bring that problem under better control. However, our earliest simple views of the mechanisms underlying the biochemical defects of diabetes mellitus have, of course, turned out to be far from complete, and we now know that not all diabetes is the result of insulin deficiency. One of the most burdensome disorders in our sedentary, overfed, and underexercised population is non-insulin-dependent diabetes mellitus, a disorder that represents one of our principal killers. It is a difficult disease, genetically complex, and influenced by a multitude of environmental and "lifestyle" influences yet to be identified. Despite these problems, a great deal has been learned recently regarding the genetic and cell biological features of this disease, as described by Tahir Pillay, John Langlois, and Jerrold Olefsky of the University of California, San Diego.

Finally, one of the quasi-success stories of traditional human biochemical genetics is that of phenylketonuria, a devastating disorder characterized by liver dysfunction and severe mental retardation. It is a classical inborn error of metabolism, a disorder of amino acid metabolism in the liver which results in accumulation of metabolites toxic to the developing brain. Through the discovery of the several enzymatic defects responsible for versions of this disease and their resulting metabolic aberrations, it was possible, long before the application of molecular techniques to the disorder, to develop a rather effective therapy, i.e., dietary restriction of phenylalanine. In most cases, this stratagem effectively leads to a reduction in the levels of the toxic metabolites that ordinarily accumulate proximal to the enzymatic block and permits almost, but not quite, normal CNS development. However, the treatment is not perfect, and it has recently become evident that patients who have had what would have seemed to be optimal control of their phenylalanine levels nevertheless suffered reductions in their IQ and other evidence of brain damage. This inadequacy of current therapy has catalyzed a great deal of additional work on the molecular genetics of this disease; this progress toward an understanding and more definitive treatment of the disorder is reviewed by Randy Eisensmith and Savio Woo from Baylor College of Medicine, Houston.

The successes of human molecular genetics are really just beginning. Progress toward a thorough characterization of the human genome is stunningly rapid and exceeding many of the earliest expectations. Disease-related genes will be falling from the skies faster than we can understand them, and mechanisms responsible for the pathogenesis of disease will be illuminated more quickly and readily than ever before. Words such as "revolutionary," "unprecedented," and "epochal" have been used to describe the opportunities and the pace of progress in molecular genetic medicine. It seems to us that these words will be found to have been appropriate.

However, there is one troublesome factor that has already put a small dent in the pace of that progress: the willingness or the ability of our traditional funding mechanisms to keep pace with the scientific opportunities. At a time when 10% or fewer of approved grant applications are being funded at the National Institutes of Health in the United States, when the general funding rate is in the vicinity of 12–15% of investigator-initiated applications, when innovation is vulnerable in favor of targeted and programmed research, and when young and eager new investigators are being repulsed by the funding problems, it is clear that, as well as we are doing in battling human disease at this "revolutionary" new level, we need to be doing even better.

Theodore Friedmann
Francesco Giannelli

# 1

# After Gene Therapy: Issues in Long-Term Clinical Follow-up and Care

**Fred D. Ledley**
GeneMedicine, Inc.
The Woodlands, Texas 77054
and Departments of Cell Biology and Pediatrics
Baylor College of Medicine
Houston, Texas 77030

## I. INTRODUCTION

The initial clinical trials of somatic gene therapy began in 1989 after several decades of debate concerning the biological, ethical, and social implications of genetic engineering (Nirenberg, 1967; Anderson and Fletcher, 1980; Fletcher, 1990; Friedmann and Roblin, 1972; Nelson, 1990; Walters, 1986; Murray, 1990; Miller, 1990, 1992; OTA, 1984; Surgeon General, 1983; Friedmann, 1983; Anderson, 1992; Ledley, 1991). Much of this analysis was conceptual in nature, predicated on the theoretical risks of gene transfer as well as first principles of ethics and social policy. To a large extent, this analysis arose independent of the traditions of clinical investigation and the practical problems associated with clinical research and clinical practice (Ledley, 1991). Furthermore, relatively little consideration was given to anticipating how somatic gene therapy would develop from an academic, scientific enterprise toward common clinical practice and commercialization. While the enormous promise of gene therapy was widely heralded, much of the analysis focused on issues related to treating rare genetic diseases rather than providing therapeutic products and clinical care for common disorders.

By the end of 1994, more than 80 clinical trials involving gene transfer into human subjects will be approved by the NIH through its Recombinant DNA Advisory Committee and more than 200 patients will have participated in these trials. These trials not only deal with rare genetic diseases, but also common diseases such as cancer, AIDS, and even arthritis. At the same time, more than 40 early stage biotechnology companies and more than 15 established pharmaceutical or biological companies have declared an intention to develop products for somatic gene therapy. It may be expected that many of these companies will enter clinical trials within the next several years in anticipation of developing marketable products in the next decade. This growth represents an enormous challenge for the understandings that have come to guide somatic gene therapy and the regulatory structures established to ensure that somatic gene therapy is developed in a fair and effective manner. In particular, this growth of clinical activity involving gene transfer requires that increased attention be focused on how clinical investigations are performed and how clinical care is provided.

Some of the most difficult questions relate to the need to provide follow-up and sometimes ongoing clinical care to patients who receive gene therapy. There are several distinct issues. The first is the need to have effective follow-up for patients who receive somatic gene therapy to fully assess potential adverse experiences resulting from these therapies. The second is the need to provide continuous clinical care to patients who are treated with somatic gene therapy. This relates to both the responsibility of investigators and institutions to ensure that patients with adverse experiences receive appropriate care for these complications and to their responsibility to provide continuing therapy to patients who may benefit from clinical trials.

While the scientific and clinical necessity of adequate follow-up and ethical imperative to provide ongoing care may be intuitive, in practice, these are difficult issues. It is extremely difficult in clinical practice to achieve effective long-term follow-up. Voluntary participation in clinical follow-up is notoriously poor, and many investigators do not pursue it for a variety of reasons (Ledley *et al.*, 1992). Furthermore, institutions may not be in a position to assume responsibility for ongoing care, particularly if gene therapy is administered as a clinical research project.

Several characteristics of the current clinical activity in gene therapy make follow-up and the commitment to long-term care particularly difficult. One characteristic is that many current clinical trials represent basic clinical research or "proof-of-principle" studies in which the results of the gene transfer will be measured against specific technical end points rather than the therapeutic effect of the gene. This is particularly true for marker gene studies and phase I studies designed to administer subtherapeutic doses of gene therapy. Such trials may be academic enterprises in which there may be little "research" interest in collecting long-term data on the consequences of gene transfer. Moreover, the patient populations participating in these initial clinical trials may not represent an ideal cohort in which to obtain prospective, controlled, and statistically significant information about the long-term consequences of gene therapy.

Thus, despite the importance of research on the clinical sequelae of gene therapy, the clinical follow-up of such patients may not be perceived to be quality clinical research. Another issue is that many clinical trials may not lead to effective therapies or commercial products. If a particular therapy is found to be ineffective in clinical trials, and no further clinical development or investigation is planned, then there is little incentive for investigators to continue tracking the consequences of such therapy. Moreover, many clinical trials are being conduced in academic centers whose commitment to somatic gene therapy may not transcend the tenure of the individual investigator or by early stage companies that may not survive. In such instances, it may not be clear who is responsible for ensuring that adequate follow-up is achieved. These issues are particularly complicated for therapies which are intended to be permanent, where even discontinuation of the therapy and termination of the study may not eliminate the risk of exposure to the therapy.

The issue of follow-up has begun to be addressed by the scientific community and agencies that regulate somatic gene therapy. In May, 1991, a proposal was submitted to the Recombinant DNA Advisory Committee (RAC) proposing that a registry of patients participating in studies be established to achieve long-term tracking of patients, consistent reporting of a critical data set, and statistical analysis of potential adverse experiences (Ledley *et al.*, 1991). The RAC passed a resolution recognizing the potential usefulness of such a registry. In several meetings the RAC has also addressed the issue of institutional responsibility for patient care after completion of clinical trials and in June, 1993, the RAC passed a resolution emphasizing the importance of continuing care. This article reviews the clinical, technical, and social issues involved in providing proper clinical follow-up to initial trials of somatic gene therapy.

## II. ASSESSING THE LONG-TERM CONSEQUENCES OF SOMATIC GENE THERAPY

### A. Theoretical risks of gene therapy

Considerable attention has focused on the theoretical risks of gene transfer methods and safety modifications that can be built into gene therapies to minimize these risks. Despite the best intentions of molecular biology, it may be expected that such precautions will never completely eliminate these risks. Clinical experience teaches that it is generally difficult to predict clinical risks *a priori* or even from animal experiments. The long-term evaluation of patients for possible adverse experiences will be essential to assess the risks of gene therapy and their incidence.

In considering the problems of follow-up, two forms of somatic gene therapy may be distinguished. The first is that which is intended to permanently insert a recombinant gene into the patient's cells. The second is that in which the recombinant gene is used like a conventional medicine with the intent that it will be eliminated from the body by metabolism, degradation, or excretion. These two approaches to therapy raise distinct issues and share certain issues in common.

Retroviral vectors based on components of Murine Leukemia Viruses (Miller, 1990) are currently being used in the majority of clinical trials. These vectors are intended to permanently integrate into populations of stable cells, such as bone marrow progenitors or hepatocytes, which may be expected to survive indefinitely within the individual with the recombinant gene in place. Gene therapies using adeno-associated virus or cells stably transfected with DNA are similarly intended to be permanent. Such permanent therapies raise difficult issues for clinical follow-up since the clinical evaluation of the safety of such therapy must continue indefinitely. Animal studies designed to assess the lifetime implications of permanent therapy will, necessarily, be incomplete when clinical trials are performed and even when products are approved for general clinical use. Patients having a recombinant gene in their body may be continually at risk for adverse experiences resulting from the presence of the gene in their cells, the attendant risk of recombination or rescue, as well as risks related to expression of the recombinant gene product. Moreover, many forms of permanent gene therapy that are currently being investigated in clinical trials are not reversible, meaning that it may be difficult, or impossible, to eliminate the recombinant gene or gene product in response to adverse experiences. The use of "suicide" vectors (which include a sequence that allows cells containing the recombinant gene to be killed by administration of a nontoxic prodrug) does little to eliminate this problem since the effectiveness and risks associated with the suicide gene and its function must themselves be established in clinical trials.

Gene therapies which are designed to function like conventional medicines for a discrete period of time before being eliminated from the body, rather than being integrated into the chromosomes of the cell, fit a more conventional paradigm for follow-up. Such therapies can be discontinued if adverse experiences are encountered. If the recombinant gene and gene product are completely eliminated from the body, then many of the direct risks of this therapy will be similarly eliminated. While there may be late adverse effects from therapy which are not reversed by elimination of the gene, for example sensitization to antigens or damage to target cells, such therapies embody a higher margin of safety than those whose persistence in the body is indefinite.

The theoretical risks of gene therapy fall into several classes. Some concerns relate explicitly to the use of attenuated or defective viral vectors in human subjects. Is there a possibility of replication-competent viruses arising *in vivo* by recombination between the attenuated or defective viruses used for therapy and wild-type viruses in the environment? This concern was heightened by evidence of replication-competent retrovirus (RCR) arising during production of replication-defective retrovirus for clinical trials, as well as the observation that RCR could be pathogenic in nonhuman primates (testimony to FDA Vaccine Advisory Committee, October 25, 1993). Some concerns relate to the use of transplanted cells. Will cells subjected to *in vitro* cultivation, gene transfer, selection, and expansion exhibit enhanced proliferation or be prone to malignancy? Will these cells spread beyond the site of implantation or induce untoward effects at the site of implantation over time? Other concerns relate generally to all methods of gene therapy. Some of these concerns are technical. Will methods for gene delivery or the therapeutic gene products themselves exhibit immunogenicity or toxicity? Could malignancies result from insertional mutagenesis by viral vectors, DNA rearrangements, or nondisjunction events caused by the presence of extraneous genetic material in cells? Could inheritable genetic damage result from inadvertent incorporation of recombinant genes into the germline? Some of these concerns relate to the social and personal consequences of gene therapy. Will genetic engineering be associated with social problems affecting the individual, family, or its community? Will it be possible to retain confidentiality and privacy? Will gene therapy lead to problems such as altered self-image or "non-disease?"

## B. Identifying adverse experiences in clinical trials

Clinical trials will need to assess both the incidence of these theoretical risks and whether adverse experiences that are reported represent "true" adverse experiences in a statistically significant manner. It is relatively easy to establish the probability and incidence of adverse experiences that are explicitly related to gene therapy as shown in Table 1.1. The confidence interval reflecting the probability of observing (or not observing) a true adverse experience is a function of the frequency of the adverse experience and the size of the patient population studied. There are two ways to consider this calculation. First, to the extent that adverse experiences are rare events, their occurrence can be assumed to follow a Poisson distribution where the likelihood ($P$) of observing an adverse experiences is given by: $P = 1 - e^{xy}$, where $e$ is the natural log, $x$ is the probability of an adverse experience in an individual patient, and $y$ is the number of patients who have been observed for adverse experiences. This relationship is put in practical form in Table 1.1. The top portion of the table shows the likelihood of observing one or more adverse experiences, whose true frequencies are shown at the top of each column, when various numbers of patients (by rows) are studied. The bottom portion of the table presents the same relationship in a different fashion by showing the number of patients who have to be

**Table 1.1.** Likelihood of Observing One or More
Occurrences of an Adverse Experience as a
Function of the Incidence of the Adverse
Experience and the Number of Patients
in the Study Population

| Patients | Incidence of adverse experience | | | | |
|---|---|---|---|---|---|
| | 0.1 | 0.3 | 0.01 | 0.003 | 0.001 |
| 10 | 0.63 | 0.26 | 0.10 | 0.03 | 0.01 |
| 30 | 0.95 | 0.59 | 0.26 | 0.09 | 0.03 |
| 100 | 0.99 | 0.95 | 0.63 | 0.26 | 0.10 |
| 300 | 0.99 | 0.99 | 0.95 | 0.59 | 0.26 |
| 1,000 | 0.99 | 0.99 | 0.99 | 0.95 | 0.63 |
| 3,000 | 0.99 | 0.99 | 0.99 | 0.99 | 0.95 |
| 10,000 | 0.99 | 0.99 | 0.99 | 0.99 | 0.99 |

observed to have a 95% probability of identifying adverse experiences having various true frequencies. Essentially, the required number of patients is equal to three times the reciprocal of the true adverse experience rate. In this regard it should be emphasized that the relatively small number of patients in clinical trials (to date no more than 30 patients have been enrolled in any one trial) does not allow any meaningful conclusion to be reached for adverse experiences having an incidence of <10%.

In assessing potential adverse experiences from gene transfer, it must also be remembered that survivors of clinical trials involving gene transfer will predictably suffer from a variety of common disease processes during the course of their lives that will be unrelated to gene transfer. Individuals will predictably suffer from diseases such as cancer, imflammatory disease, senescence, and even bear children with birth defects at an incidence equivalent to that in the general population. Many of these patients will also be treated with a variety of other established or experimental drugs which themselves may have side effects. While many reported conditions may not be related to gene therapy, the question will always be asked whether the appearance of new symptoms represents a late consequence of gene therapy. It is difficult to design clinical studies with sufficient power to make significant statistical inferences about the incidence of adverse experiences against a background incidence of similar disorders in the general population. It has proved difficult, for example, to establish whether there is, in fact, a causal relationship between breast implants and inflammatory diseases given the relatively high incidence of such diseases among women in general. As shown in Table 1.2, the number of patients required to make a

**Table 1.2.** Number of Patients Required in Study Population to Detect One Adverse Experience with Background Incidence of a Similar Event (95% Likelihood)

| | Incidence of adverse experience | |
|---|---|---|
| Background incidence | 0.1 | 0.01 |
| 0.1 | 10,000 | 980,000 |
| 0.01 | 1,600 | 110,000 |
| 0.001 | 500 | 16,000 |

statistically significant conclusion about an increase in the frequency of common diseases may be prohibitively large.

It is unlikely that any one investigator or institution will have a sufficiently large patient population to make statistically significant inferences about the incidence of most adverse experiences. The absence of such a formal assessment will make it impossible both to know the true incidence of adverse experiences and to rule out gene therapy as a cause of adverse experiences that may be reported but may, in fact, be unrelated to the genetic therapy.

Several highly public problems over the past decade, including the contamination of cadaver-derived growth hormone with Creutzfield Jacob Virus, the failure of some Bjork Shiley valves, and the confusion concerning the safety of breast implants, have demonstrated the difficulty of performing a retrospective analysis when putative adverse experiences are reported. The difficulty in identifying teratogenic effects is also evident in the slow recognition of maternal phenylketonuria, fetal hydantoin syndrome, and fetal alcohol syndrome during the 1970s.

A proper assessment of the risks of gene transfer is an important research problem and should be funded as such. This research will require prospective tracking of patients who participate in clinical trials in order to collect a consistent set of data concerning patient well-being and potential adverse experiences in the largest possible cohort of individuals. Patient tracking networks or registries are complex and require sophisticated methodological design. It is necessary to have databases that are secure and flexible as well as algorithms for analysis of the data that allow statistically significant conclusions to be drawn. It may be necessary to track control groups or employ meta-analysis to draw meaningful conclusions from a large number of small clinical trials. The design of a tracking network must establish a critical balance between protecting the confidentiality and privacy of patients and communicating useful information to patients and their health care providers. A balance must be obtained between collecting sufficient data to draw meaningful conclusions and requesting so much data from reporting physicians that responses lag. Finally, the design must be sufficiently robust as to remain valid in the fact of incomplete or incorrect reporting as well as unexpected findings. There are many examples of registries which have failed to achieve these standards, but also precedents for registries which have made a significant impact in the understanding and management of disease (Emery and Miller, 1976; Holton, 1987; Hubbard *et al.*, 1987; Mize and Pratt, 1989). The importance of registries in the long-term management of patients was emphasized in the 1989 Report of The U.S. National Commission on Orphan Diseases which reported that ". . . registries of scientific and technical data have been shown to both stimulate research and improve patients access to treatment" (Thoene and Crooks, 1989).

The role of registries is also evident in regulations promulgated by the FDA which mandate patient tracking for permanently implanted devises (FDA, 1992). The role of registries in the follow-up for gene therapy was considered by the RAC in May, 1991, in response to a proposal for establishing such a registry on a prospective basis. The RAC resolved to ". . . recognize the potential usefulness of an established registry of patients, clinicians, and scientists involved in human gene transfer trials." (RAC, 1991).

Following this resolution, the RAC has increased the surveillance of clinical protocols based on data submitted by clinical investigators, but no mechanism has been established for long-term patient or physician tracking, consistent data collection from the patient or primary health care providers, or statistical analysis.

It should be emphasized that effective patient tracking would provide substantial benefits to individuals who participate in clinical trials by enhancing their long-term care. One benefit is that a tracking network may facilitate early identification of any adverse experiences. A tracking network will also facilitate the dissemination of information about adverse experiences to patients who may require certain diagnostic tests or therapeutic actions. In contrast, it should be equally emphasized that if patient tracking is performed in a nonrigorous fashion, this may increase the risk to patients by providing a false assurance as to the quality of clinical surveillance for adverse experiences or an inadequate or inaccurate analysis of potential adverse experiences.

## III. ONGOING THERAPY FOR ADVERSE EXPERIENCES

There has been considerable debate concerning the responsibility of institutions and individual investigators to provide continuing care to patients who may suffer adverse experiences during clinical trials of gene therapy. At most research institutions in the United States, study subjects are asked to release investigators and the institution from liability for adverse experiences resulting from the research procedure as an element of the informed consent. This practice supposes that the risks of clinical research are analogous to the risks of conventional therapeutic or diagnostic procedures which are assumed by the patient if informed consent is provided or if commercial pharmaceuticals, biologicals, or devices are used in accordance with the product label. The implicit or explicit acceptance of risk by the subject in these instances does not extend to adverse experiences arising from inappropriate medical practice in which the patient is exposed to unanticipated risks and the provider is thus liable for the consequences of their malpractice. There are, however, other precedents. In the United Kingdom the majority of research ethics committees require "no fault" compensation for adverse experiences related to clinical research (Harvey and Chadwick, 1992). In Canada, a recent judgment against an institution and investigator in the case of *Weiss vs Solomon* (Glass and Freedman, 1991) has raised questions about the privileged nature of clinical investigation even under informed consent.

It is reasonable to argue that informed consent for a research procedure involves a contract between the patient and provider/investigator analogous to that associated with common clinical procedures. The vital essence of the informed consent for research is that the patient is provided with information about the possible risks and benefits and chooses to accept the risks in anticipation of the benefits. There is a significant difference, however, between the consent given for established therapeutic modalities and consent for clinical research in that the risks of participating in a research trial may be substantially greater than those of approved therapies, and the nature of the benefits may be related less to the individual's well-being and more to the concern of society for improved therapeutic agents. The sense that the subject of a clinical research trial is assuming risk on behalf of society leads many to argue that society, as represented by the investigator or the institution, has a special responsibility to the patient if adverse experiences are encountered. This is true not only for gene therapy but clinical research in general (Levine, 1986).

This issue has been addressed in detail by several federal commissions. In 1977 the Health, Education, and Welfare (HEW) Secretary's Task Force on Compensation of Injured Research Subjects recommended that

> Human subjects who suffer physical, psychological, or social injury in the course of research conducted or supported by the U.S. Public Health Service should be compensated if 1) the injury is proximately caused by such research, and 2) the injury on balance exceeds that reasonably associated with such illnesses from which the subject may be suffering as well as with treatment usually associated with such illnesses

at the time the subject began participation in the research. (Health, Education, and Welfare Secretary's Task Force on Compensation of Injured Research Subjects, 1977)

In 1982 the President's Commission for the Study of Ethical Problems in Medicine and Biomedical and Behavioral Research completed a report entitled "Compensation for Research Injuries." This commission concluded that while

> . . . it would be ethically desirable for compensation to be provided, it does not follow that the Federal government has an ethical obligation to establish or require a formal compensation program in all research projects supported or regulated by the government. (President's Commission for the Study of Ethical Problems in Medicine and Biomedical and Behavioral Research, 1982)

The commission suggested that "the Secretary of Health and Human Services conduct a small-scale experiment in which several institutions would receive Federal support over 3–5 years for the administrative and insurance costs of providing compensation on a non-fault basis to injured research subjects." No action was taken on these recommendations.

This issue has also been addressed by the Royal College of Physicians in a report entitled "Research Involving Patients." Among the conclusions of this report are

> 58. Though the chances of harm coming to patients in the course of carefully conducted research are very small, it is important that proper arrangements are made to compensate patients in the event of such harm occurring.
> 59. Bodies that sponsor research, including both publicly funded bodies and the pharmaceutical industry, should now so arrange their affairs as to implement the principle that injury due to participation in research, sponsored by them or conducted by their staff with the approval of a Research Ethics Committee shall be compensated by a simple, informal, and expeditious procedure.
> 60. In the event of any significant injury, the patient must be entitled to receive compensation regardless of whether there may or may not have been negligence or legal liability on any other basis. (Royal College of Physicians, 1990)

The Recombinant DNA Advisory Committee has extensively considered the obligation of investigators and institutions to provide long-term care for patients who may suffer adverse experiences related to clinical trials of gene therapy. A June, 1993, statement approved by the Recombinant DNA Advisory Committee cited the President's Commission report and added

> The RAC supports the inclusion in basic universal health care of coverage for injury received through participation in approved clinical research. It is the strong desire of the RAC that the NIH Director would call this issue to the attention of the individuals formulating the President's health care reform proposal. (Recombinant DNA Advisory Committee, 1993)

There is a growing consensus among ethicists and investigators that subjects should receive "no-fault" compensation for injuries associated with clinical trials. Despite the good intentions of such as policy, however, difficult issues remain. One issue is that it may often be difficult to determine whether certain adverse experiences are, in fact, the consequence of gene transfer. It is common, for example, to make a distinction between the costs of therapy related to the patient's underlying condition and those which are related to the experimental therapy. Patients and third party payers are commonly asked to pay the costs of therapeutic measures and diagnostic tests which are indicated for their underlying disease, while they are not charged for the costs of the research procedure and its evaluation. Similarly, if an adverse experience is related to gene therapy, responsibility for care might be assumed by the institution, while if a complication arises from the underlying disease, other therapeutic measures, or the normal course of health and disease, then responsibility should be borne by the patient or insurers. As discussed above, it may be impossible to make a critical and correct assessment of the origin of many potential adverse experiences and their relationship to the gene therapy.

The second issue is to identify the individual or institution responsible and capable of assuming the costs or compensation and follow-up care. Clinical investigators themselves may not be in a position to provide long-term follow-up for adverse experiences which may involve other specialties or expertise. Moreover, clinical investigators may move among various institutions, often taking with them research teams and expertise required for such continuing care. Research grants from the government, foundations, or even pharmaceutical companies are not structured to provide funding for care on an ongoing basis. Academic institutions or contract research organizations may have a limited research budget and may be unable to assume open-ended responsibility for such care or to procure insurance against such as eventuality. Such an obligation might inhibit institutional involvement in clinical research in general.

In the absence of an agreed social policy on care for research risks, it is important to focus on the fact that by providing no-fault compensation for research-related risks, institutions and investigators may significantly alter the balance of risk and benefit in favor of participation in clinical trials. Thus, the commitment to provide ongoing care may be seen simply as a component of good medical practice.

## IV. ONGOING THERAPY FOR CLINICAL BENEFIT

Difficult issues may also arise concerning the care of patients who may benefit from an experimental therapy. It is axiomatic that in providing informed consent for an experimental therapy patients are accepting risk in the interest of personal benefit. Thus, patients who experience significant benefit from the experimental therapy may expect to receive such therapy on a continuing basis. If the clinical trial leads to licensing and marketing of a therapeutic product, then the product will continue to be available to the patient and these expectations will be fulfilled. If, however, the clinical trial terminates without a therapeutic product being approved for clinical use, then the patients who participated in the clinical trial will no longer be able to receive this therapy. It is not difficult to envision clinical settings in which a gene therapy could become a critical, even life-sustaining, therapy for a patient with a severe genetic or infectious disease. While institutions do not assume responsibility for the ongoing clinical care of underlying conditions in patients participating in therapeutic clinical trials, the fact that clinical trials of gene therapy may be performed in patients with no other therapeutic options may make the cessation of care untenable. It may be difficult to provide such care, however, since this could also necessitate not only providing routine clinical care, but also producing the materials required for gene therapy and performing complex *ex vivo* procedures.

It is likely that only a small fraction of clinical trials will lead to products which are licensed by the FDA for clinical use. Many clinical trials represent basic clinical research or proof-of-principle studies aimed at a discrete, technical end-point. Such studies may be performed by investigators and institutions who will not have the commitment or capacity to continue such therapy once the research plan is completed. This is particularly true for trials performed in academic settings in which the therapeutic measure being studied may have limited commercial potential, and the academic interests of individual investigators may move to new research problems. In addition, clinical trials that are designed to evaluate potential commercial products may be found to provide benefit only to a limited number of study participants. Such trials may be considered unsuccessful if they do not provide significant benefits to large numbers of patients and the development of these products may be terminated, thus eliminating the source of material for those patients who may have found benefit. Finally, many clinical trials will be initiated by early-stage biotechnology companies that may have limited resources and will be able to bring only a limited number of products to market.

It is unrealistic to expect clinical investigators or institutions to maintain dedicated manufacturing, quality control, and clinical facilities. Moreover, there may be no way to fund such activity. Such activity would quickly cease to be "research," and thus would not be fundable from research grants. Pharmaceutical and biotechnology companies will continue to expend resources only on programs likely to provide marketable products. It may also be impossible to bill the costs of such efforts to the patient or even third party payers unless the product receives marketing or premarketing approval.

If it is not realistic to offer continuing therapy to patients who participate in clinical trials, then the risks associated with discontinuation of therapy must be clearly addressed in the informed consent. In some situations such consent may seem to be inadequate, particularly if there is a likelihood that the patient's life or well-being may become dependent upon the availability of gene therapy, and such clinical trials would need to be designed with particular caution.

## V. INFORMED CONSENT FOR FOLLOW-UP

The essence of research is that it involves unknown risks and results. This is certainly true for research involving gene therapy. Clinical research is never completely free of risk, nor it is possible to unequivocally or quantitatively assess the risks and benefits of a clinical trial before it begins. Rather, risks must be honestly and intelligibly described, and the patient must be the ultimate arbiter of the value of the potential benefits and anticipated risks through the process of providing informed consent (Surgeon General, 1966; Code of Federal Regulations, 1983; Andrews, 1987). A Lancet editorial emphasized ". . . the individual's responsibility and choice, provided it harms no one else, is fundamental to a democratic society" (Anon, 1989).

What is essential is that the patient's consent be based on a full exposition of the risks and limitations of the proposed therapy, not on false expectations of cure.

The issues of long-term care and follow-up must be part of the informed consent process. Prospective study participants must understand the importance of follow-up and the limits to the investigator's or institution's commitment to providing long-term care. While informed consent does not obviate the investigator's responsibility in any area, it must indicate the practical limitations that constrain the investigator in providing follow-up or clinical care.

Appropriate informed consent will be particularly important if the investigators intend to track patients who participate in clinical trials. It is a first principle of clinical research that study participants have the option of withdrawing at any time during the study. Patients cannot be obligated to continue their participation in clinical trials, even to the extent of allowing follow-up studies or making their medical records available to investigators performing follow-up studies. Follow-up studies, in and of themselves, can be performed only with informed consent and must offer the subject a favorable balance of risk and benefit. Many subjects may see substantial risks to their confidentiality or privacy in studies that track their medical history. These risks may be balanced by ensuring that patient tracking also provides information to patients concerning potential adverse experiences and recommended diagnostic or therapeutic interventions.

Considerable safeguards are required in designing protocols for follow-up to ensure that the individual's confidentiality and privacy are not violated, that the benefits of follow-up balance the risks, and that appropriate informed consent is obtained. Mechanisms should also be established which allow individuals to withdraw from follow-up at any time, and if children participate in follow-up studies, it will be necessary not only to obtain their assent at the time the study begins, but also their consent upon reaching the age of majority.

## VI. CONCLUSION

The clinical investigator and institution conducting clinical trials have a certain responsibility to provide follow-up care for patients who participate in clinical trials. These issues are no different in clinical trials of gene therapies than in trials of conventional pharmaceuticals, biologicals, and devises. Research involving gene therapy is different only because of the continuing public surveillance of this research and the fragile nature of public acceptance of genetic engineering in general. The two fundamental tenets of the social discourse concerning gene therapy are that the critical and multidisciplinary assessment gene transfer technologies should take place in a public forum and that the highest possible standards of the scientific method, clinical design, and ethical analysis must be followed. A commitment to long-term follow-up and clinical care must meet these standards. This commitment must adhere not only to the regulatory requirements of the FDA and the needs of patients who participate in clinical trials, but also to the sensitivities of society which has endorsed these clinical trials with studied caution.

In all clinical trials, proper informed consent represents the principal instrument for assuring that the procedures and practices are fair. In the face of practical limitations in the ability of investigators and institutions to provide follow-up care, these limitations must be explicitly described to subjects as part of obtaining informed consent.

The NIH should be encouraged to solicit and support research projects to track patients to study the long-term consequences of gene therapy. Such studies must themselves exhibit scientific excellence in design, execution, analysis, and reporting. If these studies do not meet these standards, then the risk to the patients will be compounded, and the adequacy of informed consent which states a commitment to follow-up will be compromised. The present efforts by the RAC to monitor the course of clinical trials through data reported by clinical investigators may not be sufficient to meet the needs of either society or the individuals who contribute as subjects of clinical trials.

Even with careful attention to the issues involved in the long-term care of patients, it is likely that follow-up for many patients will remain imperfect. Poor patient compliance and follow-up is a real element of clinical practice. This may, in fact, be one of the lessons which will be learned from clinical trials: that medicine is, indeed, imperfect, and that gene therapies and their application must be ethical and effective in the face of this reality.

## Acknowledgments

The author thanks Bridgette Brown, R.N., for her assistance in preparing the manuscript as well as the thoughtful insights provided by Dr. W. French Anderson, Susan Mize, Dr. Baruch Brody, Dr. Henry Miller, Dr. Nelson Wivel, and Dr. Claudia Kozinetz. Dr. Ledley is a founder with equity interest in GeneMedicine, Inc. This work was supported in part by the ACTA Foundation.

## References

Anderson, W. F., and Fletcher, J. C. (1980). Gene therapy in human beings, when is it ethical to begin? *N. Engl. J. Med.* **303**:1293–1297.

Anderson, W. F. (1992). Human gene therapy. *Science* **256**:808–813.

Andrews, L. B. (1987). Medical Genetics, a Legal Frontier. American Bar Foundation, Chicago.

Anon. (1989). Gene Therapy (editorial). Lancet i:193.

Emery, A. E. H., and Miller, J. R. (1976). Registers for the Prevention of Genetic Disease. Stratton Intercontinental Medical Book Corp., New York.

Fletcher, J. C. (1990). Evolution of ethical debate about human gene therapy. *Hum. Gene Ther.* **1**:55–68.

Food and Drug Administration (FDA) (1992). Medical Devises; Devise Tracking, vol. 57, pp. 10702–10717. Fed. Reg., Friday, March 27, 1992.

Friedmann, T. (1983). Gene Therapy, Fact and Fiction. Cold Spring Harbor Laboratory Press, Cold Spring Harbor, NY.

Friedmann, T. A., and Roblin, R. (1972). Gene therapy for human genetic disease. *Science* **175**:949–952.

Glass, K. C., and Freedman, B. (1991). Legal liability for injury to research subjects. *Clin. Invest. Med.* **14**:176–180.

Harvey, I., and Chadwick, R. (1992). Compensation for harm: The implications for medical research. *Soc. Sci. Med.* **34**:1399–1404.

Holton, J. B. (1987). Registers for inherited metabolic diseases. *J. Inher. Metab. Dis.* **10**:309–316.

Hubbard, S. M., Henney, J. E., and Devita, V. T. (1987). A computer data base for information on cancer treatment. *N. Engl. J. Med.* **316**:315–318.

Ledley, F. D. (1991). Clinical considerations in the design of protocols for somatic gene therapy. *Hum. Gene Ther.* **2**:77–84.

Ledley, F. D., Kozinetz, C. A., Brody, B., Mize, S., and Anderson, W. F. (1991). Gene Transfer Patient and Provider Network (GENTRANET). Submission to Recombinant DNA Advisory Committee, May, 1992.

Ledley, F. D., Brody, B., Kozinetz, C., and Mize, M. (1992). The challenge of follow-up for clinical trials of somatic gene therapy. *Hum. Gene Ther.* **3**:657–664.

Levine, R. J. (1986). Ethics and Regulation of Clinical Research.

Miller, A. D. (1990). Human gene therapy: Part of a therapeutic continuum. *Hum. Gene Ther.* **1**:3–4.

Miller, A. D. (1992). Human gene therapy comes of age. *Nature* **357**:1455–1460.

Mize, S. G., and Pratt, J. H. (1989). Register of inherited metabolic disorders. In "Proceedings of the 7th National Neonatal Screening Symposium, New Orleans, La.," pp. 87–90.

Murray, T. H. (1990). Human gene therapy, the public, and the public policy. *Hum. Gene Ther.* **1**:49–54.

Nelson, J. R. (1990). The role of religions in the analysis of the ethical issues of human gene therapy. *Hum. Gene Ther.* **1**:43–48.

Nirenberg, M. (1967). Will society be prepared? *Science* **157**:633–635.

Office of Technology Assessment (OTA) (1984). Human Gene Therapy. Background Paper. (U.S. Government Printing Office, Washington, DC.)

Royal College of Physicians (1990). Research involving patients. Summary and recommendations of a report of the Royal College of Physicians. *J. R. Col. Phys.* **24**:10–14.

Surgeon General (1983). President's Commission for the Study of Ethical Problems in Medicine and Biomedical and Behavioral Research. Splicing Life. U.S. Government Printing Office, Washington DC.

Thoene, J., and Crooks, G. M. (1989). Report of the National Commission on Orphan Diseases. USDHHS, Public Health Service, Washington, DC.

Walters, L. (1986). The ethics of human gene therapy. *Nature* **320**:225–226.

# 2

# Gaucher Disease[1]

**Ernest Beutler**
Department of Molecular and Experimental Medicine
The Scripps Research Institute
La Jolla, California 92037

## I. INTRODUCTION

In 1882 Phillipe Charles Ernest Gaucher in his thesis, *De l'epithelioma primitif de la rate, hypertrophie idiopathique del la rate san leucemie,* described the disease that now bears his name. Believing that he had described an epithelioma of the spleen, he hardly could have anticipated the real nature of the disease, and the extensive information that would become available a century later.

We now know that the disease is caused by hereditary deficiency of glucocerebrosidase resulting from mutations in the glucocerebrosidase gene, GBA, or, very rarely, to mutations of the prosaposin gene, PSAP, which directs production of a cofactor that is needed for functioning of glucocerebrosidase.

Investigations of the molecular biology of the disease have disclosed important information about the mutations that cause Gaucher disease, their frequency in various populations, and their phenotypic effect. This disease has become one of the relatively few enzyme deficiencies that can be treated effectively by replacement of the missing enzyme. The new findings about Gaucher disease have, as is usually the case, raised as many questions as they have answered.

In this paper I shall review what has become known about Gaucher disease, particularly as a result of the study of its molecular biology, and discuss some of the questions that have remained unanswered. Finally, I shall address

[1]This is manuscript 8587-MEM from The Scripps Research Institute. Supported by National Institutes of Health Grants DK36639 and RR00833 and the Sam Stein and Rose Stein Charitable Trust Fund.

some of the criticisms that have been made of my findings concerning the treatment of Gaucher disease, findings that have an enormous financial impact on the treatment of patients with this disorder.

## II. CLINICAL MANIFESTATIONS OF DISEASE

Gaucher disease is a lysosomal storage disorder characterized by the accumulation of the glycolipid glucocerebroside. The manifestations of this disorder can vary enormously. Many of the patients are totally asymptomatic, quite unaware of the presence of storage cells in their marrow, liver, and spleen, and living out a normal life. At the other extreme are patients with the rare neuronopathic forms of the disease, usually fatal by the age of 18 months. Accordingly, the disease has been classified into three major types.

### A. Type I Gaucher disease

Type I disease is by far the most common. It is defined as the form of Gaucher disease in which there is no primary involvement of the central nervous system. Occasionally, neurologic disturbances may occur in type I disease, but when they do, they are secondary to complications such as collapse of a vertebra or a stroke (Grewal et al., 1991).

The predominant manifestations of this type of disease are visceral, with enlargement and dysfunction of the liver and spleen, on the one hand, and skeletal disease on the other. Even within a given genotype of the disease manifestations may be entirely skeletal, entirely visceral, or a combination of both. The reason for the markedly different sites of involvement is not understood and will be discussed below. Occasionally, pulmonary function is impaired in type I disease. In published cases this has generally represented involvement by the direct presence of Gaucher cells in pulmonary macrophages (Schneider et al., 1977; Smith et al., 1978; Roberts and Fredrickson, 1967; Myers, 1937). However, there also seem to be patients who have serious hemodynamic changes in the lung without major direct pulmonary involvement, as determined either by X-ray or transbronchial biopsy. Both intrapulmonary shunting and pulmonary hypertension occur. In such patients without frank pulmonary involvement with Gaucher cells this may be secondary to severe liver disease.

Even more remarkable than the differences in areas of involvement, and as poorly understood, are the differences in severity of the disease. As will be pointed out below, the mutations that a patient has inherited play a distinct role in the severity of clinical manifestations, but provide far from the whole answer.

## B. Type II Gaucher disease

Type II disease is the most devastating form of this disorder. Patients have severe neurological involvement. Extensive visceral involvement with hepato-splenomegaly is usually present, but it is the early onset of neurologic disease that characterizes type II Gaucher disease. Oculomotor abnormalities are often the first manifestations, with the appearance of bilateral fixed strabismus (Kolodny et al., 1982) or of oculomotor apraxia. Rapid head thrusts are a characteristic manifestation, presenting as an attempt to compensate when trying to follow a moving object (Conradi et al., 1991). Hypertonia of the neck muscles with extreme arching of the neck, bulbar signs, limb rigidity, seizures, and sometimes choreoathetoid movements occur. Gaucher disease may manifest as hydrops fetalis (Sun et al., 1984), and a severe ichthyosis-like skin disorder may exist at birth; these infants have been referred to as "collodion babies" (Commens et al., 1988; Lipson et al., 1991).

It is likely that a total absence of glucocerebrosidase is not compatible with extrauterine life. No patient has been found with two mutations both of which prevent formation of any glucocerebrosidase. This is true despite the fact that one of the more common Jewish mutations, 84GG, produces a frameshift in the far 5' end of the molecule. Moreover, mice with targeted disruption of the gene die within 24 hr of birth and manifest extensive lysosomal glucocerebroside storage (Tybulewicz et al., 1992).

## C. Type III Gaucher disease

Intermediate in severity between type I and type II disease, type III disease is characterized by a later onset of neurological symptoms. The prototype of type III disease is a form of Gaucher disease found at high frequencies in certain population isolates in northern Sweden. This disease, known as the Norrbott-nian form after the district in which it is found, has been studied in considerable detail (Svennerholm et al., 1991; Erikson, 1986). The median age of onset of symptoms is at 1 year, with a range in 22 patients of 0.1–14.2 years. The first symptoms are usually the results of visceral involvement with neurologic findings developing in about one-half of the children during the first decade of life. As in type II disease, disorders of eye movement are the usual first symptoms with the subsequent development of other neurologic manifestations, such as ataxia. Further subdivision of type III disease into type IIIa and IIIb has been proposed (Patterson et al., 1993). The former is characterized by a clinical picture dominated by slowly progressive dementia, ataxia, and spasticity as well as horizontal supranuclear gaze palsy. The patients with type IIIb disease have horizontal supranuclear gaze palsy without other major neurologic signs but have prominent visceral manifestations.

## III. MOLECULAR BIOLOGY

With the availability of only scanty amino acid sequence, the glucocerebrosidase gene was cloned and sequenced after screening several fibroblast cDNA libraries with antibody that had been prepared against purified glucocerebrosidase (Sorge *et al.*, 1985b,a). The cDNA was cloned independently (Tsuji *et al.*, 1986) and a very similar sequence obtained. Subsequently, the entire gene and a pseudogene were cloned (Reiner *et al.*, 1988; Reiner and Horowitz, 1988) and sequenced (Horowitz *et al.*, 1989; Beutler *et al.*, 1992b).

### A. The gene and pseudogene

Classical cell hybridization techniques demonstrated that the glucocerebrosidase gene is on chromosome I (Shafit-Zagardo *et al.*, 1981). Subsequently it was mapped to band q21 (Ginns *et al.*, 1985). The gene itself is about 7 kb in length, while the pseudogene is only 5 kb long. A particularly informative patient with Gaucher disease was found to have a crossover between the gene and the pseudogene. The 5′ sequence was that of the gene and the 3′ sequence that of the pseudogene. Using *Sac*II, a restriction endonuclease that cuts outside of both genes, a shortened fragment was present in the digested genomic DNA of the patient. From its length it was possible to deduce that the pseudogene is located about 16 kb downstream from the functioning gene (Zimran *et al.*, 1990).

Both the gene and the pseudogene contain *Alu* sequences, but the gene contains several such sequences that are missing from the pseudogene, accounting for its greater length. Apparently the pseudogene was formed by tandem duplication during evolution and subsequent to its formation *Alu* sequences were inserted into introns of the functioning gene but not the pseudogene. The pseudogene has undergone a number of mutations including a 55-bp deletion from the coding region and numerous point mutations.

An unusual feature of the pseudogene is that it is transcribed. Attachment of a reporter to the putative promotor region of the gene and pseudogene showed that both promoters had some activity, although the activity of the functioning gene promotor was much higher than the activity of the pseudogene promotor (Reiner and Horowitz, 1988). Indeed, in cultured fibroblasts and lymphoblasts it was possible to find ample amounts of mRNA with the pseudogene sequence (Sorge *et al.*, 1990). Because two of the point mutations in the pseudogene involve splice sites, processing of the primary transcript of the pseudogene does not proceed in the same fashion as processing of the functioning gene. Exon 2, the 5′ portion of exon 3, and exon 4 are missing from the processed transcripts. In each case, the exon upstream of the destroyed donor site has been deleted. Apparently the next upstream donor site does not use its own normal acceptor, but rather one downstream from the lost donor site. In the case of the intron 3 donor site the normal acceptor site at the 3' end of intron 4 is used, deleting exon 3. In the case of the lost intron 2 donor site the intron 1 donor does not use the acceptor at the 3' end of intron 2, but rather another site in the middle of pseudogene exon 3. This site does not exist in the normal gene, but the mutation of a TpG at nt 1724–1725 of the functional gene to an ApG creates a new splicing acceptor that is used in preference to the intact site at the 3' end of intron 3. Similar exon skipping has been found in other genes: for example, the collagen gene (Weil *et al.*, 1988).

## B. The complementary DNA

The sequence of the cDNA revealed the presence of two upstream in-frame ATGs. Both of these serve as starting points for translation: in a cell-free system two translation products of appropriate length are obtained (Sorge *et al.*, 1987). A hydrophobic leader sequence is present (Ginns *et al.*, 1985).

The presence of two functional start ATGs poses an interesting problem. Have these evolved because they provide a functional advantage or is their presence merely happenstance, conferring no selective advantage or disadvantage? Intuitively it seems that the start point for translation would not be a neutral event. Certainly, starting translation at the upstream ATG results in the incorporation of 20 additional amino acids into the leader sequence, which is then quickly destroyed. For this reason the function of the two start ATGs has been investigated. Mutagenesis of either site with subsequent retrovirus-mediated transduction of murine fibroblasts has allowed us to follow the fate of glucocerebrosidase synthesized *in vivo* utilizing either start site. With either the upstream or the downstream site acting alone, the enzyme is made and targeted to the lysosomal fraction of the cell.

One may postulate that two start sites could provide two different leader sequences. The normal donor site at genomic nt 611 (Horowitz *et al.*, 1989) could use a cryptic recipient site at genomic nt 1047. This would not produce a frameshift. The result would be to splice out the hydrophobic leader, leaving a hydrophilic leader, and this could markedly influence the subsequent processing of the enzyme. Such differential splicing could take place in different tissues, but reverse transcription of mRNA from a variety of adult human tissues has failed to identify the postulated processed mRNA. It is possible, of course, that such differential splicing takes place only at different stages of embryonic development. However, if there is an advantage to the two ATGs it is not one that exists throughout phylogeny. The mouse cDNA has been cloned and appears to have only a single start codon, homologous with the human downstream start codon (O'Neill *et al.*, 1989).

**Table 2.1.** Mutations Known to Cause Gaucher Disease

| | cDNA from 5' ATG | Amino acid mature protein | Genomic (Horowitz et al., 1989) | Exon | Base substitution | Amino acid substitution | Rapid detection method | Disease type | Reference |
|---|---|---|---|---|---|---|---|---|---|
| 1 | 72 | None | 1023 | 2 | C→Del | Stop | +AluI | I | Beutler et al., 1993a |
| 2 | 84 | None | 1035 | 2 | G→GG | Stop | | I | Beutler et al., 1991a; Beutler, 1991b |
| 3 | IVS2 | IVS2 | 1067 | Ivs2(+1) | g→a* | Splice | −HphI | I | Beutler et al., 1992a |
| 4 | 203 | None | 1707 | 3 | C→Del | Stop | ASOH | I | Beutler et al., 1994 |
| 5 | 476 | 120 | 3060 | 5 | G→A | Arg→Gln | +BspNI EcoRII | I | Graves et al., 1988 |
| 6 | 481 | 122 | 3065 | 5 | C→T | Pro→Ser | −KpnI | | Beutler et al., 1992a, 1993a |
| 7 | 535+ | 140 | 3119 | 5 | G→C | Asp→His | +BstHI + NlaIII + MnlI | I | Eyal et al., 1991 |
| 8 | 586 | 157 | 3170 | 5 | A→C | Lys→Gln | +ScrFI | II | Latham et al., 1991; Eyal et al., 1991 |
| 9 | 644 | 176 | 3438 | 6 | C→A | Ala→Asp | −PflMI | I | Beutler et al., 1994 |
| 10 | 661 | 182 | 3455 | 6 | C→A | Pro→Thr | −HphI | I | Beutler et al., 1994 |
| 11 | 721 | 202 | 3515 | 6 | G→A* | Gly→Arg | −NciI | I | Beutler et al., 1994 |
| 12 | 751 | 212 | 3545 | 6 | T→C | Tyr→His | +DraIII | I | Beutler et al., 1992a, 1993a |
| 13 | 754 | 213 | 3548 | 6 | T→A* | Phe→Ile | | III | Kawame and Eto, 1991 |
| 14 | 764 | 216 | 4113 | 7 | T→A | Phe→Tyr | +KpnI | I | Beutler and Gelbart, 1990 |
| 15 | 983 | 289 | 4332 | 7 | C→T | Pro→Leu | | I | He et al., 1992 |
| 16 | 1043 | 309 | 5259 | 8 | C→T | Ala→Val | +MaeIII −BanI | I | Latham et al., 1991 |
| 17 | 1053 | 312 | 5269 | 8 | G→T | Trp→Cys | −BanI −NlaIV −KpnI | I | Latham et al., 1991 |
| 18 | 1085 | 323 | 5301 | 8 | C→T | Thr→Ile | +FokI | I | He et al., 1992 |
| 19 | 1090 | 325 | 5306 | 8 | G→A* | Gly→Arg | +Bsu36I | II | Eyal et al., 1990 |
| 20 | 1093+ | 326 | 5309 | 8 | G→A | Glu→Lys | +BbvII +MboII −BsmaI | I | Eyal et al., 1991 |
| 21 | 1141 | 342 | 5357 | 8 | T→G | Cys→Gly | −StuI | II | Eyal et al., 1990 |
| 22 | 1192 | 359 | 5408 | 8 | C→T | Arg→Stop | −TaqI −Sau3AI | I,II | Beutler and Gelbart, 1994 |

| 23 | 1193 | 359 | 5409 | 8 | G→A | Arg→Gln | −TaqI | I | Kawame et al., 1992 |
|----|------|------|------|----|------|---------|-------|-----|---------------------|
| 24 | 1208 | 364 | 5424 | 8 | G→C | Ser→Thr | −CviJI | I | Latham et al., 1991 |
| 25 | 1226 | 370 | 5841 | 9 | A→G | Asn→Ser | +CviJI | I | Tsuji et al., 1988 |
| 26 | 1246 | 377 | 5861 | 9 | G→A | Gly→Ser | | I | Laubscher et al., 1993 |
| 27 | 1249 | 378 | 5864 | 9 | T→G | Trp→Gly | ASOH | I | Beutler et al., 1994 |
| 28 | 1255 | 380 | 5870 | 9 | G→A | Asp→Asn | | I | Beutler et al., 1994 |
| 29 | 1256 | 380 | 5871 | 9 | A→C | Asp→Ala | +ScrFI | I | Walley and Harris, 1993 |
| 30 | 1263 | None | 5878 | 9 | 55Del | Stop | −SalI | I | Beutler et al., 1992a, 1993a |
| 31 | 1297 | 394 | 5912 | 9 | G→T | Val→Leu | −HgiEII | I,III | Theophilus et al., 1989 |
| 32 | 1312 | 399 | 5927 | 9 | G→A | Asp→Asn | −TaqI | II | Beutler and Gelbart, 1994 |
| 33 | 1342 | 409 | 5957 | 9 | G→C* | Asp→His | −StyI | I,III | Eyal et al., 1990; Theophilus et al., 1989 |
| 34 | 1343 | 409 | 5958 | 9 | A→T | Asp→Val | −AflIII +MaeIII | III | Theophilus et al., 1989 |
| 35 | 1361 | 415 | 5976 | 9 | C→G | Pro→Arg | +HhaI | II | Wigderson et al., 1989 |
| 36 | 1390 | 425 | 6375 | 10 | A→G | Lys→Glu | | III | Kawame et al., 1992 |
| 37 | 1448 | 444 | 6433 | 10 | T→C* | Leu→Pro | +NciI | I,II,III | Tsuji et al., 1987 |
| 38 | 1504 | 463 | 6489 | 10 | C→T | Arg→Cys | +BsrI −MspI | I,III | Hong et al., 1990 |
| 39 | 1505 | IVS463 | 6490 | 10 | G→A | Arg→His | −MspI | III | Ohshima et al., 1993 |
| 40 | 1549 | 478 | 6628 | 11 | G→A | Gly→Ser | +AluI | I | Beutler et al., 1993a |
| 41 | 1603 | 496 | 6682 | 11 | C→T | Arg→Cys | −BsaHI | I | Kawame et al., 1992 |
| 42 | 1604 | 496 | 6683 | 11 | G→A | Arg→His | +HphI | I | Beutler et al., 1992a, 1993a |

*Mutation identical to pseudogene sequence
+ Same allele
ASOH = Allele specific oligonucleotide hybridization

23

## C. Mutations that cause Gaucher disease

More than 30 different mutations that cause Gaucher disease have been identified. These are summarized in Table 2.1. Included are single base pair changes in the coding region and at splice sites, single nucleotide insertions and deletions, deletions of a 55-bp segment, homologous with a deletion that occurs normally in the pseudogene, deletion of the entire gene, gene fusions, and gene conversions. No promotor mutations or mutations involving the 5' or 3' noncoding region have been identified so far.

A relationship exists between the nature of the mutation and the disease phenotype that is observed. To a large extent, the severity of the phenotype can be predicted from the mutations. Thus, the 84GG mutation and the IVS2(+1) mutation, both of which preclude the formation of any enzyme at all, have never been seen in the homozygous state. Like the mouse with targeted disruption of the glucocerebrosidase gene (Tybulewicz et al., 1992), these mutations are presumably fatal before birth. At the other extreme is the 1226G mutation, a relatively conservative change from asparagine to serine. Homozygotes for this mutation have, on the average, a late onset of disease and a mild course. They never suffer the neuronopathic form of the disease (Zimran et al., 1989, 1992; Beutler et al., 1992a; Sibille et al., 1993). Intermediate in severity is the 1448C mutation and some of the recombination events in which this mutation is present. Here, the homozygous state is compatible with life, but is usually associated with neuronopathic disease, either the fulminating type II or the more chronic type III disease. The relationship between severity of the disease and the age of patients with the more common genotypes is summarized in Figure 2.1. It is obvious that although on the average, the phenotype does predict the genotype, there is much variation within any given genotype and therefore considerable overlap between different genotypes. The possible causes of this variability are discussed below.

## IV. POPULATION GENETICS

The introns of the glucocerebrosidase gene contain 11 polymorphic sites, all of which are in linkage disequilibrium (Beutler et al., 1992b). The two haplotypes that are produced are designated + and − based on the polymorphic PvuII site that was the first polymorphism to be found (Sorge et al., 1985a). In Western populations, the gene frequency of the − haplotype is approximately 0.65, while that of the + haplotype is 0.35. In oriental populations the frequencies are reversed: the frequency of the − haplotype is 0.35 and that of the + haplotype is 0.65. The liver type pyruvate kinase gene (PKLR), is also known to reside on band q21 of chromosome 1 (Tani et al., 1987). A polymorphic site on this gene

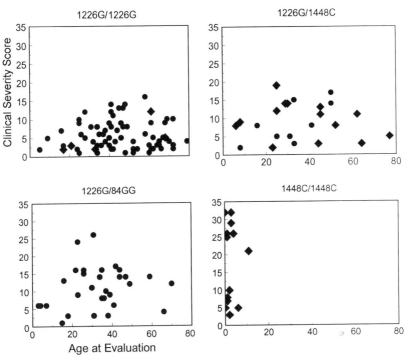

**Figure 2.1.** The relationship between age and disease severity in patients with the four most common Gaucher disease genotypes. Circles denote Jewish patients. Non-Jewish patients are represented by diamonds. The severity score is calculated from arbitrary values assigned to various disease manifestations in different organ systems as presented previously (Zimran *et al.*, 1992). Patients with the 1226G/1226G genotype tend to be older, the disease is somewhat milder, and there is no tendency for disease manifestations to be greater in the older patients. Patients with the 1226G/1448C and 1226G/84GG genotypes have more severe disease, which is to some extent age related. The most severe disease is present in the patients with the 1448C/1448C genotype.

(Kanno *et al.*, 1993) is in marked linkage disequilibrium with the polymorphic sites of the glucocerebrosidase gene (Figure 2.2).

Certain mutations are particularly common causes of Gaucher disease. In Jewish patients the most common two mutations are 1226G and 84GG accounting for about 77 and 13% of Gaucher disease mutations in this population. Two additional mutations, found almost exclusively in the Jewish population, IVS2(+1) and 1297T, each account for an additional 2% of the disease-producing alleles in the Jewish populations (Beutler *et al.*, 1992a; Beutler and Gelbart, 1993). Without exception, each of these four "Jewish" mutations has

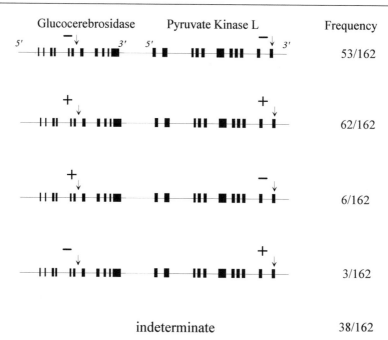

**Figure 2.2.** Diagrammatic presentation of linkage disequilibrium between polymorphic sites in the liver-type pyruvate kinase gene (*PKLR*) and the glucocerebrosidase gene (*GBA*) in 81 normal individuals (Glenn *et al.*, 1994). The arrows show the locations of the polymorphic sites, viz. the *Pvu*II polymorphic site in the glucocerebrosidase gene and the A→C transition that destroys a *Bsp*HI restriction site at cDNA nt1705 of the pyruvate kinase gene (Kanno *et al.*, 1992). According to previous convention we have designated the Pv1.1 polymorphism + for the nt 3931G allele and − for the 3931A allele. In the case of the *PKLR* polymorphism nt1705A is arbitrarily designated as + and nt 1705C as −.

been found in its own haplotype, a finding that is consistent with a single origin of each mutation.

In contrast to the situation with the Jewish 1226, 84GG, and IVS2(+1) mutations, the panethnic 1448C mutation is found in the context of both the + and − haplotypes (Beutler and Gelbart, 1993), indicating that it has repeatedly arisen independently. It is the 1448C mutation that is most common in non-Jews from all over the world. The substitution at this nucleotide creates the sequence that is normally present in the pseudogene and it may be presumed that it has repeatedly been created by gene conversion. Indeed, often other mutations also representing the pseudogene are present downstream from the 1448C mutation (Horowitz *et al.*, 1993), even in the absence of evidence for a physical crossover between the glucocerebrosidase gene and its pseudogene. In non-Jewish patients with Gaucher disease many different mutations are encountered; they presumably represent the background "noise" in this gene. In Europe, however, but not in Asia, the 1226G gene is often found in patients with Gaucher disease (Beutler and Gelbart, 1993), probably as a result of ethnic admixture.

## A. Gaucher disease as a balanced polymorphism in the Jewish population

It has sometimes been suggested that the high prevalence of Gaucher disease in the Jewish population might be due to a founder effect or to genetic drift. If this were the case the high gene frequency would be mere happenstance and there would be no necessity for heterozygotes to have a survival advantage. There are, however, two circumstances that compel me to conclude that the heterozygous state for Gaucher disease confers, or conferred sometime in the past, a selective advantage upon those individuals who possessed it. First, while a founder effect or drift might explain the high prevalence of a single Gaucher disease allele in the Jewish population, there are no less than four mutations that are found predominantly in the Jewish population, and two of these exist at fairly high gene frequencies, 0.034 and 0.0021 for the 1226G and 84GG mutations, respectively (Beutler et al., 1993b). Second, it is notable that there are two other disorders of glycolipid metabolism, viz., Tay–Sachs disease and Niemann–Pick disease, that are most common in the Jewish population. Thus, it seems clear that at some time in the Jewish experience a partial deficiency of one of the glycolipid metabolizing enzymes was advantageous. One may raise the question of whether it is not a *heterozygote* advantage that is enjoyed in the case of Gaucher disease, but rather a *homozygous* advantage. This might be a possibility because so many of those who inherit the 1226G/1226G genotype have no disease manifestations. If they enjoyed a selective advantage, this might counterbalance the relatively few with this genotype who have reduced fitness. However, such a mutation would drive the much more serious 84GG mutation out of the population, since the 1226G/84GG heterozygotes have markedly reduced fitness.

## V. TREATMENT

Patients with Gaucher disease are often highly productive members of society, and in the milder forms of the disease, the life span of the patient seems to be compromised very little if at all. However, the symptoms of the disease may be quite disabling and, in patients more severely affected, life threatening. Thus, a considerable amount of attention has been paid to attempting to improve the quality of life of patients with Gaucher disease and, if possible, to reverse the signs and symptoms of the disorder.

## A. Symptomatic management

### 1. Skeletal symptoms

Painful crises are usually managed with analgesics and bed rest.

The quality of life of patients with Gaucher disease may be greatly enhanced by appropriate orthopedic surgical intervention. Hip replacement has been particularly useful (Goldblatt et al., 1988) and, in some cases, successful replacement of knee joints has also been accomplished.

Aminohydroxypropylidine biphosphonate administraiton was reported in 1984 to be effective in reversing bone disease in a patient with Gaucher disease (Harinck et al., 1984). Seven years later, improved calcium balance and bone density were reported in two additional cases (Ostlere et al., 1991). No controlled studies of the effect of this promising class of compounds have been reported.

Once they occur, pathologic fractures often heal slowly or not at all. Collapse of vertebrae or of the femoral head has permanent sequelae. To help prevent these complications, we advise patients with Gaucher disease to avoid activities that could result in bone damage, such as competitive running, tennis, skiing, or various "contact sports." Swimming is probably the safest exercise for patients with bone disease.

### 2. Splenectomy

Splenectomy is a very effective treatment of the thrombocytopenia and to a considerable extent the anemia that often occurs in the course of Gaucher disease (Medoff and Bayrd, 1954; Fleshner et al., 1991). Splenectomy is also indicated when splenomegaly is so massive as to become symptomatic and to interfere with normal growth and development. The initial response to splenectomy is virtually always highly satisfactory. However, concern has been expressed about the possible effect of the removal of the spleen on progressive deposition of glycolipid in other organs (Ashkenazi et al., 1986; Rose et al., 1982). While removing an organ that serves as an important storage site could result in accelerated deposition in other organs such as the skeleton (Rose et al., 1982; Kyllerman et al.,1990) and, in type 3 disease, the central nervous system (Conradi et al., 1984; Svennerholm et al., 1982), the evidence that this is the case is largely anecdotal. From a review of over 200 patients, Lee (1982) concluded that the presence or absence of bony disease was unrelated to splenectomy. Other investigators have reached similar conclusions based on clinical evaluations of affected type 1 patients (Goldblatt and Beighton, 1982; Beighton *et al.*, 1982). However, Norrbottnian patients have been found to accumulate larger amounts of glucosylceramide in the plasma and brains following splenectomy, which correlated with a more rapid clinical deterioration (Nilsson *et al.*, 1985, 1982; Nilsson and Svennerholm, 1982; Conradi *et al.*, 1984; Rose *et al.*, 1982). Splenectomy should be reserved for patients with platelet counts consistently under 40,000/μl, growth retardation, and/or mechanical cardiopulmonary compromise.

Partial splenectomy was introduced in an attempt to obtain the therapeutic benefits of splenectomy while avoiding the possible adverse effect on the course of the disease (Beutler, 1977, 1979). The procedure was also proposed as a means of avoiding the susceptibility to sepsis that occasionally follows total splenectomy (Guzzetta *et al.*, 1987). Since there is some regrowth of the splenic remnant (Kyllerman *et al.*, 1990; Guzzetta *et al.*, 1987; Rodgers *et al.*, 1987; Guzzetta *et al.*, 1990), it is indeed possible that allowing some of the spleen to remain could prevent progression of the disease. However, death from bleeding of a regrown splenic remnant has been reported (Holcomb and Greene, 1993), and there are no controlled studies that provide clear-cut justification for this procedure. Splenic embolization as been advocated either as an adjunct to or a substitute for splenectomy in the treatment of Gaucher disease (Thanopoulos *et al.*, 1987; Samama *et al.*, 1989), but the amount of experience with this approach is apparently quite meager.

## B. Specific therapy

### 1. Enzyme replacement

In 1964, DeDuve (1964) first suggested that replacement of the missing enzyme with exogenous enzyme might be a successful approach to the treatment of lysosomal storage diseases. Type I Gaucher disease is a particularly attractive candidate for such an approach because there is no primary central nervous system involvement and the target cell is the macrophage. Early attempts to treat patients with administration of glucocerebrosidase were disappointing. Brady *et al.* administered unmodified placental acid β-glucosidase intravenously to two patients with Gaucher disease (Brady *et al.* 1974a,b, 1975; Pentchev *et al.*, 1975). Liver biopsies before and after enzyme administration appeared to show clearance of glycolipid. However, in view of the heterogeneous distribution of acid β-glucosidase in liver (Beutler *et al.*, 1977a), the biopsy results were difficult to interpret. In any case, no clinical response was reported. We administered smaller amounts of enzyme, but targeted it to the reticuloendothelial system by encapsulation in erythrocyte ghosts that were coated with immunoglobulin (Beutler *et al.*, 1977a). This resulted in the reduction of liver size as judged by technetium scanning. However, treatment of six additional patients using this technique (Beutler, 1981; Beutler et al., 1980; Beutler and Dale, 1979; Dale and Beutler, 1982; Beutler et al., 1977b) gave disappointing results. Trials of enzyme replacement were also undertaken by Belchetz and Gregoriadis (Belchetz et al., 1977) using liposome-entrapped enzyme. Alarming symptoms occurred after some liposome infusions, and the therapeutic results were equivocal (Gregoriadis et al., 1982).

With the realization that the uptake of glycoproteins occurred by way of carbohydrate-specific receptors (Ashwell and Morell, 1974) and, in particular, that macrophages contained mannose-specific receptors (Achord et al., 1977), attempts were made to target acid β-glucosidase to macrophages by modifying the oligosaccharides of the enzyme (Furbish et al., 1978; Pentchev et al., 1978; Steer et al., 1978; Furbish et al., 1981; Doebber et al., 1982). That paved the way for developing macrophage-directed enzyme therapy for Gaucher disease. Industrial-scale production of human placental enzyme, modified to expose covered N-acetylglucosamine and mannose residues (Ceredase; alglucerase) has made sufficient material available for prolonged clinical administration of large amounts of acid β-glucosidase. Clear-cut clinical responses were demonstrated (Barton et al., 1990, 1991a) and readily confirmed (Beutler et al., 1991b; Kay et al., 1991; Fallet et al., 1992). In anemic patients the hemoglobin concentration of the blood began to rise within the first few months of enzyme therapy. Similarly, rapid rises in platelet counts were observed in splenectomized patients with thrombocytopenia. The platelet response is slower in patients whose spleen has not been removed. Regression of organomegaly was generally evident within the first 6 months of treatment, with an average decrease of excess liver volume by about 20 to 25%. Bone pain gradually decreased, but the X-ray abnormalities of bone appeared to respond very slowly. Magnetic resonance imaging showed a decrease in bone involvement. Few untoward reactions have been reported (Pastores et al., 1993). Processing of the placental extract from which the enzyme is purified is such that known viruses, including the human immunodeficiency virus, are destroyed. Recombinant enzyme is not yet available commercially, but clinical trials with this enzyme indicated efficacy (Grabowski et al., 1993). Ten to 15% of patients treated with alglucerase develop antibodies against the enzyme protein. A few of these develop pruritus with enzyme infusion, but it has been possible to continue therapy in such patients without the occurrence of serious reactions (Pastores et al., 1993). One patient with antibodies developed anaphylaxis during an enzyme infusion.

Studies of the binding of alglucerase to murine macrophages and to human monocyte-derived macrophages have shown that only a very small amount binds to the classical mannose receptor. Most of the mannose-terminated enzyme is bound by another mannose-dependent but calcium-independent receptor that has a lower affinity and much higher copy number and that is present in many cells, including endothelial cells (Sato and Beutler,

**Table 2.2.** Binding of Alglucerase to Two Different
Mannose-Dependent Receptors

|                               | Classical | New     |
| ----------------------------- | --------- | ------- |
| Distribution                  | Limited   | Broad   |
| Number/macrophage             | 20,000    | 500,000 |
| Affinity for alglucerase      | High?     | Low     |
| Affinity for mannose–albumen  | High      | None    |
| Calcium requirement           | +         | 0       |
| Inhibition by fucose          | +         | 0       |
| Inhibition by α-glucose       | 0         | +       |

1993). The properties of this receptor are compared with those of the classical mannose receptor in Table 2.2. Although in one study remarkable amounts of infused enzyme were found in the liver, very little of the administered enzyme was detected in the lung, which was severely involved with Gaucher disease (Fallet *et al.*, 1992). We have studied the appearance of β-glucosidase activity in hip bone removed surgically from patients with Gaucher disease, and remarkably little enzyme was found, even immediately after enzyme infusion (Beutler *et al.*, 1995) (Table 2.3). A maximum increase of about 2 μUnits of enzyme per microgram of bone DNA was detected in this marrow-rich bone after 60 units of enzyme per kilogram body weight had been infused. Since the average DNA content of

**Table 2.3.** β-Glucosidase Activity in Bones of Normal
Subjects, Untreated Patients with Gaucher
Disease, and Patients after Enzyme Infusion

| Subject                             | β-Glucosidase activity (μUnits/μg DNA) |
| ----------------------------------- | -------------------------------------- |
| Normal controls (4)                 | 9.3 ± 5.3                              |
| Untreated Gaucher disease patient   |                                        |
| 1                                   | 0.8                                    |
| 2                                   | 1.5                                    |
| 3                                   | 0.8                                    |
| 4                                   | 0.7                                    |
| 5                                   | 1.5                                    |
| 5[a]                                | 2.6                                    |
| 6[b]                                | 2.0                                    |

[a]Bone taken at end of 2-hr infusion of 60 U/kg body wt.
[b]Bone taken 2 days after infusion of 60 U/kg body wt.

human tissues is 100 mg DNA/kg body wt even distribution of the infused enzyme should have increased the enzyme activity of the bone 60 units of enzyme per 100 mg of DNA, or 600 units of enzyme per gram of DNA even if the enzyme had been distributed uniformly to all body cells, let alone been "targeted" to the macrophage-rich marrow. Only about 0.3% of the amount of enzyme reached the bone that would have been found had it been evenly distributed to all body cells. We have made similar measurements on marrow aspirates obtained from a patient who had just received alglucerase infusions.

Clearly, alglucerase is not efficiently macrophage targeted. The vast majority of the infused enzyme disappears to other tissues, very likely endothelium, which is richly endowed with receptors for the mannose-terminated enzyme. Thus, the reason for the very satisfactory clinical response to alglucerase is not clear. It may be that the small amount that is taken up by macrophages is sufficient to decrease the glucosyl ceramide burden or that the enzyme removes the glycolipid from areas other than macrophages.

The dose of alglucerase used in the early studies (Barton et al., 1991a) and recommended by the commercial producer and the Food and Drug Administration (Anonymous, 1991) is 60 units per kilogram every 2 weeks. Treatment of a 70-kg patient according to this schedule at the current market price of alglucerase is $380,000 per year for enzyme alone. Thus, the preparation is extremely costly and well out of the price range of many patients, particularly in developing countries (Zimran et al., 1991). Initial doses of only 30 U/kg, even when given at the very inefficient interval of every 2 weeks (see below), have an effect that is not clearly distinguishable from twice that dose (Pastores et al., 1993).

Because of the high cost of alglucerase, it is desirable to administer the preparation in as efficient a manner as possible. It is difficult to carry out clinical studies to define the most efficient mode of administration because of the great variability of both disease severity and response to treatment. However, the short intracellular half-life of exogenous acid β-glucosidase (Pentchev et al., 1978) suggests that frequent administration of the enzyme would be more effective than infrequent dosing. A considerable number of studies have now been carried out to compare the efficacy of frequent, low doses of alglucerase with infrequent, much higher doses. The effect of giving the drug every 2 weeks has been documented by Barton et al. (1991b, 1993) and by Fallet, Pastore, Grabowski, and their collaborators (Pastores et al., 1993). The effect of administration of enzyme three times a week has been studied by our group (Figueroa et al., 1992; Zimran et al., 1993a; Beutler, 1993; Beutler et al., 1991b; Kay et al., 1991; Beutler and Garber, 1993), by Zimran et al., (1993a,b), and by Hollak et al. (1993). The results obtained with respect to changes in liver size and spleen size at an end point of approximately 6 months are summarized in Figures 2.3 and 2.4. Since the effect of treatment on skeletal lesions occurs very slowly, if at all, it has not been documented in any of the published studies.

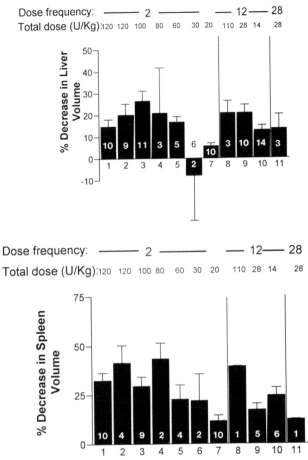

**Figure 2.3.** The change in liver and spleen size in patients with Gaucher disease being treated with different dosage schedules. The numbers 1 through 11 denote the source of the data: 1, Barton et al. (1991a) as corrected in (Barton et al., 1993); 2–5, Fallet et al., (1992) and Pastores et al., (1993); 6, Beutler, (1994); 7, Barton et al., (1993); 8, 9, 11 Figueroa et al. (Figueroa et al., 1992) and Beutler (Beutler, 1994); 10, Hollak et al., (1993). The number on the bars represents the number of patients in each group, and one standard error is shown. The dose frequencies and total dose are given for a 4-week period. It is apparent that the response to doses as low as 15 U per kilogram is no different than the response to 130 U per kilogram, provided that the dose is fractionated to three times weekly. When the dose is not fractionated, decreasing doses result in decreased response of liver and spleen size. The difference in cost of drug for these dosages is over $350,000 per year for a 70-kg patient.

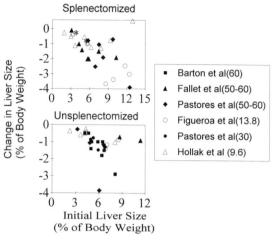

**Figure 2.4.** The response of liver size to treatment with different schedules of alglucerase compared with initial liver size as percentage of body weight. The dose given each 2 weeks is given in parentheses. Data from Barton *et al.*, 1991a, Fallet *et al.*, 1992, Pastores *et al.*, 1993, Figueroa *et al.*, 1992, Hollak *et al.*, 1993, and Hollak, 1994 are from the cited references. The responses calculated for Barton *et al.* may be somewhat exaggerated since 9- to 12-month weights rather than 6-month weights were used for the calculation (Barton *et al.*, 1993). The response of patients with larger livers is clearly greater than that of those who have only slight hepatomegaly. There is no difference between the response of patients receiving therapy every 2 weeks with large, bolus doses (closed symbols) and that of those receiving the much smaller, less costly, fractionated doses (open symbols).

## 2. Marrow transplantation

Abnormalities of macrophages in various organs are the cause of the phenotypic expression of Gaucher disease. Since the macrophage descends from the hematopoietic stem cell, it was to be expected that stem cell transplantation would cure Gaucher disease. The first patient transplanted in the treatment of Gaucher disease showed clearance of Gaucher cells from the marrow in about 6 months (Rappeport and Ginns, 1984) but died of infection before 1 year. Subsequently, a number of other patients with type 1 Gaucher disease (Hobbs, 1987; Hobbs *et al.*, 1987; August *et al.*, 1984) and type 3 disease (Erikson *et al.*, 1990; Ringdén *et al.*, 1988; Lonnqvist *et al.*, 1984; Tsai *et al.*, 1992; Svennerholm *et al.*, 1991) have undergone transplantation. Regression of disease has been observed uniformly in those patients with type 1 disease who survived, and although there are suggestions that the progression of neurological disease is arrested by the procedure, the long-term effects are by no means certain.

    The indications for marrow transplantation are uncertain. Transplantation costs less than enzyme replacement, and may therefore be available to some patients who would not have access to enzyme augmentation therapy. Moreover, it provides a permanent cure when successful. However, it is difficult to recommend transplantation to patients with type 1 disease in most circumstances because of the 10% mortality incident to tranplantation, even in the best circumstances, and because of adverse long-term effects on growth and development (Sullivan and Reid, 1991; Storb *et al.*, 1991; Beutler, 1991a). The risk is even higher in patients with advanced organ dysfunction, those who most need treatment.

## 3. Gene transfer

Since Gaucher disease can be cured by transplantation of allogeneic stem cells, correction of the defect in the patient's own hematopoietic stem cells should be an effective way to treat this disease. Stem cells that produce glucocerebrosidase would not have a proliferative advantage over those that do not. The marrow of a 70-kg human is estimated to have a volume of about 1.3 liters (Davidson and Wells, 1962). Based on the volume of marrow cells and their representation in the marrow (Williams and Nelson, 1990) one may calculate that there are about $1.4 \times 10^{12}$ nucleated cells in the marrow. A good yield of cells is considered to be about $0.03 \times 10^{12}$ nucleated cells, usually obtained as about 1 liter of liquid marrow. Thus, the infused marrow contains only about 2% as many stem cells as are endogenous to the patient, and cure would be expected only if the patient's untransformed cells were at least partially ablated by chemotherapy or irradiation. A rational strategy for the treatment of Gaucher disease, then, would be marrow ablation followed by autologous transplantation with transformed hematopoietic stem cells. The transfer of stable, functional acid β-glucosidase into cultured fibroblasts and transformed lymphoblasts using a retroviral vector was readily accomplished (Sorge *et al.* 1987a; Choudary *et al.* 1986a,b), but transfer into hematopoietic stem cells is much more difficult (Beutler *et al.*, 1988). However, with the use of modern vectors, relatively high efficiency transfer of the human acid β-glucosidase cDNA into murine and human hematopoietic stem cells or progenitors has been accomplished (Nolta *et al.*, 1990; Fink *et al.*, 1990; Kohn *et al.*, 1991) with evidence of sustained long-term expression of β-glucosidase in transplanted mice (Weinthal *et al.*, 1991; Correll *et al.*, 1992; Ohashi *et al.*, 1992). Some of the factors that must be taken into account in developing a successful strategy have been reviewed (Beutler and Sorge, 1990; Karlsson, 1991).

Transplantation of transduced stem cells is not the only possible therapeutic approach, however. Vectors that introduce an episome that functions for days or weeks could be the basis of effective correction of the disease phenotype, if not permanently then for many months or years. Adenoviruses might serve as such vectors (Stratford-Perricaudet et al., 1992), but the immune response that would develop to the viral proteins might preclude more than a single administration of the agent. Liposomal vectors are less likely to be antigenic and have shown the ability to transfect cells in intact mice with great efficiency (Zhu et al., 1993). Such vectors have the potential of forming an effective basis for the treatment of Gaucher disease.

# VI.  UNRESOLVED ISSUES

## A.  Variability of the Gaucher disease genotype

There is no doubt that a relationship exists between the clinical phenotype of patients with Gaucher disease and the mutations that they carry. However, there is considerable variability in the manifestations that occur in patients with the identical phenotype (Figure 2.1). Such variability can, theoretically, be due to environmental factors, linked genetic variability, or unlinked genetic variability.

   *Environmental variability* could include infections. For example, a patient with Gaucher disease might become infected with the cytomegalovirus or the Epstein–Barr virus and develop increased splenomegaly. This, in turn, would shorten white cell lifespan and therefore increase the inflow of globosides and gangliosides, the precursors of glucocerebrosidase. The increased sequestration of glucocerebroside would further increase the splenomegaly and set up a vicious cycle. On the other hand, dietary factors are unlikely to be important environmental factors in influencing the course of Gaucher disease; the source of the storage glycolipid is endogenous, not exogenous.

   *Linked genetic variability* could include additional mutations or polymorphisms in the glucocerebrosidase gene.

   *Unlinked genetic variability* could include many other genes. For example, the hydrolysis of glucocerebroside by glucocerebrosidase requires the presence of another protein, a saposin (Levy et al., 1991). Conceivably, variation in the amount or structure of the saposin available might influence the course of the disease. There are many other more subtle causes of genetic variability such as the intensity of the tissue reaction to store glucocerebroside or the presence of alternative excretory or catabolic pathways.

   There are some observations of the population genetics that can help distinguish between these three possible causes for variability of the Gaucher disease phenotype. If the variability is observed in the general population any of these three causes could be responsible. If, on the other hand, variability was observed between sibs, linked genetic differences are unlikely to play a major role. Finally, if identical twins show markedly discordant expression, then environmental causes are most likely responsible.

To my knowledge, no data have been published that analyze the degree of variability between individuals with the same genotype among the general population, sibs, and identical twins. My own impression of variability among sibs is that it is almost as great as that in the general population. I know of two identical twin pairs with Gaucher disease, and here too the phenotype is moderately discordant. This leads me to believe that environmental factors may be quite important in determining the disease phenotype.

One aspect of the variability of Gaucher disease phenotypes that requires special attention is the difference between patients in which tissues are involved. In the case of involvement of the central nervous system it seems quite clear that the nature of the mutations is of primary importance. Some mutations, notably 1226G, are never associated with neurologic disease, one report of such disease not withstanding (Sidransky et al., 1992). Others, particularly 1448C, are associated with neurologic disease and apparently regularly so when inherited in the homozygous state. Why this is the case in not entirely clear. The quantitative deficiency produced by the alleles that are associated with neuronopathic disease when they are expressed in tissue culture cells is more severe, in general, than the deficiency produced by alleles that produce only nonneuronopathic disease (Grace et al., 1990; Beutler et al., 1984; van Weely et al., 1993). It may be that the severity of the deficiency in general, is the most important factor. However, it is most likely that the severity of the deficiency in the brain would be the critical feature that differentiates neuronopathic from nonneuronopathic disease, and this may depend on factors other than the severity of the defect in leukocytes or cultured fibroblasts. For example, it is possible that some mutant proteins are more likely to undergo proteolysis in the brain than are others (Beutler, 1983).

It is much more difficult to explain why some patients have extensive bone disease and no visceral disease, others have marked hepatosplenomegaly and no bone disease, and still others have involvement of both viscera and skeleton. Such differences are often evident even among sibs and do not seem to be related to the type of Gaucher disease mutation (Figure 2.5). In some respects this seems similar to the tissue trophism that is sometimes observed with neoplastic disease. Certain tumors characteristically involve particular tissues. Bony metastases are characteristic of breast and prostate cancers, while liver involvement is more characteristic of colon and pancreatic cancer. Such differences are not due to the circulation draining the tumor bed alone. They are probably related to cell adhesion molecules on the surface of the cell. Indeed, neoplasms from the same tissue behave very differently in different patients. It may be that an inherited difference in the affinity of various cell adhesion molecules may play a role in determining where Gaucher cells "home," but there is no evidence regarding the cause of differences in tissue distribution of disease manifestations.

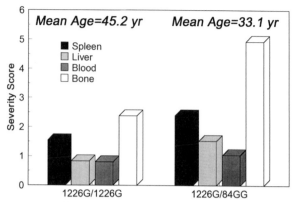

**Figure 2.5.** Severity scores relevant to spleen size (or splenectomy), liver size, cytopenias, and bone involvement in 74 patients with the 1226G/1226G genotype and 31 patients with the 1226G/84GG genotype. Even though the latter patients were younger at the time of evaluation, disease manifestations are more severe. However, the relative severity of the involvement of different organ systems is the same.

## B.  Selective advantage of heterozygotes for Gaucher disease

The compelling reasons for believing that the heterozygous state for Gaucher disease confers a selective advantage have been summarized above. The nature of the advantage is much less clear.

When seeking the advantage conferred by a single dose of the gene that produces a loss of fitness in the double dose it is logical to consider protection against those disorders that are particularly dangerous to the individual in the ethnic group that is being studied. Infectious diseases are particularly important in this regard, and historically malaria has been a great killer in the tropics. Thus, high gene frequencies for several red cell abnormalities, e.g., glucose-6-phosphate dehydrogenase deficiency and sickle cell disease, seem to have developed because of protection offered against that infection (Luzzatto, 1979). It has been proposed that the gene for cystic fibrosis may protect against cholera (Bijman *et al.*, 1988). Following this lead, it has been proposed that deficiencies in the enzymes of glycolipid metabolism might have conferred protection against infection with tuberculosis, certainly likely to have been an important cause of disability and death in the crowded medieval Jewish ghettos in Europe (Rotter and Diamond, 1987). Although an attractive idea, there is no evidence in its favor. Indeed, a study of the frequency of the heterozygous state in Jewish tuberculosis patients in Israel showed no deviation from that of the general population (Zimran and Beutler, 1994). It has been suggested that, since glycolipids are of critical importance in functioning of the brain, a deficiency of one of these enzymes might improve intelligence, an alteration that would probably be reflected in greater fertility in the context of the Jewish society. However, comparison of psychologic traits and intelligence in sibs consisting of a normal individual and one heterozygous for Gaucher disease failed to show any difference (Zimran and Beutler, 1994). Cardiovascular disease is common in the Jewish population and perhaps heterozygotes for Gaucher disease have a lower incidence of coronary artery disease. No studies to test this hypothesis have been performed, but it is notable that plasma cholesterol levels of patients with Gaucher disease are quite uniformly less than normal (Ginsberg et al., 1984; Zimran et al., 1992).

## VII. CONTROVERSIES

There has been considerable controversy about the natural course of Gaucher disease and about the best dose and schedule for the administration of alglucerase. The intensity of the controversy is probably partly a function of the high economic stakes. If the disease is indolent in most adults, as we have shown, treatment is not required. If doses much smaller than those recommended by the manufacturer can be given with the same therapeutic effect, as we and others have shown, the profits from the sale of Ceredase would shrink markedly. Moreover, if the doses must be given very frequently the patient will be inconvenienced, and patient advocacy groups have therefore been reluctant to accept the fact that frequent infusions are, indeed, required.

In this section I will review some of the assertions that have been made questioning some of the data presented in this review.

*Low doses of enzyme may be effective in some patients with mild disease but large doses should be given to patients with severe disease manifestations.*

Numerous studies show that patients with severe disease, as defined by massive organomegaly, respond as well to small doses of alglucerase as they do to large doses. Indeed, with both large and small doses the *least* response occurs in patients who only have slight enlargement of liver or spleen (see Figure 2.3). If one were to base dosage on severity of disease, the data would lead to the conclusion that patients with mild disease, if treated at all, should receive the largest doses.

*The response to small doses has been exaggerated by calculation of a decrease in excess liver size rather than in a decrease in absolute or relative liver size* (Grabowski, 1993).

It is implied in this criticism that the response has been measured in one way using small doses and in another, less favorable way when analyzing data using large doses. In the benchmark paper (Figueroa et al., 1992) in which small doses were compared with large doses all changes in liver and spleen size were reported in three different ways: (1) absolute decrease in liver size, (2) relative decrease in organ size, and (3) decrease in excess organ size. In all comparisons the same method of calculation was used, of course, at each dosage level and schedule. Change in excess organ size was not calculated to distort the results, but rather to normalize them. The desirability of normalizing the data is apparent from careful consideration of Figure 2.3. Thus, calculating the change in excess organ size tends to lower the apparent exaggerated effect of treatment in patients with massive organomegaly and increase the apparent effect in patients with mild organomegaly. This serves to eliminate the perturbing effect of differences in degrees of hepato- or splenomegaly in different series of patients and is not designed nor does it serve to make one treatment appear better than another.

*There is only a single class of receptors for alglucerase* (Barton *et al.*, 1991b).

The extraordinary claim that all of the administered enzyme binds to the same class of receptors has been contrived to imply that giving small fractionated doses would have no theoretical advantage in the treatment of Gaucher disease. The idea that there is only a single class of binding sites for *any* large glycoprotein ligand is without precedent. Obviously, proteins bind with more or less avidity to any cell surface. The "evidence" that there is only a single binding site consists of nothing more than an interpretation of the rates of clearance of alglucerase from plasma after infusion at various rates (Barton *et al.*, 1991b). This obviously tells us next to nothing about the number, distribution, or avidity of binding sites. Such information can only be obtained by performing classical binding studies, and the situation has been clarified by performing such studies which show unequivocally that there are at least two classes of mannose-inhibitable binding sites and additional mannose noninhibitable binding sites (Sato and Beutler, 1993). Interestingly, one of the mannose-inhibitable binding sites exists in high copy numbers on all cells that we have tested, including endothelium. Their avidity for alglucerase is one order of magnitude less than that of the classic binding sites that are largely limited to macrophages.

*Home therapy for Gaucher disease should not be started until the patient has received therapy for 9 to 12 months in a medical setting.*

This warning, issued by Genzyme Corporation in a letter circulated to physicians, is cloaked with concern for patient safety. It is perhaps not coincidental that adherence to this warning would serve to make it very difficult for most patients to receive the three infusions weekly that we have shown represents a more cost-effective way to treat Gaucher disease with alglucerase. Thus, the warning turns out to be quite self-serving for the manufacturer, who can increase the amount of the product sold if the less-efficient bolus schedules are used.

What are the facts? There have been two instances in which patients have developed apparently life-threatening anaphylactic reactions to the infusion of alglucerase. Clearly, this is a cause for some concern. However, it is notable that both of these episodes occurred in patients receiving large doses of alglucerase on the every-other-week schedule. Untoward reactions have not occurred to date with the small doses that are being given as home therapy (Zimran *et al.*, 1993a). It is notable that even though over a thousand such infusions are being given every month worldwide, not a single anaphylactic reaction has been observed. This is not to suggest that such reactions could not occur. No treatment is 100% safe for every patient in every circumstance. Moreover, in some of our patients catheter infections have occurred, necessitating antibiotic therapy and removal of the catheter. But economic considerations as well as considerations of quality of life have made home therapy the standard of care with a variety of intravenous medications, including antihemophilic globulin and antibiotics. Decisions about the relative safety of home therapy should be based upon data when the recommended therapy is used, not when some other therapy in which massive doses of the drug are being used, and these data provide no support for deferring such treatment for 6 to 12 months.

## References

Achord, D., Brot, F., Gonzalez-Noriega, A., Sly, W., and Stahl, P. (1977). Human beta-glucuronidase. II. Fate of infused human placental beta-glucuronidase in the rat. *Pediatr. Res.* **11**:816–822.

Anonymous (1991). Alglucerase approved for Gaucher disease. *FDA Med. Bull.* **21**:6–7.

Ashkenazi, A., Zaizov, R., and Matoth, Y. (1986). Effect of splenectomy on destructive bone changes in children with chronic (Type I) Gaucher disease. *Eur. J. Pediatr.* **145**:138–141.

Ashwell, G., and Morell, A. G. (1974). The role of surface carbohydrate in the hepatic recognition and transport or circulating glycoproteins. *Adv. Enzymol.* **41**:99–128.

August, C., Palmieri, M., Nowell, P., Elkins, W., D'Angelo, G., Glew, R., and Daniels, L. (1984). Bone marrow transplantation (BMT) in Gaucher's disease. *Pediatr. Res.* **18**:236a.

Barton, N. W., Furbish, F. S., Murray, G. J., Garfield, M., and Brady, R. O. (1990). Therapeutic response to intravenous infusions of glucocerebrosidase in a patient with Gaucher disease. *Proc. Natl. Acad. Sci. USA* **87**:1913–1916.

Barton, N. W., Brady, R. O., Dambrosia, J. M., Di Bisceglie, A. M., Doppelt, S. H., Hill, S. C., Mankin, H. J., Murray, G. J., Parker, R. I., Argoff, C. E., Grewal, R. P., and Yu, K.-T. (1991a). Replacement therapy for inherited enzyme deficiency—Macrophage-targeted glucocerebrosidase for Gaucher's disease. *N. Engl. J. Med.* **324**:1464–1470.

Barton, N. W., Brady, R. O., Murray, G. J., Argoff, C. E., Grewal, R. P., Yu, K.-T., Dambrosia, J. M., DiBisceglie, A. M., Hill, S. C., Parker, R. I., Doppelt, S. H., and Mankin, H. J. (1991b). Enzyme-replacement therapy for Gaucher's disease: Reply. *N. Engl. J. Med.* **325**:1811.

Barton, N. W., Brady, R. O., and Dambrosia, J. M. (1993). Treatment of Gaucher's disease. *N. Engl. J. Med.* **328**:1564–1565.

Beighton, P., Goldblatt, J., and Sacks, S. (1982). Bone involvement in Gaucher disease. *In* "Gaucher Disease: A Century of Delineation and Research" (R. J. Desnick, S. Gatt, and G. A. Grabowski, Eds.), pp. 107–129. A. R. Liss, New York.

Belchetz, P. E., Crawley, J. C. W., Braidman, I. P., and Gregoriadis, G. (1977). Treatment of Gaucher's disease with liposome-entrapped glucocerebroside: Beta-glucosidase. *Lancet* **2**:116–117.

Beutler, E. (1977). Newer aspects of some interesting lipid storage diseases: Tay–Sachs and Gaucher's diseases. *West. J. Med.* **126**:46–54.

Beutler, E., Dale, G. L., Guinto, E., and Kuhl, W. (1977a). Enzyme replacement therapy in Gaucher's disease: Preliminary clinical trial of a new enzyme preparation. *Proc. Natl. Acad. Sci. USA* **74**:4620–4623.

Beutler, E., Dale, G. L., and Kuhl, W. (1977b). Enzyme replacement with red cells. *N. Engl. J. Med.* **296**:942–943.

Beutler, E. (1979). Gaucher disease. *In* "Genetic Diseases Among Ashkenazi Jews" (R. M. Goodman and A. G. Motulsky, Eds.), pp. 157–169. Raven Press, New York.

Beutler, E., Dale, G. L., and Kuhl, W. (1980). Replacement therapy in Gaucher's disease. *In* "Enzyme Therapy in Genetic Diseases: 2" (R. J. Desnick, Ed.), pp. 369–381. A. R. Liss, New York.

Beutler, E. (1981). Enzyme replacement therapy. *TIBS Rev.* **6**:95–97.

Beutler, E. (1983). Selectivity of proteases as a basis for tissue distribution of enzymes in hereditary deficiencies. *Proc. Natl. Acad. Sci. USA* **80**:3767–3768.

Beutler, E., Kuhl, W., and Sorge, J. (1984). Cross-reacting material in Gaucher disease fibroblasts. *Proc. Natl. Acad. Sci. USA* **81**:6506–6510.

Beutler, E., Sorge, J. A., Zimran, A., West, C., Kuhl, W., Westwood, B., and Gelbart, T. (1988). The molecular biology of Gaucher disease. *In* "Lipid Storage Disorders. Biological and Medical Aspects" (R. Salvayre, L. Douste-Blazy, and S. Gatt, Eds.), pp. 19–27. Plenum Press, New York.

Beutler, E. (1991a). Bone marrow transplantation for sickle cell anemia: Summarizing comments. *Semin. Hematol.* **28**:263–267.

Beutler, E. (1991b). Gaucher's disease. *N. Engl. J. Med.* **325**:1354–1360.

Beutler, E., Gelbart, T., Kuhl, W., Sorge, J., and West, C. (1991a). Identification of the second common Jewish Gaucher disease mutation makes possible population based screening for the heterozygous state. *Proc. Natl. Acad. Sci. USA* **88**:10544–10547.

Beutler, E., Kay, A., Saven, A., Garver, P., Thurston, D., Dawson, A., and Rosenbloom, B. (1991b). Enzyme replacement therapy for Gaucher disease. *Blood* **78**:1183–1189.

Beutler, E. (1992). Gaucher disease: New molecular approaches to diagnosis and treatment. *Science* **256**:794–799.

Beutler, E., Gelbart, T., Kuhl, W., Zimran, A., and West, C. (1992a). Mutations in Jewish patients with Gaucher disease. *Blood* **79**:1662–1666.

Beutler, E., West, C., and Gelbart, T. (1992b). Polymorphisms in the human glucocerebrosidase gene. *Genomics* **12**:795–800.

Beutler, E. (1993). Modern diagnosis and treatment of Gaucher's disease. *Am. J. Dis. Child.* **147**:1175–1183.

Beutler, E., Gelbart, T., and West, C. (1993a). Identification of six new Gaucher disease mutations. *Genomics* **15**:203–205.

Beutler, E., Nguyen, N. J., Henneberger, M. W., Smolec, J. M., McPherson, R. A., West, C., and Gelbart, T. (1993b). Gaucher disease: Gene frequencies in the Ashkenazi Jewish population. *Am. J. Hum. Genet.* **52**:85–88.

Beutler, E. (1994). Unpublished.

Beutler, E., Gelbart, T., and Demina, A. (1994). Glucocerebrosidase Mutations in Gaucher disease. *Mol. Med.* **1**:82–92.

Beutler, E., Kuhl, W., and Vaughan, L. M. (1995). Failure of alglucerase infused into Gaucher disease patients to localize in marrow macrophages. *Mol. Med.*, in press.

Beutler, E., and Dale, G. L. (1979). Enzyme replacement therapy. *In* "Covalent and Non-covalent Modulation of Protein Function" (D. Atkinson and C. F. Fox, Eds.), pp. 449–461. Academic Press, New York.

Beutler, E., and Garber, A. M. (1994). Alglucerase for Gaucher disease: Dose, costs and benefits. *PharmacoEconomics* **5**:453–459.

Beutler, E., and Gelbart, T. (1990). Gaucher disease associated with a unique KpnI restriction site: Identification of the amino acid substitution. *Ann. Hum. Genet.* **54**:149–153.

Beutler, E., and Gelbart, T. (1993). Gaucher disease mutations in non-Jewish patients. *Br. J. Haematol.* **85**:401–405.

Beutler, E., and Gelbart, T. (1994). Two new Gaucher disease mutations. *Hum. Genet.* **93**:209–210.

Beutler, E., and Sorge, J. (1990). Gene transfer in the treatment of hematologic disease. *Exp. Hematol.* **18**:857–860.

Bijman, J., De Jonge, H., and Wine, J. (1988). Cystic fibrosis advantage. *Nature* **336**:430.

Brady, R. O., Gal, A. E., and Pentchev, P. G. (1974a). Evolution of enzyme replacement therapy for lipid storage diseases. *Life Sci.* **7**:1235–1248.

Brady, R. O., Pentchev, P. G., Gal, A. E., Hibbert, S. R., and Dekaban, A. S. (1974b). Replacement therapy for inherited enzyme deficiency. Use of purified glucocerebrosidase in Gaucher's disease. *N. Engl. J. Med.* **291**:989–993.

Brady, R. O., Pentchev, P. G., and Gal, A. E. (1975). Investigations in enzyme replacement therapy in lipid storage diseases. *Fed. Proc.* **34**:1310–1315.

Choudary, P. V., Barranger, J. A., Tsuji, S., Mayor, J., LaMarca, M. E., Cepko, C. L., Mulligan, R. C., and Ginns, E. I. (1986a). Retrovirus-mediated transfer of the human glucocerebrosidase gene to Gaucher fibroblasts. *Mol. Biol. Med.* **3**:293–299.

Choudary, P. V., Horowitz, M., Barranger, J. A., and Ginns, E. I. (1986b). Gene transfer and expression of active human glucocerebrosidase in mammalian cell cultures. *DNA* **5**:78.

Commens, K. L., Choong, R., and Jaworski, R. (1988). Collodion babies with Gaucher's disease. *Arch. Dis. Child.* **63**:854–865.

Conradi, N., Kyllerman, M., Månsson, J.-E., Percy, A. K., and Svennerholm, L. (1991). Late-infantile Gaucher disease in a child with myoclonus and bulbar signs: Neuropathological and neurochemical findings. *Acta Neuropathol. (Berlin)* **82**:152–157.

Conradi, N. G., Sourander, P., Nilsson, O., Svennerholm, L., and Erikson, A. (1984). Neuropathology of the Norbottnian type of Gaucher disease: Morphological and biochemical studies. *Acta Neuropathol. (Belrin)* **65**:99–100.

Correll, P. H., Colilla, S., Dave, H. P. G., and Karlsson, S. (1992). High levels of human glucocerebrosidase activity in macrophages of long-term reconstituted mice after retroviral infection of hematopoietic stem cells. *Blood* **80**:331–336.

Dale, G. L., and Beutler, E. (1982). Enzyme therapy in Gaucher disease: Clinical trials and model system studies. *In* "Advances in the Treatment of Inborn Errors of Metabolism" (M. A. Crawford, D. A. Gibbs, and R. W. E. Watts, Eds.), pp. 77–91. Wiley, New York.

Davidson, I., and Wells, B. B. (1962). *Clinical Diagnosis by Laboratory Methods*. W. B. Saunders, Philadelphia.

De Duve, C. (1964). From cytases to lysosomes. *Fed. Proc.* **23**:1045–1049.

Doebber, T. W., Wu, M. S., Bugianesi, R. L., Ponpipom, M. M., Furbish, F. S., Barranger, J. A., Brady, R. O., and Shen, T. Y. (1982). Enhanced macrophage uptake of synthetically glycosylated human placental beta-glucocerebrosidase. *J. Biol. Chem.* **257**:2193–2199.

Erikson, A. (1986). Gaucher disease. Norrbottnian Type (III). Neuropaediatric and neurobiological aspects of clinical patterns and treatment. *Acta Paediatr. Scand.* **326**:7–42.

Erikson, A., Groth, C. G., Månsson, J.-E., Percy, A., Ringdén, O., and Svennerholm, L. (1990). Clinical and biochemical outcome of marrow transplantation for Gaucher disease of the Norrbottnian type. *Acta Paediatr. Scand.* **79**:680–685.

Eyal, N., Wilder, S., and Horowitz, M. (1990). Prevalent and rare mutations among Gaucher patients. *Gene* **96**:277–283.

Eyal, N., Firon, N., Wilder, S., Kolodny, E. H., and Horowitz, M. (1991). Three unique base pair changes in a family with Gaucher disease. *Hum. Genet.* **87**:328–332.

Fallet, S., Sibille, A., Mendlson, R., Shapiro, D., Hermann, G., and Grabowski, G. A. (1992). Enzyme augmentation in moderate to life-threatening Gaucher disease. *Pediatr. Res.* **31**:496–502.

Figueroa, M. L., Rosenbloom, B. E., Kay, A. C., Garver, P., Thurston, D. W., Koziol, J. A., Gelbart, T., and Beutler, E. (1992). A less costly regimen of alglucerase to treat Gaucher's disease. *N. Engl. J. Med.* **327**:1632–1636.

Fink, J. K., Correll, P. H., Perry, L. K., Brady, R. O., and Karlsson, S. (1990). Correction of glucocerebrosidase deficiency after retroviral-mediated gene transfer into hematopoietic progenitor cells from patients with Gaucher disease. *Proc. Natl. Acad. Sci. USA* **87**:2334–2338.

Fleshner, P. R., Aufses, A. H., Jr., Grabowski, G. A., and Elias, R. (1991). A 27-year experience with splenectomy for Gaucher's disease. *Am J. Surg.* **161**:69–75.

Furbish, F. S., Steer, C. J., Barranger, J. A., Jones, E. A., and Brady, R. O. (1978). The uptake of native and desialylated glucocerebrosidase by rat hepatocytes and Kupffer cells. *Biochem. Biophys. Res. Commun.* **81**:1047–1053.

Furbish, F. S., Steer, C. J., Krett, N. L., and Barranger, J. A. (1981). Uptake and distribution of placental glucocerebrosidase in rat hepatic cells and effects of sequential deglycosylation. *Biochim. Biophys. Acta* **673**:425–434.

Ginns, E. I., Choudary, P. V., Tsuji, S., Martin, B., Stubblefield, B., Sawyer, J., Hozier, J., and Barranger, J. A. (1985). Gene mapping and leader polypeptide sequence of human glucocerebrosidase: Implications for Gaucher disease. *Proc. Natl. Acad. Sci. USA* **82**:7101–7105.

Ginsberg, H., Grabowski, G. A., Gibson, J. C., Fagerstrom, R., Goldblatt, J., Gilbert, H. S., and Desnick, R. J. (1984). Reduced plasma concentrations of total, low density lipoprotein and high density lipoprotein cholesterol in patients with Gaucher type I disease. *Clin. Genet.* **26**:109–116.

Glenn, D., Gelbart, T., and Beutler, E. (1994). Tight linkage of pyruvate kinase (*PKLR*) and glucocerebrosidase (*GBA*) genes. *Hum. Genet.* **93**:635–638.

Goldblatt, J., Sacks, S., Dall, D., and Beigton, P. (1988). Total hip arthoplasty in Gaucher's disease. Long-term prognosis. *Clin. Orthop.* **228**:94–98.

Goldblatt, J., and Beighton, P. (1982). South African variants of Gaucher disease. *In* "Gaucher Disease: A Century of Delineation and Research" (R. J. Desnick, S. Gatt, and G. A. Grabowski, Eds.), pp. 95–106. A. R. Liss, New York.

Grabowski, G. A. (1993). Treatment of Gaucher's disease. *N. Engl. J. Med.* **328**:1565.

Grabowski, G. A., Pastores, G., Brady, R. O., and Barton, N. (1993). Gaucher disease type I: Safety and efficacy of macrophage targeted recombinant glucocerebrosidase therapy. *Clin. Res.* **41**:390a.

Grace, M. E., Graves, P. N., Smith, F. I., and Grabowski, G. A. (1990). Analyses of catalytic activity and inhibitor binding of human acid β-glucosidase by site-directed mutagenesis. Identification of residues critical to catalysis and evidence for causality of two Ashkenazi Jewish Gaucher disease Type 1 mutations. *J. Biol. Chem.* **265**:6827–6835.

Graves, P. N., Grabowski, G. A., Eisner, R., Palese, P., and Smith, F. I. (1988). Gaucher disease type 1: Cloning and characterization of a cDNA encoding acid β-glucosidase from an Ashkenazi Jewish patient. *DNA* **7**:521–528.

Gregoriadis, G., Weereratne, H., Blair, H., and Bull, G. M. (1982). Liposomes in Gaucher type I disease: Use in enzyme therapy and the creation of an animal model. *In* "Gaucher Disease: A Century of Delineation and Research" (R. J. Desnick, S. Gatt, and G. A. Grabowski, Eds.), pp. 681–701. A. R. Liss, New York.

Grewal, R. P., Doppelt, S. H., Thompson, M. A., Katz, D., Brady, R. O., and Barton, N. W.

(1991). Neurologic complications of nonneuronopathic Gaucher's disease. *Arch. Neurol.* **48**:1271–1272.

Guzzetta, P. C., Connors, R. H., Fink, J., and Barranger, J. A. (1987). Operative technique and results of subtotal splenectomy for Gaucher disease. *Surg. Gynecol. Obstet.* **164**:359–362.

Guzzetta, P. C., Ruley, E. J., Merrick, H. F. W., Verderese, C., and Barton, N. (1990). Elective subtotal splenectomy. Indications and results in 33 patients. *Ann. Surg.* **211**:34–42.

Harinck, H. I. J., Bijvoet, O. L. M., van der Meer, J. W. H., Jones, B., and Onvlee, G. J. (1984). Regression of bone lesions in Gaucher's disease during treatment with aminohydroxypropylidene bisphosphonate. *Lancet* **2**:513.

He, G.-S., Grace, M. E., and Grabowski, G. A. (1992). Gaucher disease: Four rare missense mutations encoding F213I, F289Y, T323I and R463C in type I variants. *Hum. Mutat.* **1**:423–427.

Hobbs, J. R. (1987). Experience with bone marrow transplantation for inborn errors of metabolism. *Enzyme* **38**:194–206.

Hobbs, J. R., Shaw, P. J., Jones, K. H., Lindsay, I., and Hancock, M. (1987). Beneficial effect of pre-transplant splenectomy on displacement bone marrow transplantation for Gaucher's syndrome. *Lancet* **1**:1111–1115.

Holcomb, G. W., III, and Greene, H. L. (1993). Fatal hemorrhage caused by disease progression after partial splenectomy for type III Gaucher's disease. *J. Pediatr. Surg.* **28**:1572–1574.

Hollak, C. E. M., Aerts, J. M. F. G., van Weely, S., Phoa, S. S. K. S., Goudsmit, R., von dem Borne, A. E. G. K., and van Oers, M. H. J. (1993). Enzyme supplementation therapy for type I Gaucher disease. Efficacy of very low dose alglucerase in 12 patients. *Blood* **82**:33a.

Hollak, C. E. M. (1994). Personal communication.

Hong, C. M., Ohashi, T., Yu, X. J., Weiler, S., and Barranger, J. A. (1990). Sequence of two alleles responsible for Gaucher disease. *DNA Cell Biol.* **9**:233–241.

Horowitz, M., Wilder, S., Horowitz, Z., Reiner, O., Gelbart, T., and Beutler, E. (1989). The human glucocerebrosidase gene and pseudogene: Structure and evolution. *Genomics* **4**:87–96.

Horowitz, M., Tzuri, G., Eyal, N., Berebi, A., Kolodny, E. H., Brady, R. O., Barton, N. W., Abrahamov, A., and Zimran, A. (1993). Prevalence of nine mutations among Jewish and non-Jewish Gaucher disease patients. *Am. J. Hum. Genet.* **53**:921–930.

Kanno, H., Fujii, H., Hirono, A., Omine, M., and Miwa, S. (1992). Identical point mutations of the R-type pyruvate kinase (PK) cDNA found in unrelated PK variants associated with hereditary hemolytic anemia. *Blood* **79**:1347–1350.

Kanno, H., Fujii, H., and Miwa, S. (1993). Low substrate affinity of pyruvate kinase variant (PK Sapporo) caused by a single amino acid substitution (426 Arg → Gln) associated with hereditary hemolytic anemia. *Blood* **81**:2439–2441.

Karlsson, S. (1991). Treatment of genetic defects in hematopoietic cell function by gene transfer. *Blood* **78**:2481–2492.

Kawame, H., Hasegawa, Y., Eto, Y., and Maekawa, K. (1992). Rapid identification of mutations in the glucocerebrosidase gene of Gaucher disease patients by analysis of single-strand conformation polymorphisms. *Hum. Genet.* **90**:294–296.

Kawame, H., and Eto, Y. (1991). A new glucocerebrosidase-gene missense mutation responsible for neuronopathic Gaucher disease in Japanese patients. *Am. J. Hum. Genet.* **49**:1378–1380.

Kay, A. C., Saven, A., Garver, P., Thurston, D. W., Rosenbloom, B. F., and Beutler, E. (1991). Enzyme replacement therapy in type I Gaucher disease. *Trans. Assoc. Am. Phys.* **104**:258–264.

Kohn, D. B., Nolta, J. A., Weinthal, J., Bahner, I., Yu, X. J., Lilley, J., and Crooks, G. M. (1991). Toward gene therapy for Gaucher disease. *Hum. Gene Ther.* **2**:101–105.

Kolodny, E. H., Ullman, M. D., Mankin, H. J., Raghavan, S. S., Topol, J., and Sullivan, J. L. (1982). Phenotypic manifestations of Gaucher disease: Clinical features in 48 biochemically verified Type I patients and comment on Type II patients. *In* "Gaucher Disease: A Century of Delineation and Research" (R. J. Desnick, S. Gatt, and G. A. Grabowski, Eds.), pp. 33–65. A. R. Liss, New York.

Kyllerman, M., Conradi, N., Månsson, J.-E., Percy, A. K., and Svennerholm, L. (1990). Rapidly progressive type III Gaucher disease: Deterioration following partial splenectomy. *Acta Paediatr. Scand.* **79:**448–453.

Latham, T. E., Theophilus, B. D. M., Grabowski, G. A., and Smith, F. I. (1991). Heterogeneity of mutations in the acid β-glucosidase gene of Gaucher disease patients. *DNA Cell Biol.* **10:**15–21.

Laubscher, K. H., Glew, R. H., Lee, R. E., and Okinaka, R. T. (1994). Use of denaturing gradient gel electrophoresis to identify mutant sequences in the β-glucosidase gene. *Hum. Mutat.* **3:**411–415.

Lee, R. E. (1982). The pathology of Gaucher disease. *In* "Gaucher Disease: A Century of Delineation and Research" (R. J. Desnick, S. Gatt, and G. A. Grabowski, Eds.), pp. 177–217. A. R. Liss, New York.

Levy, H., Or, A., Eyal, N., Wilder, S., Widgerson, M., Kolodny, E. H., Zimran, A., and Horowitz, M. (1991). Molecular aspects of Gaucher disease. *Dev. Neurosci.* **13:**352–362.

Lipson, A. H., Rogers, M., and Berry, A. (1991). Collodion babies with Gaucher's disease—A further case. *Arch. Dis. Child.* **66:**667.

Lonnqvist, B., Ringden, O., Wahren, B., Gahrton, G., and Lundgren, G. (1984). Cytomegalovirus infection associated with and preceding chronic graft-versus-host disease. *Transplantation* **38:** 465–468.

Luzzatto, L. (1979). Genetics of red cells and susceptibility to malaria. *Blood* **54:**961–976.

Medoff, A. S., and Bayrd, E. D. (1954). Gauchers disease in 29 cases: Hematologic complications and effect of splenectomy. *Ann. Intern. Med.* **40:**481–492.

Myers, B. (1937). Gaucher's disease of the lung. *BMJ* **X:**8–10.

Nilsson, O., Håkansson, G., Dreborg, S., Groth, C. G., and Svennerholm, L. (1982). Increased cerebroside concentration in plasma and erythrocytes in Gaucher disease: Significant differences between type I and type III. *Clin. Genet.* **22:**274–279.

Nilsson, O., Grabowski, G. A., Ludman, M. D., Desnick, R. J., and Svennerholm, L. (1985). Glycosphingolipid studies of visceral tissues and brain from type 1 Gaucher disease variants. *Clin. Genet.* **27:**443–450.

Nilsson, O., and Svennerholm, L. (1982). Accumulation of glucosylceramide and glucosylsphingosine (psychosine) in cerebrum and cerebellum in infantile and juvenile Gaucher disease. *J. Neurochem.* **39:**709–718.

Nolta, J. A., Sender, L. S., Barranger, J. A., and Kohn, D. B. (1990). Expression of human glucocerebrosidase in murine long-term bone marrow cultures after retroviral vector-mediated transfer. *Blood* **75:**787–797.

Ohashi, T., Boggs, S., Robbins, P., Bahnson, A., Patrene, K., Wei, F.-S., Wei, J.-F., Li, J., Lucht, L., Fei, Y., Clark, S., Kimak, M., He, H., Mowery-Rushton, P., and Barranger, J. A. (1992). Efficient transfer and sustained high expression of the human glucocerebrosidase gene in mice and their functional macrophages following transplantation of bone marrow transduced by a retroviral vector. *Proc. Natl. Acad. Sci. USA* **89:**11332–11336.

Ohshima, T., Sasaki, M., Matsuzaka, T., and Sakuragawa, N. (1993). A novel splicing abnormality in a Japanese patient with Gaucher's disease. *Hum. Mol. Genet.* **2:**1497–1498.

O'Neill, R. R., Tokoro, T., Kozak, C. A., and Brady, R. O. (1989). Comparison of the chromosomal localization of murine and human glucocerebrosidase genes and of the deduced amino acid sequences. *Proc. Natl. Acad. Sci. USA* **86:**5049–5053.

Ostlere, L., Warner, T., Meunier, P. J., Hulme, P., Hesp, R., Watts, R. W. E., and Reeve, J. (1991). Treatment of type 1 Gaucher's disease affecting bone with aminohydroxypropylidene bisphosphonate (pamidronate). *Q. J. Med.* **79:**503–515.

Pastores, G. M., Sibille, A. R., and Grabowski, G. A. (1993). Enzyme therapy in Gaucher disease type 1: Dosage efficacy and adverse effects in thirty-three patients treated for 6 to 24 months. *Blood* **82**:408–416.

Patterson, M. C., Horowitz, M., Abel, R. B., Currie, J. N., Yu, K.-T., Kaneski, C., Higgins, J. J., O'Neill, R. R., Fedio, P., Pikus, A., Brady, R. O., and Barton, N. W. (1993). Isolated horizontal supranuclear gaze palsy as a marker of severe systemic involvement in Gaucher's disease. *Neurology* **43**:1993–1997.

Pentchev, P. G., Brady, R. O., Gal, A. E., and Hibbert, S. R. (1975). Replacement therapy for inherited enzyme deficiency. Sustained clearance of accumulated glucocerebroside in Gaucher's disease following infusion of purified glucocerebrosidase. *J. Mol. Med.* **1**:73–78.

Pentchev, P. G., Kusiak, J. W., Barranger, J. A., Furbish, F. S., Rapoport, S. I., Massey, J. M., and Brady, R. O. (1978). Factors that influence the uptake and turnover of glucocerebrosidase and alpha-galactosidase in mammalian liver. *Adv. Exp. Med. Biol.* **101**:745–752.

Rappeport, J. M., and Ginns, E. I. (1984). Bone-marrow transplantation in severe Gaucher Disease. *N. Engl. J. Med.* **311**:84–88.

Reiner, O., Wigderson, M., and Horowitz, M. (1988). Structural analysis of the human glucocerebrosidase genes. *DNA* **7**:107–116.

Reiner, O., and Horowitz, M. (1988). Differential expression of the human glucocerebrosidase-coding gene. *Gene* **73**:469–478.

Ringdén, O., Groth, C.-G., Erikson, A., Bäckman, L., Granqvist, S., Måansson, J.-E., and Svennerholm, L. (1988). Long-term follow-up of the first successful bone marrow transplantation in Gaucher disease. *Transplantation* **46**:66–70.

Roberts, W. C., and Fredrickson, D. S. (1967). Gaucher's disease of the lung causing severe pulmonary hypertension with associated acute recurrent pericarditis.*Criculation* **35**:783–789.

Rodgers, B. M., Tribble, C., and Joob, A. (1987). Partial splenectomy for Gaucher's disease. *Ann. Surg.* **205**:693–698.

Rose, J. S., Grabowski, G. A., Barnett, S. H., and Desnick, R. J. (1982). Accelerated skeletal deterioration after splenectomy in Gaucher type 1 disease. *Am. J. Roentgenol.* **139**:1202–1204.

Rotter, J. I., and Diamond, J. M. (1987). What maintains the frequencies of human genetic diseases? *Nature* **329**:289.

Samama, G., Brefort, J. L., Dolley, M., and Leporrier, M. (1989). Huge splenomegaly in Gaucher's disease. Treatment by embolization before splenectomy. *Presse Med.* **18**:1078–1079.

Sato, Y., and Beutler, E. (1993). Binding, internalization, and degradation of mannose-terminated glucocerebrosidase by macrophages. *J. Clin. Invest.* **91**:1909–1917.

Schneider, E. L., Epstein, C. J., Kaback, M. J., and Brandes, D. (1977). Severe pulmonary involvement in adult Gaucher's disease. *Am. J. Med.* **63**:475–480.

Shafit-Zagardo, B., Devine, E. A., Smith, M., Arredondo, G. A. F., and Desnick, R. J. (1981). Assignment of the gene for acid beta-glucosidase to human chromosome 1. *Am. J. Hum. Genet.* **33**:564–575.

Sibille, A., Eng, C. M., Kim, S.-J., Pastores, G., and Grabowski, G. A. (1993). Phenotype/genotype correlations in Gaucher disease type I: Clinical and therapeutic implications. *Am. J. Hum. Genet.* **52**:1094–1101.

Sidransky, E., Tsuji, S., Martin, B. M., Stubblefield, B., and Ginns, E. I. (1992). DNA mutation analysis of Gaucher patients. *Am. J. Med. Genet.* **42**:331–336.

Smith, R. L., Hutchins, G. M., Sack, G. H., Jr., and Ridolfi, R. L. (1978). Unusual cardiac, renal and pulmonary involvement in Gaucher's disease. Interstitial glucocerebroside accumulation, pulmonary hypertension and fatal bone marrow embolization. *Am. J. Med.* **65**:352–360.

Sorge, J., Gelbart, T., West, C., Westwood, B., and Beutler, E. (1985a). Heterogeneity in type I Gaucher disease demonstrated by restriction mapping of the gene. *Proc. Natl. Acad. Sci. USA* **82**:5442–5445.

Sorge, J., West, C., Westwood, B., and Beutler, E. (1985b). Molecular cloning and nucleotide sequence of the human glucocerebrosidase gene. *Proc. Natl. Acad. Sci. USA* **82**:7289–7293.

Sorge, J., Kuhl, W., West, C., and Beutler, E. (1987a). Complete correction of the enzymatic defect of type I Gaucher disease fibroblasts by retroviral-mediated gene transfer. *Proc. Natl. Acad. Sci. USA* **84**:906–909.

Sorge, J. A., West, C., Kuhl, W., Treger, L., and Beutler, E. (1987b). The human glucocerebrosidase gene has two functional ATG initiator codons. *Am. J. Hum. Genet.* **41**:1016–1024.

Sorge, J., Gross, E., West, C., and Beutler, E. (1990). High level transcription of the glucocerebrosidase pseudogene in normal subjects and patients with Gaucher disease. *J. Clin. Invest.* **86**:1137–1141.

Steer, C. J., Furbish, F. S., Barranger, J. A., Brady, R. O., and Jones, E. A. (1978). The uptake of agalacto-glucocerebroside by rat hepatocytes and Kupffer cells. *FEBS Lett.* **91**:202–205.

Storb, R., Anasetti, C., Appelbaum, F., Bensinger, W., Buckner, C. D., Clift, R., Deeg, H. J., Doney, K., Hansen, J., Loughran, T., Martin, P., Pepe, M., Petersen, F., Sanders, J., Singer, J., Stewart, P., Sullivan, K. M., Witherspoon, R., and Thomas, E. D. (1991). Marrow transplantation for severe aplastic anemia and thalassemia major. *Semin. Hematol.* **28**:235–239.

Stratford-Perricaudet, L. D., Makeh, I., Perricaudet, M., and Briand, P. (1992). Widespread long-term gene transfer to mouse skeletal muscles and heart. *J. Clin. Invest.* **90**:626–630.

Sullivan, K. M., and Reid, C. D. (1991). Introduction to a symposium on sickle cell anemia: Current results of comprehensive care and the evolving role of bone marrow transplantation. *Semin. Hematol.* **28**:177–179.

Sun, C. C., Panny, S., Combs, J., and Gutberlett, R. (1984). Hydrops fetalis associated with Gaucher disease. *Pathol. Res. Pract.* **179**:101–104.

Svennerholm, L., Dreborg, S., Erikson, A., Groth, C. G., Hillborg, P. O., Håkansson, G., Nilsson, O., and Tibblin, E. (1982). Gaucher disease of the Norrbottnian type (type III). Phenotypic manifestations. *In* "Gaucher Disease: A Century of Delineation and Research" (R. J. Desnick, S. Gatt, and G. A. Grabowski, Eds.), pp. 67–94. A. R. Liss, New York.

Svennerholm, L., Erikson, A., Groth, C. G., Ringdén, O., and Månsson, J.-E. (1991). Norrbottnian type of Gaucher disease—Clinical, biochemical and molecular biology aspects: Successful treatment with bone marrow transplantation. *Dev. Neurosci.* **13**:345–351.

Tani, K., Fujii, H., Tsutsumi, H., Sukegawa, J., Toyoshima, K., Yoshida, M. C., Noguchi, T., Tanaka, T., and Miwa, S. (1987). Human liver type-pyruvate kinase: cDNA cloning and chromosomal assignment. *Biochem. Biophys. Res. Commun.* **143**:431–438.

Thanopoulos, B. D., Frimas, C. A., Mantagos, S. P., and Beratis, N. G. (1987). Gaucher disease: Treatment of hypersplenism with splenic embolization. *Acta Paediatr. Scand.* **76**:1003–1007.

Theophilus, B. D. M., Latham, T., Grabowski, G. A., and Smith, F. I. (1989). Comparison of RNase A, chemical cleavage, and GC-clamped denaturing gradient gel electrophoresis for the detection of mutations in exon 9 of the human acid β-glucosidase gene. *Nucleic Acids Res.* **17**:7707–7722.

Tsai, P., Lipton, J. M., Sahdev, I., Najfeld, V., Rankin, L. R., Slyper, A. H., Ludman, M., and Grabowski, G. A. (1992). Allogenic bone marrow transplantation in severe Gaucher disease. *Pediatr. Res.* **31**:503–507.

Tsuji, S., Choudary, P. V., Martin, B. M., Winfield, S., Barranger, J. A., and Ginns, E. I. (1986). Nucleotide sequence of cDNA containing the complete coding sequence for human lysosomal glucocerebrosidase. *J. Biol. Chem.* **261**:50–53.

Tsuji, S., Choudary, P. V., Martin, B. M., Stubblefield, B. K., Mayor, J. A., Barranger, J. A., and Ginns, E. I. (1987). A mutation in the human glucocerebrosidase gene in neuronopathic Gaucher's disease. *N. Engl. J. Med.* **316**:570–575.

Tsuji, S., Martin, B. M., Barranger, J. A., Stubblefield, B. K., LaMarca, M. E., and Ginns, E. I. (1988). Genetic heterogeneity in type 1 Gaucher disease: Multiple genotypes in Ashkenazic and non-Ashkenazic individuals. *Proc. Natl. Acad. Sci. USA* **85**:2349–2352, 5708.

Tybulewicz, V. L. J., Tremblay, M. L., LaMarca, M. E., Willemsen, R., Stubblefield, B. K., Winfield, S., Zablocka, B., Sidransky, E., Martin, B. M., Huang, S. P., Mintzer, K. A., Westphal, H., Mulligan, R. C., and Ginns, E. I. (1992). Animal model of Gaucher's disease from targeted disruption of the mouse glucocerebrosidase gene. *Nature* **357**:407–410.

van Weely, S., Van den Berg, M., Barranger, J. A., Sa Miranda, M. C., Tager, J. M., and Aerts, J. M. F. G. (1993). Role of pH in determining the cell-type-specific residual activity of glucocerebrosidase in type 1 Gaucher disease. *J. Clin. Invest.* **91**:1167–1175.

Walley, A. J., and Harris, A. (1993). A novel point mutation (D380A) and a rare deletion (1255del55) in the glucocerebrosidase gene causing Gaucher's disease. *Hum. Mol. Genet.* **2**: 1737–1738.

Weil, D., Bernard, M., Combates, N., Wirtz, M. K., Hollister, D. W., Steinmann, B., and Ramirez, F. (1988). Identification of a mutation that causes exon skipping during collagen pre-mRNA splicing in an Ehlers–Danlos syndrome variant. *J. Biol. Chem.* **263**:8561–8564.

Weinthal, J., Nolta, J. A., Yu, X. J., Lilley, J., Uribe, L., and Kohn, D. B. (1991). Expression of human glucocerebrosidase following retroviral vector-mediated transduction of murine hematopoietic stem cells. *Bone Marrow Transplant.* **8**:403–412.

Wigderson, M., Firon, N., Horowitz, Z., Wilder, S., Frishberg, Y., Reiner, O., and Horowitz, M. (1989). Characterization of mutations in Gaucher patients by cDNA cloning. *Am. J. Hum. Genet.* **44**:365–377.

Williams, W. J., and Nelson, D. A. (1990). Examination of the marrow. In "Hematology" (W. J. Williams, E. Beutler, A. J. Erslev, and M. A. Lichtman, Eds.), pp. 24–31. McGraw–Hill, New York.

Zhu, N., Liggitt, D., Liu, Y., and Debs, R. (1993). Systemic gene expression after intravenous DNA delivery into adult mice. *Science* **261**:209–211.

Zimran, A., Sorge, J., Gross, E., Kubitz, M., West, C., and Beutler, E. (1989). Prediction of severity of Gaucher's disease by identification of mutations at DNA level. *Lancet* **2**:349–352.

Zimran, A., Sorge, J., Gross, E., Kubitz, M., West, C., and Beutler, E. (1990). A glucocerebrosidase fusion gene in Gaucher disease. Implications for the molecular anatomy, pathogenesis and diagnosis of this disorder. *J. Clin. Invest.* **85**:219–222.

Zimran, A., Hadas-Halpern, I., and Abrahamov, A. (1991). Enzyme replacement therapy for Gaucher's disease. *N. Engl. J. Med.* **325**:1810–1811.

Zimran, A., Kay, A. C., Gelbart, T., Garver, P., Saven, A., and Beutler, E. (1992). Gaucher disease: Clinical, laboratory, radiologic and genetic features of 53 patients. *Medicine (Baltimore)* **71**:337–353.

Zimran, A., Hollak, C. E. M., Abrahamov, A., van Oers, M. H. J., Kelly, M., and Beutler, E. (1993a). Home treatment with intravenous enzyme replacement therapy for Gaucher disease: An international collaborative study of 33 patients. *Blood* **82**:1107–1109.

Zimran, A., Kannai, R., Cohen, Y., Zevin, S., Hadas-Halpern, I., and Abrahamov, A. (1993b). Low dose enzyme replacement therapy for patients with Gaucher disease: Effects of age, sex, genotype and splenectomy on response to treatment in 29 patients. *Blood* **82**:33a.

Zimran, A., and Beutler, E. (1994). Unpublished.

# 3

# The Genetics of Non-insulin-Dependent Diabetes Mellitus

**T. S. Pillay, W. J. Langlois, and J. M. Olefsky**
Division of Endocrinology & Metabolism
Department of Medicine
University of California, San Diego
La Jolla, California 92093
and the Veteran's Administration Medical Center
Research Service
San Diego, California 92161

## I. INTRODUCTION

Non-insulin-dependent diabetes (NIDDM) is a genetically heterogeneous disorder of glucose homeostasis that affects approximately 5% of people in Westernized countries. A variety of biochemical abnormalities have been identified in NIDDM, and the relative contribution of these different physiologic or cellular defects differs among patient groups. However, regardless of the pathophysiologic sequence in an individual patient, once the full-blown hyperglycemic situation develops, a characteristic set of metabolic derangements can be identified in the great majority of NIDDM patients (Olefsky, 1989; DeFronzo et al., 1992). These consist of abnormalities in the liver, peripheral insulin target tissues, and the pancreatic islets, which in aggregate represent the final common metabolic pathway for the pathogenesis of hyperglycemia in NIDDM (Olefsky, 1991). Figure 3.1 summarizes these abnormalities. The role of the liver in the pathogenesis of NIDDM is overproduction of glucose. Increased basal hepatic glucose production is a characteristic feature of essentially all NIDDM patients with fasting hyperglycemia (Olefsky, 1991). The figure depicts skeletal muscle as the prototypical peripheral insulin target tissue, since in the *in vivo* insulin-stimulated state, 80–90% of all glucose uptake is into skeletal muscle. Target tissues are insulin resistant in NIDDM, and this has been described in many

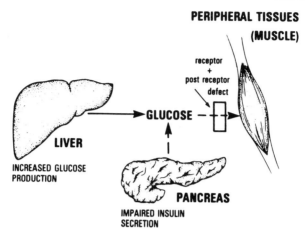

**PERIPHERAL TISSUES (MUSCLE)**

**Figure 3.1.** The pathogenesis of NIDDM. Summary of the metabolic abnormalities which contribute to hyperglycemia. Increased hepatic glucose production, impaired insulin secretion, and insulin resistance due to receptor and postreceptor defects all combine to generate the hyperglycemic state. Reproduced with permission from Olefsky (1989).

different populations. Finally, abnormal islet cell function plays a central role in the development of hyperglycemia during the natural history of NIDDM (Porte, 1991). Decreased β-cell function and increased glucagon secretion are frequent concomitants of the NIDDM state. In aggregate, these metabolic abnormalities are responsible for the hyperglycemic condition in patients with NIDDM.

## II. NATURAL HISTORY OF THE DEVELOPMENT OF NIDDM

The metabolic abnormalities depicted in Figure 3.1 represent a single point in time, following development of the full NIDDM syndrome. However, this analysis does not elucidate the progressive development of this disease. The latter is a subject which has received considerable attention, and Figure 3.2 represents a schematic representation of the natural history, or progression, toward NIDDM. One begins with insulin resistance, which can be genetic or acquired, or both. Many lines of evidence demonstrate that those individuals who evolve to overt NIDDM manifest only insulin resistance in the prediabetic state (Olefsky, 1991). Acquired factors, such as obesity and a sedentary life style, may be additive, but insulin resistance is considered to be a primary inherited component of the disease. This concept comes from a verity of studies, all demonstrating that prediabetic patients are characterized by insulin resistance with normal hepatic glucose metabolism and β-cell function many years prior to the development of

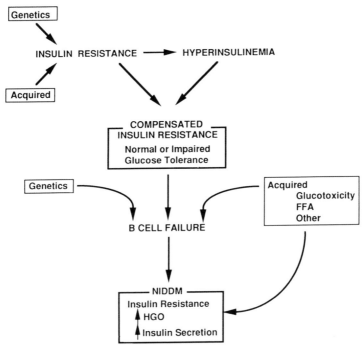

**Figure 3.2.** The etiology of NIDDM. Proposed etiology for the development of NIDDM. Reproduced with permission from Olefsky (1989).

NIDDM. Given a primary state of insulin resistance, if β-cell function is normal, this will lead to compensatory hyperinsulinemia which, in turn, maintains relatively normal glucose homeostasis. Thus, in the compensated insulin-resistant state, one has either normal glucose tolerance or impaired glucose tolerance (IGT), but not diabetes. A subpopulation of individuals with this compensated insulin-resistant state then goes on to develop NIDDM. The magnitude of this subpopulation depends on the particular ethnic groups studied.

Following the transition from the compensated state to frank NIDDM, at least three pathophysiologic changes can be observed (Weir and Leahy, 1994). First, there is a marked decrease in β-cell function and insulin secretion. Whether this is caused by preprogrammed genetic abnormalities in β-cell function and/or by acquired insults, such as that due to the chronic effects of mild hyperglycemia referred to as "glucotoxicity," or both, remains to be elucidated. Regardless, a marked decrease in insulin secretion accompanies this transition and is clearly a major contributor to the development of the hyperglycemic state. In other words, the β-cell can no longer compensate for this insulin resistance, and as the plasma insulin levels decline, plasma glucose levels increase.

A second metabolic abnormality occurs in the liver. Patients with IGT have normal basal rates of hepatic glucose production, whereas patients with fasting hyperglycemia have increased rates. The exact mechanisms of this increase in hepatic glucose output remain to be fully elucidated, but the majority of this increased glucose production is accounted for by an increase in gluconeogenesis (Weir and Leahy, 1994). Whatever the exact mechanisms, it is quite clear that this hepatic abnormality is a secondary, rather than a primary, event.

The third metabolic abnormality which occurs following the transition to NIDDM is worsening of the insulin-resistant state. Whether this increment in insulin resistance is secondary to glucose toxicity or other acquired factors remains to be determined.

## III. THE ROLE OF GENETIC FACTORS IN THE ETIOLOGY OF NIDDM

The contribution of genetic abnormalities to the etiology of NIDDM is best illustrated by studies of identical twins. In twin pairs with one affected individual, over 90% of the cotwins developed NIDDM (Barnett et al., 1981). Although the discordant twins did not have NIDDM, they had mild glucose intolerance and an abnormal insulin response, suggesting the possible eventual development of overt NIDDM, which would then drive the concordance rate to 100%. Despite the high concordance rate in twins, it is obvious that NIDDM is not simply the result of a single gene defect, since the incidence of NIDDM in first and second degree relatives is lower than one would expect in such a case (Barnett et al., 1981; Warram et al., 1990). In a 13 year follow-up of 155 offspring of parents who both had NIDDM, only 16% of the offspring developed NIDDM (Warram et al., 1990), which is eight times the risk for a corresponding Caucasian population of similar age, but less than would be expected given a single gene defect (Zimmet, 1992).

Studies in populations with a high prevalence of NIDDM have also provided relevant data to support the genetic basis and primary role of insulin resistance in the etiology of this disease. The prevalence of NIDDM in the United States is 2–4% for Caucasians, but is 4–6% for African-Americans (Harris et al., 1987), 10–15% for Mexican-Americans (Haffner et al., 1990), and 35% for the Pima native Americans in Arizona (Knowler et al., 1981), the group with the highest incidence of NIDDM in the world. Similarly, in Micronesians from the island of Nauru, familial clustering of age-adjusted glucose tolerance is seen (Serjeantson et al., 1991). Seventy-nine percent of 43 offspring were hyperglycemic when both parents were hyperglycemic, 48% of 77 offspring were hyperglycemic when one parent was diabetic, and when neither was diabetic, only 5% of 20 offspring were hyperglycemic (Serjeantson *et al.*, 1987). Further study of this population demonstrated that those with hyperinsulinemia at baseline were most likely to develop NIDDM (Zimmet, 1992). These data have been confirmed in other population studies, including the study of Pima native Americans in which 80% of the 35–44 year old offspring of Pima native American parents with early onset NIDDM also had diabetes (Knowler *et al.*, 1981; Kadowaki *et al.*, 1984; Haffner *et al.*, 1990; Charles *et al.*, 1991). Decreased insulin-mediated glucose disposal (insulin resistance) in Pima native Americans is associated with hyperinsulinemia and predicts the development of both IGT and NIDDM (Lillioja *et al.*, 1988; Saad *et al.*, 1991).

In summary, the phenotypic manifestations of classical NIDDM involve elevated hepatic glucose output, impaired pancreatic insulin secretion, and peripheral insulin resistance. NIDDM has a strong genetic component, and studies of prediabetic subjects indicate that insulin resistance, accompanied by hyperinsulinemia, exists prior to any deterioration of glucose homeostasis. After a period of compensatory hyperinsulinemia with normal glucose tolerance or IGT, β-cell insulin secretion declines, and overt NIDDM results.

## IV. STRATEGIES FOR IDENTIFYING DIABETES-SUSCEPTIBILITY GENES IN NIDDM

Diabetes-susceptibility genes have been sought using molecular genetic techniques (Weatherall, 1991). In general, two methods have been used. The first is the candidate gene approach and the second involves various techniques of gene mapping (Figure 3.3).

### A. The candidate gene approach

Many proteins are involved in the overall regulation of glucose homeostasis, and the genes for a number of these proteins have been cloned and sequenced. Based on what is known about the pathophysiology of NIDDM, one can hypothesize that genetic defects exist in these proteins, and many studies have been undertaken to search for inherited abnormalities in these "candidate genes" (Table 3.1). Some of these studies have involved direct sequencing of genes, or portions thereof, in NIDDM compared to normal subjects. Although laborious, such studies are rather definitive, in so far as the numbers of subjects studied are sufficient.

Molecular scanning approaches are available (Cotton, 1989) which allow examination of a much larger population base. For example, the technique of single-stranded conformational polymorphism (SSCP) (Orita *et al.*, 1989) has

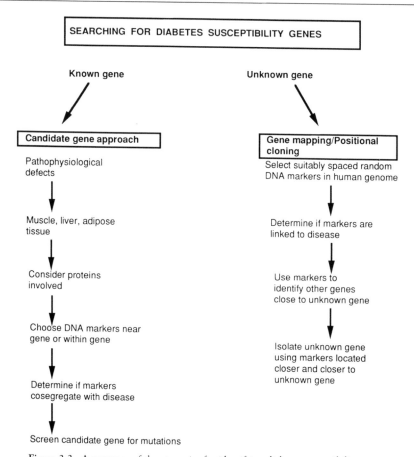

**Figure 3.3.** A summary of the strategies for identifying diabetes-susceptibility genes.

been widely employed as a useful method to detect single nucleotide changes without sequencing the entire gene. With this technique, labeled primers are incorporated into polymerase chain reaction (PCR) products and the strands are denatured and separated by electrophoresis on nondenaturing gels. The method is extremely sensitive and changes in a single base will alter the mobility of a DNA fragment allowing the detection of mutations. Ultimately, the importance of any mutation must be proven by tracking its segregation through affected and nonaffected members of well-defined pedigrees and by expressing the mutant protein *in vitro* and studying its biologic function.

An additional, but less definitive method of studying candidate genes is to evaluate linkage between polymorphic markers in the gene of interest and the

**Table 3.1.** Candidate Genes for NIDDM

Pancreatic
   Insulin
   Islet amyloid polypeptide (IAPP)
   Glucokinase
   Glucokinase regulatory protein
   Glucose transporter 2
   Glucagon-like peptide 1 receptor
Sites of insulin action (liver and muscle)
   Insulin receptor
   Insulin receptor substrate 1
   Glucose transporter 4
   Hexokinase 1 and 2
   Glycogen synthase
   Glucagon receptor
   Phosphoenolpyruvate carboxykinase

NIDDM phenotype (Weatherall, 1991; Cox *et al.*, 1992). If there is linkage between a marker and diabetes, this indicates that the marker and the disease are coinherited. The likelihood of linkage is estimated from the concordance of the distribution of the markers and the phenotype (NIDDM) and is expressed as the logarithm of the odds (LOD) score (Weatherall, 1991). Such linkage studies have been conducted by examining the relationship between polymorphic markers and NIDDM within affected and unaffected members of well-defined multigeneration families, or by studying linkage between the polymorphic genotype and phenotype in a general population of NIDDM and nondiabetic subjects (Cox and Bell, 1989).

## B. Genome mapping approaches

In most multifactorial polygenic diseases, the molecular defects are not understood and, hence, little or nothing is known about the gene(s) involved. It then becomes necessary to use alternative approaches including linkage analysis and "positional cloning" (Wicking and Williamson, 1994). In order to isolate a gene without any information about its protein product, one must determine the chromosomal location of the unknown gene. The large amount of nucleotide variation found in regions of DNA between genes and exons has yielded a substantial amount of useful DNA markers to facilitate the study of the human genome. These random polymorphic genetic markers have been used to "map" or assign the location of disease phenotypes to a particular chromosome in a collection of individuals, either a pedigree or a population. They are scattered throughout the genome and generally occur at a frequency of 1/500 base pairs (bp) (Weatherall, 1991). This frequency implies that several polymorphic loci occur near or within every single gene. For a monogenic disease, about 150 markers separated by approximately 20 million bp (20 centimorgans) are needed to cover the whole genome (Botstein *et al.*, 1980). If the disease is inherited with the DNA marker, then it suggests that the causative agent and the marker are physically close together on a chromosome. Such markers have been used to identify genes which confer susceptibility to NIDDM (Permutt, 1991).

There are three main types of DNA polymorphisms in use and these are restriction fragment length polymorphisms (RFLPs), variable number tandem repeats (VNTRs), and $(CA)_n$ minirepeats.

RFLPs were the first type of DNA polymorphism to be used in molecular genetics (Botstein *et al.*, 1980). RFLPs occur when changes in nucleotide bases abolish or create restriction enzyme cutting sites. This means that only nucleotide changes which affect the action of a restriction endonuclease can be detected by this method and this limits its sensitivity. Other forms of nucleotide variation that do not alter restriction sites will not be detected. This form of polymorphism produces two alleles and it is not possible to identify the parental chromosome. Haplotypes can be constructed using multiple RFLPs found in the same gene. This is more useful than single-site polymorphisms.

Repetitive DNA sequences, such as VNTRs and minirepeats, have been found to be more useful for linkage analysis as they tend to be multiallellic and randomly distributed. VNTRs are short elements of DNA made up of four to six nucleotides repeated in variably sized blocks of DNA. These are often G–C rich and provide useful markers. One such example is the VNTR located at the 5'-end of the human insulin gene on chromosome 11. Most individuals are heterozygous for length and DNA sequence at this locus. In contrast to RFLPs, VNTRs can often be used to determine unambiguous transmission of the parental chromosome. VNTRs are produced by unequal recombination in meiosis and, hence, are unstable. Such instability makes it difficult to track a mutation through generations.

$(CA)_n$ minirepeats are another class of DNA polymorphisms scattered throughout the human genome. The value of $n$ is between 4 and 40 and usually $<25$. Along with their complement $(GT)_n$, they are spaced approximately 50–100 kb apart, generally do not form clusters, and are uniformly dispersed throughout the human genome. Repeats that are close to candidate genes provide useful markers. If the flanking sequence of a minirepeat is known, it can then be amplified using PCR. The size of the allelic fragment can then be determined and these can be used as linkage markers. Minisatellite repeats provide linkage markers for any candidate gene.

*Alu* variable polyA DNA polymorphisms have also been used (Weissenbach *et al.*, 1992). These contain the recognition sequence for the restriction enzyme *Alu*I (AGCT) and range in length from 130 to 300 bp. In contrast to other simple sequence elements, they are generally not tandemly repeated, but exist as isolated elements. About a million copies are present in the human genome.

The mapping of the disease gene is then accomplished in the following way: several generations of a pedigree are assembled with sufficient numbers of affected individuals. One then determines whether the DNA markers cosegregate with the disease. The strength of association is defined by the LOD score. One then attempts to map the disease gene with markers located even closer to the gene. This approach is based on the assumption that only a single or a few major genes cause the disease, while some minor genes may exert a background effect. One may use a population or a pedigree to study gene–disease association. Population analysis requires careful ethnic matching of control groups and may be complicated by heterogeneity.

Pedigree analysis can be accomplished using sib pairs or a combined segregation-linkage analysis approach. For sib-pair analysis only affected sib pairs in a pedigree are evaluated and this is useful with late-onset diseases where it is not known if the unaffected relative will develop the disease. Consequently, the unaffected relatives need not be considered in the calculations. Furthermore, it is not necessary to consider the mode of transmission as is the case with segregation-linkage analysis. However, despite its apparent simplicity, problems arise with sib-pair analysis. These include the possibility that the marker alleles cannot be identified with certainty and a homozygous parent may be typed as a heterozygote. Furthermore, the phenotype of the parents is ignored.

Linkage studies in families provide a generally more powerful approach than population association studies. Linkage disequilibrium is maintained to a larger extent in families than within populations. Hence, a diabetes-susceptibility gene located 15 megabases away from the marker will be detected. In contrast, using the population approach one would need a marker at least 0.1 megabases away to detect a diabetes-susceptibility gene. For family based linkage analysis, only 100–200 polymorphic markers are needed to cover the human genome to search for an unknown gene if one does not have a candidate gene. With the population approach many more markers are needed. As will be discussed later, using linkage analysis and mapping approaches in an early onset subtype of NIDDM, diabetes-susceptibility genes were localized to chromosomes 20 (Bell *et al.*, 1991) and chromosome 7 (Froguel *et al.*, 1992).

If a chromosomal assignment of a disease phenotype has been accomplished, one can then use consecutive polymorphic DNA markers to exclude regions of the chromosome in a sequential manner to map the region of the chromosome where the disease gene is located. Once a gene has been localized using these DNA markers, it can then be cloned and studied in greater detail. The DNA markers used in the initial localization can then be used to find markers which are closer to the gene and this can be repeated for each set of markers, a process referred to as "chromosome walking" (Bellanné-Chantelot et al., 1992). This approach, which was originally called "reverse genetics," and is now termed "positional cloning," involves the manipulation of large genomic DNA fragments which have been cloned into yeast artificial chromosomes. This is because the initial DNA markers may be located millions of bases away and in order to "walk" toward the gene, it is necessary to isolate large DNA fragments sequentially until one reaches the gene.

Linkage analysis in diabetes has not been as rewarding as initially hoped, owing to the complexity of NIDDM. Current methods in molecular genetics make it possible to clone any single-locus human disease gene using established strategies for locating and cloning genes. This approach is clearly not the case with the common forms of NIDDM. The unclear mode of inheritance, polygenic etiology, and genetic heterogeneity coupled with unclear definitions of the affected phenotype have all contributed to making linkage analysis less useful (Hitman and McCarthy, 1991). Furthermore, other genetic and environmental factors may determine whether an individual carrying a disease-susceptibility allele expresses NIDDM. Manifestations of the disease are determined by variable age of onset and penetrance (proportion of individuals carrying the affected allele who display the phenotype). Patients with an affected gene may be normoglycemic and the development of diabetes may depend on other factors including ageing and obesity.

A number of approaches may be used to circumvent these difficulties (Froguel et al., 1993a): (1) phenotypic definitions need to be broadened so that subclinical disease can be detected. The WHO criteria for diabetes may not detect early disease and indeed groups investigating the molecular genetics of diabetes have considered a fasting glucose level of 6.1 mmol/liter as a cutoff, rather than the WHO cutoff level of 7.8 mmol/liter (WHO, 1985). Phenotypic characterization must be detailed. (2) Subphenotypic classification may be necessary (Froguel et al., 1993a). The common diabetes endpoint of hyperglycemia is probably the result of several gene defects. A more accurate assessment of the genes involved may arise from subclassifying NIDDM patients into groups with predominant insulin secretion defects or predominant insulin resistance. Furthermore, careful metabolic staging may be necessary (Granner and O'Brien, 1992) in order to define the role of individual diabetes-susceptibility genes. (3) Large pedigrees must be assembled—this is especially true for a genetically heterogeneous disorder, such as diabetes, since the cause of NIDDM in a pedigree is likely to be more homogeneous than in the population at large.

Despite the shortcomings discussed above, linkage analysis is still a valuable tool for elucidating the genetics of NIDDM if the difficulties and assumptions underlying it are recognized. While a LOD score of 3 may be taken to indicate linkage in single gene disorders, more stringent values are required for polygenic, multifactorial diseases.

## C. Subtraction or differential cloning

Subtraction cloning is used to identify genes that are expressed in one cell population and not in another. These two cell populations can be different tissues or cells at a different developmental stage. In its application to NIDDM, these two tissues can be the same cell type derived from a normal and a diabetic patient, respectively. A number of strategies are available to isolate differentially expressed genes. Differential hybridization, although tedious and labor intensive, is widely established in a number of different adaptations. All are based on the hybridization of the mRNA species from one cell type with cDNA from another. The remaining enriched mRNA is then separated and used as the starting point to construct a cDNA library.

Other recently developed and less well-established techniques include representational difference analysis (RDA) (Lisitsyn et al., 1993) and PCR-based mRNA differential display (Liang and Pardee, 1992). In RDA restriction fragments present in one population of DNA (e.g., diabetic or "driver" DNA) but not another (e.g., normal or "tester" DNA) are purified and used as probes to detect polymorphism between individuals. Driver and tester genomic DNA is initially cleaved with restriction enzymes. One set of DNA (tester) is then ligated to oligonucleotide adaptors. The two populations of DNA are mixed and then subjected to PCR amplification. Common sequences between the two DNA populations will hybridize. The mixture is then subjected to PCR using the oligonucleotide adaptors as primers. This results in exponential amplification of target sequences that are unique to the tester DNA population, i.e., sequences or polymorphisms present in normal DNA as opposed to diabetic DNA and vice versa.

PCR-based mRNA differential display uses a set of oligonucleotide primers, one arbitrary in sequence and another (polyT) which hybridizes to the polyadenylate (polyA) mRNA tails to initially prime a reverse-transcriptase reaction and then PCR amplification. The arbitrary primer will anneal to different positions relative to the polyT primer depending on the mRNA species (e.g., normal vs diabetic tissue). Hence, mRNA from different tissues will generate fragments of different sizes resulting in different patterns on electrophoresis. The bands which appear to be unique to one tissue can then be isolated, cloned, and sequenced to determine their identities. Thus far, only subtraction cloning (Reynet and Kahn, 1993) and differential display methods (Nishio et al., 1994) have been applied to NIDDM.

Subtraction or differential cloning has been used to identify genes associated with NIDDM (Reynet and Kahn, 1993). Messenger RNA and cDNA were prepared from diabetic and normal muscle. Hybridization of mRNA from one tissue with cDNA from the other tissue, followed by separation of the unhybridized DNA, allowed the construction of a cDNA library enriched in mRNA species preferentially expressed in an NIDDM patient. The enriched mRNA species was also used to synthesize radioactively labeled probes ("subtracted probes") which were then used to screen the subtracted cDNA library. In this way, a novel clone, *Ras* associated with *diabetes* (Rad), was identified. The 29-kDa Rad protein belongs to the Ras-GTPase superfamily. It is primarily expressed in skeletal and cardiac muscle. Its overexpression in skeletal muscle of NIDDM patients suggests a pathophysiological role but the precise function and role of Rad in normal and diabetic skeletal muscle remains to be elucidated. Overexpression of Rad may be compensatory, or an epiphenomenon, or it may act to inhibit Ras in the insulin-initiated signaling pathway.

## D. Animal models for NIDDM

A number of animal models have been used to investigate the genetics of NIDDM (Karasik *et al.*, 1994). These have important advantages for the genetic analysis of diabetes-susceptibility genes in that environmental factors can be controlled and detailed tissue analysis is possible. Furthermore, the strains will be genetically homogeneous and one will have unlimited access to several generations of offspring. Such animal studies make it possible to understand the effects of a single mutation or polygenic factors in the development of diabetes. In the *db* mouse, a mutation on chromosome 4 causes spontaneous obesity in the homozygous state (Karasik *et al.*, 1994). The *db/db* mouse develops different degrees of diabetes depending on the genetic background into which it is bred. Obesity results in insulin resistance and islet hyperplasia and marked hyperinsulinemia in inbred strains. Obesity-induced insulin resistance in other strains of mice results in a milder diabetes similar to NIDDM. The obesity-induced diabetes phenotype is determined by multiple genes and this has similar implications for the genetic analysis of obesity-induced diabetes in man (Karasik and Hattori, 1994). Chromosomal mapping of these genes can be accomplished by high-resolution linkage analysis. Then, homologous regions in the human genome can be identified by synteny mapping. For the *db* mouse chromosome 4 diabetes gene the equivalent flanking genes in the human genome are present on chromosome 1p31-36 (Bahary *et al.*, 1994). Genetic markers for this region can be used to investigate human NIDDM in sib-pair analysis and population association. The major shortcoming of this approach is that animal susceptibility genes are not always applicable to common NIDDM in man because no animal model has diabetic features identical to human disease.

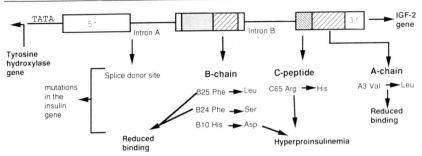

**Figure 3.4.** Intron–exon structure of the insulin gene on chromosome 11p. The insulin gene is located between the tyrosine hydroxylase and the insulin-like growth factor II genes. The positions of the various mutations and their functional effects are illustrated. Adapted with permission from Steiner *et al.* (1990).

---

## V. CANDIDATE GENES

### A. The insulin gene

The insulin gene was the first candidate gene investigated in NIDDM. The insulin gene spans 1430 bp with three exons separated by introns of 179 and 786 bp, respectively (Ullrich *et al.*, 1980). The gene is located on chromosome 11p15 (Owerbach *et al.*, 1980) and is adjacent to the insulin-like growth factor II and tyrosine hydroxylase genes (Figure 3.4). These three genes are expressed differentially in a tissue-specific manner and are not coordinately regulated despite their proximity to each other.

Defects in the coding regions of the insulin gene have been described in 11 families with diabetes (Steiner *et al.*, 1990) and are due to five different mutations. Three of the mutations occur in the mature insulin molecule: "Insulin Chicago" (a PheB25 → Leu substitution), "Insulin Los Angeles" (PheB24 → Ser), and "Insulin Wakayama" (a ValA3 → Leu substitution). These affect the binding of insulin to its receptor and result in a clinical syndrome, insulinopathy, which is characterized by hyperinsulinemia, and altered insulin/C-peptide ratio and frequently mild IGT. Sensitivity to exogenous insulin in these subjects is normal or slightly impaired, despite the exaggerated hyperinsulinemia.

In four additional families, impaired processing of proinsulin results in familial hyperproinsulinemia (Steiner *et al.*, 1990). This disorder is the result of mutation of an arginine at the Lys64–Arg65 endopeptidase cleavage site or a HisB10 → Asp mutation. Some of these patients have impaired glucose tolerance.

Polymorphisms in the insulin gene have also been investigated in

NIDDM. A VNTR hypervariable region is located 375 bp upstream from the transcription initiation site of the insulin gene (Bell *et al.*, 1981). Differences in the composition of the repeat units between individuals are observed. Usually there are 30–139 repeats of the 14- to 15-bp unit, while occasionally as many as 540 repeats have been observed. The apparent nonrandom size distribution of the hypervariable region of the insulin gene has been used to group insulin alleles into three classes (Bell *et al.*, 1981, 1982). Class 1 alleles contain 40 repeat units (570 bp), class 2 alleles contain about 95 copies of repeats and class 3 alleles contain 170 copies (2470 bp). Class 1 and class 3 alleles are most commonly found in European populations, while in populations of African descent the class 2 alleles are also common resulting in a trimodal distribution of all three alleles (Elbein *et al.*, 1985). There is extensive polymorphic variation and heterogeneity in the hypervariable region within ethnic groups beyond that indicated by allele classification (Permutt and Elbein, 1990). Linkage analysis of the insulin gene using DNA polymorphisms and the hypervariable region has provided useful information.

Initially, the large class 3 allele was found to be associated with NIDDM (Permutt and Elbein, 1990). Investigation of other racial groups did not confirm this association. In separate studies, both class 1 and class 3 alleles were found to be associated with NIDDM in Caucasians (Hitman and McCarthy, 1991) but not in seven other racial groups (Permutt, 1991). The smaller class 1 allele has been found to be associated with IDDM in Caucasians, although this was not shared in sib pairs with IDDM (Permutt, 1991).

Hypoinsulinemia in NIDDM may arise from mutations in the promoter region which could alter gene regulation and decrease transcription (Permutt, 1991). Three hundred seventy-five base pairs of the insulin gene promoter were amplified by PCR from genomic DNA of NIDDM patients and this allowed the sequences of both alleles to be determined. Extensive investigation of the promoter hypervariable region has failed to find a significant role for this region as a diabetes-susceptibility gene. However, a variant allele of the promoter has been detected in 5% of African-American NIDDM patients (Olansky *et al.*, 1992). *In vitro*, this promoter reduces transcription by 30–50% implying a possible role in this subgroup of NIDDM patients.

## B. The insulin receptor

The insulin receptor gene is a logical candidate gene to analyze as a possible cause of non-insulin-dependent diabetes mellitus. Insulin resistance clearly plays a major role in the pathophysiology of NIDDM, and, in fact, predates the onset of clinical NIDDM in patients followed over time (Olefsky, 1991).

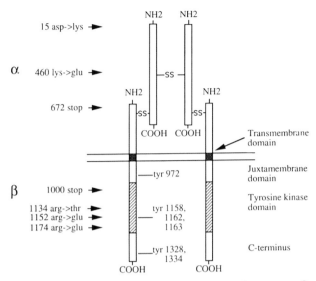

**Figure 3.5.** Schematic representation of the structure of the insulin receptor. Specific receptor mutations discussed in this review are indicated by the arrows on the left.

## 1. Structure and function of the insulin receptor

Molecular cloning of the insulin receptor (Ullrich *et al.*, 1985; Ebina *et al.*, 1985) revealed that it is structurally a member of the tyrosine kinase family of transmembrane receptors, which includes the EGF, PDGF, CSF-1, and IGF-1 receptors. The receptor is encoded by a single gene on chromosome 19 and consists of 22 exons spanning 120 kb (Seino *et al.*, 1989). The α subunit is encoded by exons 1 to 11, and the remaining 11 exons encode the β subunit (Figures 3.5 and 3.6). Translation of receptor mRNA in the RER yields a proreceptor precursor of 1382 amino acids, including a 27 residue signal peptide at the amino terminal which assists in intracellular receptor trafficking and is subsequently cleaved. During transport through the Golgi apparatus the proreceptor undergoes extensive N- and O-linked glycosylation, and two monomeric proreceptors are linked into a dimeric structure by disulfide bonds. Finally, the α and β subunits are produced by proteolytic cleavage at a tetrabasic cleavage sit, but remain joined by α–β disulfide bonds, yielding the mature $\alpha_2\beta_2$ heterotetrameric receptor (Ullrich and Schlessinger, 1990) (Figure 3.5).

The α subunit, which is entirely extracellular, forms the high-affinity insulin-binding site (Figure 3.5). Insulin-binding specificity appears to reside in amino acids 1 to 131 and 315 to 514 (Schumacher *et al.*, 1993). The intervening

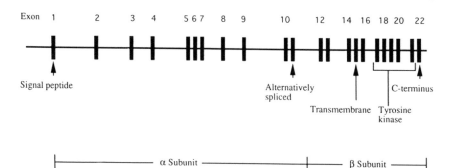

**Figure 3.6.** The insulin receptor gene. Receptor protein domains encoded by specific exons are indicated. Adapted with permission from Seino *et al.* (1989).

region, which is cysteine rich, may form a contact point (Yip, 1992) but does not seem to confer insulin-binding specificity (Schumacher *et al.*, 1993). There are in fact two different isoforms (A and B) of the receptor, which differ by alternative splicing of exon 11. This results in two mature protein products which differ by 12 amino acids at the carboxy terminal of the α subunit. This alternative splicing is tissue specific (Benecke *et al.*, 1993), and it has been shown that the exon 11⁻ (B) form has an approximately twofold higher affinity for insulin. The functional consequences of the alternative isoforms are not entirely clear. The B isoform has a slightly different autophosphorylation pattern on peptide mapping and couples more efficiently to substrate phosphorylation (Kosaki *et al.*, unpublished) (all amino acid numbering given is for the exon 11⁺ form of the receptor) (Ebina *et al.*, 1985).

   The β subunit has a number of distinct functional domains. These are the 193 amino acid extracellular domain, the 23 residue transmembrane domain, and the 402 residue intracellular domain. The latter can be further subdivided into the juxtamembrane domain, the tyrosine kinase domain, and the carboxy terminus. Little is known of the function of the extracellular domain other than the fact that it participates in the α–β disulfide bonds. The transmembrane domain plays a key role in signal transduction, linking the extracellular insulin-binding domain to the intracellular effector domain of the receptor.

   The juxtamembrane domain of the intracellular β subunit has been shown to play an important role in receptor endocytosis induced by insulin binding. The low-density lipoprotein receptor requires an NPXY motif for endocytosis, which is thought to interact with adaptor proteins in clathrin-coated pits (Brown *et al.*, 1991). The insulin receptor has two analogous motifs in the juxtamembrane region, and mutagenesis experiments have shown that the GPLY sequence at residues 962–965 plays a major role in receptor endocytosis

(Rajagopalan et al., 1991). Tyrosine 972 in this region of the receptor may also play a role in interaction with receptor substrates, including insulin receptor substrate-1 (IRS-1) (White et al., 1988).

The tyrosine kinase domain is highly conserved among the family of tyrosine kinase receptors. It contains an invariant ATP-binding consensus sequence, Gly-X-X-Gly-X-Gly-X, at residues 1003 to 1009, and a lysine residue at position 1030 which permits tight ATP binding by forming a salt bridge with the β-phosphate of ATP. Further, C-terminal in the kinase domain is a cluster of three tyrosines, at positions 1158, 1162, and 1163, which plays an important regulatory role. After insulin stimulation they undergo rapid autophosphorylation, which permits activation of the receptor as a kinase (reviewed in Olefsky, 1990).

The carboxy terminus of the insulin receptor is the domain with least sequence identity to other tyrosine kinase receptors (Ullrich et al., 1990). It contains two main targets of tyrosine phosphorylation, at positions 1328 and 1334, as well as sites for serine/threonine phosphorylation (Pillay et al., 1991; Coghlan et al., 1994; Lewis et al., 1994). There is evidence that phosphorylation of these tyrosines may inhibit receptor mediation of mitogenic effects (Takata et al., 1991). Furthermore, the rest of the C-terminus seems to be required for effective signaling of the metabolic effects of insulin (Maegawa et al., 1988). Thus, the C-terminus may participate in modulating receptor function, promoting the specific metabolic effects which make insulin unique among the growth factors.

Activation of the receptor is initiated by binding of insulin to the α subunit. This induces a conformational change in the C-terminus (Baron et al., 1992), which is followed by ATP binding. The "regulatory" tyrosines at positions 1158, 1162, and 1163 undergo autophosphorylation, and the receptor is activated as a tyrosine kinase (Wilden et al., 1992), resulting in further autophosphorylation of the receptor on tyrosines 1328, 1334, and 972.

The intracellular signaling pathways utilized by the insulin receptor are the subject of intensive study, and a great deal of information has recently been elucidated (Figure 3.8). Activation of the receptor as a kinase results in the tyrosyl phosphorylation of IRS-1. This, in turn, interacts with various SH2 domain-containing proteins, which bind to IRS-1 phosphotyrosine residues in the context of specific amino acid motifs (White and Kahn, 1994). One such protein is the p85 subunit of phosphatidylinositol 3-kinase (PI-3 kinase) which binds to IRS-1 at YMXM motifs. Binding results in the activation of the PI-3 kinase activity of the p110 subunit. Other proteins which interact with IRS-1 include Grb2, syp, and nck. Insulin stimulation also causes the tyrosyl phosphorylation of Shc, another adaptor protein. This then binds to Grb2, which is constitutively bound to the Ras-activating protein Son-of-sevenless gene product (SOS) (Egan et al., 1993). Subsequent activation of Ras results in the sequential activation of Raf, MEK, and MAP kinase, which can phosphorylate transcription factors, including Elk. However, despite the multiple signaling pathways which appear to be activated by the insulin receptor, the significance of each, and how they interact, is as yet unclear. Insulin has pleiotropic effects, causing the stimulation of glucose transport and glucose metabolic pathways, but it also has effects similar to other growth factors on gene regulation and cell proliferation. It is probable that the activation of multiple pathways permits this diversity of bioeffects, and interactions between pathways may permit careful modulation of signaling.

## 2. Leprechaunism

Leprechaunism is a congenital syndrome of severe insulin resistance associated with multiple somatic abnormalities. These include intrauterine growth retardation with subnormal postnatal growth, typical dysmorphic facies, and lipoatrophy. Patients also have acanthosis nigricans, a cutaneous manifestation seen in patients with severe insulin resistance of many causes, characterized by hyperpigmented, hyperkeratotic lesions, often with a velvety appearance, and seen most commonly in the flexural areas and the nape of the neck. Metabolically, these patients usually demonstrate fasting hypoglycemia and postprandial hyperglycemia, with insulin levels greatly above normal. They seldom survive to 1 year of age.

Leprechaunism is clearly caused by mutations in the insulin receptor. Most patients studies to date have mutations in both alleles of the receptor, being either homozygotes or compound heterozygotes. Two patients have only had a mutation discovered in one allele, but in each case the other allele can be inferred to be abnormal, possibly having mutations in regulatory domains of the gene (Taylor, 1992).

Patient leprechaun/Ark-1 has been the most thoroughly studied. Initial investigation of insulin binding to the patient's monocytes showed receptor number to be reduced to 10 to 20% of normal, and binding was also relatively insensitive to reductions in pH (Taylor *et al.*, 1981). The patient was subsequently found to have a different mutation in each insulin receptor allele (Kadowaki *et al.*, 1988). There was a nonsense mutation in one allele at codon 672 in the $\alpha$ subunit. Because this termination codon resulted in the cessation of translation before the transmembrane domain, there was no protein product expressed at the membrane. The other allele displayed a missense mutation at codon 460, resulting in the replacement of lysine by glutamine.

In order to study the effect of the $Glu^{460}$ mutant, it was expressed in Chinese hampster ovary (CHO) cells (Kadowaki *et al.*, 1988, 1990a). This is the best available way to analyze the biologic function of mutant receptors as it allows their study in isolation from any other genetic defects the patient may have, as well as freeing them from the abnormal metabolic milieu of hyperglycemia if the patient is diabetic, which itself has been shown to affect receptor function (Friedenberg et al., 1988). Normally, insulin binding is greatly decreased as the pH is reduced from physiologic levels, but in the mutant receptor this effect was significantly blunted (Kadowaki et al., 1988, 1990a). After endocytosis of the receptor–ligand complex, the acid pH in the endosome normally causes ligand dissociation. Some of the receptor is then degraded, but some is recycled to the plasma membrane. The Glu$^{460}$ mutant was found to internalize normally, but to remain associated with insulin for a prolonged time after endocytosis. There was reduced recycling of receptors to the plasma membrane, presumably a result of impaired ligand–receptor dissociation. This led to a reduction in receptor half-life. However, receptor autophosphorylation and tyrosine kinase activity were normal (Kadowaki et al., 1990a).

Thus, the insulin resistance manifested by leprechaun Ark-1 was felt to result from one mutant insulin receptor allele expressing no functional product because of premature termination of translation and one allele expressing a receptor which was degraded abnormally rapidly, with the net result being the expression of very few receptors at the plasma membrane. Family studies revealed the father to be heterozygous for the termination codon mutation. He demonstrated decreased receptor number in binding studies and was mildly insulin resistant. The mother was heterozygous for the Glu$^{460}$ allele and had normal receptor number but decreased pH sensitivity and normal insulin sensitivity (Kadowaki et al., 1988).

## 3. Rabson–Mendenhall syndrome

The Rabson–Mendenhall syndrome is the presentation in early childhood of insulin-resistant diabetes mellitus, associated with abnormal facies, dental dysplasia, thickened nails, hirsutism, precocious puberty, and acanthosis nigricans (Rabson et al., 1956). Patients are differentiated from those with leprechaunism primarily by having a less severe form of insulin resistance, with older age of presentation of hyperglycemia. The genetic etiology of the syndrome was evident from its frequent occurrence in more than one sibling in a family, suggesting an autosomal recessive form of inheritance.

Analysis of three patients with this syndrome has shown them to have a variety of insulin receptor mutations, and in all three patients both alleles were abnormal (Taylor, 1992). A representative patient, RM-1, had a nonsense mutation at codon 1000 of one allele (Kadowaki et al., 1990b,c). The predicted protein product of this allele would be truncated before the tyrosine kinase domain and would therefore be nonfunctional. Furthermore, the premature termination codon seemed to exert a cis-dominant effect, causing greatly decreased transcription of mRNA from this allele. The other allele was found to have a missense mutation, with replacement of Asp by Lys at the 15th position in the N-terminus of the $\alpha$ subunit (Kadowaki *et al.*, 1990b).

Analysis of the Lys[15] mutant by transfection into NIH 3T3 cells revealed that receptor processing was impaired (Kadowaki *et al.*, 1990b). There was evidence of impaired cleavage of the proreceptor to $\alpha$ and $\beta$ subunits and impaired processing of the high mannose form of N-linked carbohydrate to complex carbohydrate. Some mature receptor was expressed at the plasma membrane, but it was found to have a fivefold reduction in insulin-binding affinity. As it has such profound effects on both receptor processing and insulin-binding affinity, it was postulated that the point mutation affected the three-dimensional folding of the receptor (Kadowaki *et al.*, 1990b).

## 4. Type A insulin resistance

The syndrome of Type A insulin resistance was first described by Kahn and colleagues in 1976 (Kahn *et al.*, 1976). They described three young nonobese female who had acanthosis nigricans, hirsutism, amenorrhea, hyperandrogenism, and polycystic ovaries. All were severely insulin resistant, and all demonstrated decreased insulin binding to monocytes. They were designated Type A to differentiate them from Type B insulin-resistant patients who were also insulin resistant, but whose resistance resulted from an autoimmune disease characterized by anti-insulin receptor antibodies (Kahn *et al.*, 1976). Despite the insulin resistance, many patients with the Type A syndrome have normal or only mildly impaired glucose tolerance, as the insulin resistance is compensated for by hyperinsulinemia. The reason for their coming to medical attention is therefore often related to the effects of hyperandrogenism, which may explain why few affected males have been described (Taylor, 1992).

Both acanthosis nigricans and hyperandrogenism are seen in patients with any of the syndromes of severe insulin resistance, including leprechaunism, Rabson–Mendenhall, and Type A. Although the etiology is not clear, a commonly proposed mechanism is that insulin, at greatly elevated levels, may interact with insulin-like growth factor-1 receptors present in both the skin and the ovarian thecal tissue, which might cause abnormal epidermal development and drive ovarian androgen overproduction (Dunaif, 1993).

A large number of Type A patients have mutations of the insulin receptor gene. The whole spectrum of possible receptor abnormalities has been seen in these patients, including decreased receptor synthesis, impaired receptor processing and transfer to the plasma membrane, impaired insulin binding, decreased autophosphorylation and tyrosine kinase activity, and accelerated receptor degradation (Taylor, 1992).

One Type A patient has been described who had a severe reduction in erythrocyte insulin binding (Imano *et al.*, 1991). All 22 exons of both receptor alleles were sequenced and were found to be normal. However, receptor mRNA levels were so low as to be undetectable by RNase protection assay, and as the mRNA was not found to be unstable, it was concluded that it was being transcribed at a greatly reduced rate (Imano et al., 1991). From this, it was inferred that each allele must have a mutation in a regulatory domain, causing decreased expression of that allele. The regulation of receptor gene expression is only poorly understood, and there is no direct evidence supporting this contention. However, this illustrates the potential complexity of screening patients for receptor mutations, as screening only the exon sequences will not reveal this type of regulatory domain mutation.

While most patients with Type A insulin resistance due to receptor mutations seem to have two mutant receptor alleles, there is a subgroup who may demonstrate inheritance in a dominant pattern, and who are heterozygous for mutations in the tyrosine kinase domain of the receptor, with the other allele being normal. One kinship displaying this pattern has been described in which three sisters had Type A resistance, the father was quite insulin resistant although he did not have acanthosis nigricans, and the mother was normal (Moller et al., 1990). All three sisters and the father were heterozygous for a missense Thr → Ala mutation at position 1134. This residue lies in the tyrosine kinase domain of the receptor in a short region almost 100% conserved among the tyrosine kinases. It is invariant in the human, mouse, and rat insulin receptors and the IGF-1 receptor, suggesting that mutation might have substantial effects on receptor function (Moller et al., 1990). This was confirmed by transfection of the mutant into CHO cells (Moller et al., 1990, 1991), which demonstrated that, while receptor processing and insulin binding were normal, receptor autophosphorylation was impaired by more than 95% and tyrosine kinase activity toward either IRS-1 in vivo or exogenous substrates in vitro was undetectable. Furthermore, the receptor did not mediate insulin stimulation of either metabolic or mitogenic effects.

The mechanism by which a single mutant allele coding for a kinase-defective receptor can cause severe insulin resistance despite a normal second allele is not entirely clear. If it only resulted in a diminution of active receptors by 50%, with no other detrimental effect on receptor function, there would be normal or minimally decreased insulin sensitivity, as suggested both by the large variation in receptor number in normal subjects (Dons et al., 1981) as well as by the normal insulin sensitivity in many subjects heterozygous for nonsense mutations (Taylor, 1992). At least four different theories have been proposed to explain this discrepancy, including formation of inactive hybrid receptors, competition for receptor substrates, selective downregulation of normal receptors, and interference with receptor aggregation (Figure 3.7).

As discussed, the insulin receptor is initially synthesized as $\alpha$–$\beta$ pro-receptor monomers which are subsequently joined by disulfide bonds during

A.  Hybrid Formation

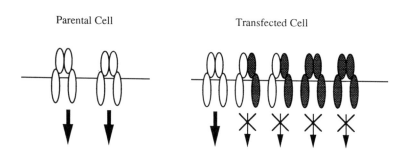

B.  Substrate Competition

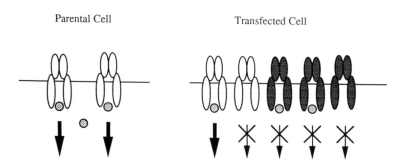

**Figure 3.7.** Mechanism of dominant negative inhibition. Two possible mechanisms of dominant negative inhibition of endogenous insulin receptors (white) by transfected kinase-inactive receptors (black). The ability of receptors to transmit insulin signaling is indicated by arrows. (A) Hybrid receptor formation. Part of the complement of endogenous receptors forms kinase-inactive hybrids with the transfected receptors, decreasing the number of active endogenous receptors capable of signaling. (B) Substrate competition. Transfected kinase-inactive receptors compete for a limiting substrate (◎), reducing the amount of substrate available to the endogenous receptors. Adapted from Olefsky (1990).

processing to form the mature $\alpha_2\beta_2$ receptor. It seems possible, therefore, that two proreceptors translated from different alleles might be combined in mature receptors. There is clear evidence that this can happen in transfected cells in which such "hybrid" receptors composed of one endogenous wild-type half-receptor and one transfected mutant half-receptor have been demonstrated (Frattali *et al.*, 1992). Furthermore, hybrid receptors of one insulin half-receptor and one IGF-1 half-receptor have been found in human placenta (Frattali *et al.*, 1992). These findings suggest that hybrids of normal and mutant insulin receptors may occur *in vivo*. Experiments with one such hybrid, composed of one normal and one kinase-inactive half-receptor, have shown that the hybrid receptor is itself kinase inactive (Treadway *et al.*, 1991). If analogous hybrid formation occurs in patients heterozygous for kinase-inactive mutant receptors, many of the half-receptors from the normal allele may be present in inactive hybrids, thus reducing active receptor expression well below 50% of normal and causing insulin resistance (Olefsky, 1990) (Figure 3.7).

There is evidence to suggest that in some cases the dominant negative phenotype is not due to hybrid receptor formation, but to other mechanisms. In one cell line transfected with a kinase-inactive mutant, the dominant negative phenotype was seen, although fewer than 10% of endogenous receptors were found to be in hybrids (Levy-Toledano *et al.*, 1994). The transfected receptors were found to bind IRS-1 after insulin stimulation, and this was associated with decreased IRS-1 phosphorylation by the wild-type receptors. Thus, inactive receptors may sequester IRS-1 and possibly other signaling molecules, reducing their availability to the normal receptors and impairing signaling (Figure 3.7).

The final two mechanisms that have been proposed include selective downregulation of the normal receptor population and impairment of receptor aggregation. It is known that kinase-inactive receptors cannot undergo ligand-induced endocytosis, which is required for receptor downregulation. In the face of equal receptor production from the normal and mutant allele, the inability of the mutant allele to be downregulated might result in a steady state in which the mutant receptors comprised more than half of the receptors on the plasma membrane. Finally, there is some evidence that aggregation of receptors on the plasma membrane after insulin stimulation may increase insulin signaling (Kubar *et al.*, 1989). If kinase-inactive receptors can participate in this aggregation, they may interfere with generation of a normal signal by the normal receptors.

Although all patients with either leprechaunism or the Rabson–Mendenhall syndrome studied to date have had insulin receptor mutations, this is not the case for those with Type A insulin resistance. Of the cases of Type A resistance found to be caused by receptor mutation, the majority had receptor defects predicted prior to genetic receptor analysis on the basis of decreased receptor expression or impaired function in patient-derived cells (Moller *et al.*, 1994). To assess the incidence of receptor mutations in patients with Type A resistance not preselected for receptor abnormalities, Moller and colleagues studied a group of 22 patients with moderate to severe insulin resistance, acanthosis nigricans, and ovarian hyperandrogenism. The insulin receptor gene was screened for gross deletions, insertions, or rearrangements by Southern blotting, and then exons 2 to 22 were analyzed for point mutations by single-stranded conformation polymorphism (Moller *et al.*, 1994). One patient was found to be heterozygous for a point mutation resulting in the replacement of Arg[1174], a conserved residue in the tyrosine kinase domain, by Gln. This mutation is likely to be the cause of the insulin resistance in this patient, and the family history is suggestive of a dominant inheritance pattern, consistent with the pattern found with kinase-defective mutants. However, no mutations were found in the remaining 21 patients, suggesting that receptor mutation is in fact an uncommon cause of this syndrome.

## 5. Insulin receptor mutations and NIDDM

Support for the possibility of receptor mutations common type NIDDM has been provided by investigation of patients with the various syndromes of severe insulin resistance as discussed above. The families of patients with defined receptor mutations have allowed the assessment of insulin resistance and glucose tolerance in individuals who do not have the phenotype of severe insulin resistance, but who nonetheless are heterozygous for mutant insulin receptors. The father of leprechaun Ark-1 is heterozygous for the nonsense mutation at codon 672. He was quite insulin resistant, although his glucose tolerance is normal (Kadowaki *et al.*, 1988). Similarly, the father of the Type A proband heterozygous for the Ala[1134] → Thr mutation has also been studied. He is heterozygous, manifesting insulin resistance, and is glucose intolerant (Moller *et al.*, 1990). In contrast, the mother of leprechaun Ark-1, who is heterozygous for the Lys[460] → Glu mutation, has normal insulin sensitivity (Kadowaki *et al.*, 1988). This suggests that some, although not all, who ae heterozygotes for receptor mutations will manifest significant insulin resistance despite not conforming to a phenotypically recognizable syndrome. It is probable, although not proven, that these people are at greater risk for the development of NIDDM than the general population. It has been estimated that, based on the incidence of leprechaunism, at least 0.1% of the population must be heterozygous for significant receptor mutations (Taylor, 1992). If these heterozygotes are at great risk of developing NIDDM, then Taylor has estimated that the incidence of receptor mutations among those with NIDDM may be between 1 and 10%.

Various approaches have been used to attempt to determine whether insulin receptor mutations are in fact implicated in common-type NIDDM. A number of studies have looked at the association of receptor RFLP with NIDDM in various populations. Results have been equivocal, with some studies finding an association of different RFLPs with NIDDM (McClain *et al.*, 1988; Rabuodi *et al.*, 1989) and others finding none (Elbein *et al.*, 1986). This is not surprising, as mutant receptors could cause, at most, a small proportion of NIDDM, and there are likely many different mutants, each with different RFLP associations. Other investigators have attempted to overcome these limitations by looking for RFLP linkage to NIDDM in kindreds. Within a family the etiology of NIDDM is likely to be more homogeneous than in the population as a whole, and linkage with the insulin receptor should be more readily demonstrated. No such linkage has been found, although the number of families studied has been low (O'Rahilly et al., 1988).

The entire insulin receptor cDNA has been sequenced in three Pima native Americans (Moller et al., 1989; Cama et al., 1990) and one Caucasian (Kusari et al., 1991c) with NIDDM and in each case was normal. The tyrosine kinase domain, exons 16–21, has been analyzed in a total of 161 patients with NIDDM using the techniques of SSCP analysis (O'Rahilly et al., 1991; Kim et al., 1992), denaturing gradient gel electrophoresis (Cocozza et al., 1992), or direct sequencing (Kusari et al., 1991a). Two patients were found to be heterozygous for potentially significant mutations. One had a missense mutation causing replacement of $Arg^{1152}$ by Gln (Cocozza et al., 1992). Transfection of this mutant into NIH 3T3 cells has confirmed that it is kinase inactive, and therefore it likely plays a causative role in the NIDDM manifested by the proband (Kusari et al., 1991a). The other mutant allele detected had a $Lys^{1068}$ to Glu mutation, which has not yet been further characterized (O'Rahilly et al., 1991). Thus, these studies suggest that no more than 1 or 2% of patients with NIDDM have significant mutations in the tyrosine kinase domain of the receptor.

## C. Insulin receptor substrate-1 (IRS-1)

The molecules involved in insulin's intracellular signaling cascade (Figure 3.8) are potential candidates for causation of NIDDM, as defects in these could cause insulin resistance at a "post-receptor" level. The potential role of IRS-1 as a causative gene was examined in a group of Danish diabetic subjects (Almino et al., 1993). The entire IRS-1 coding sequence of each patient was examined by SSCP. Two different polymorphisms were found, resulting in amino acid substitutions at residue 513 or 972. Both polymorphisms were found in both diabetic and control subjects, and although there was a trend to a higher frequency in the diabetics, it was not significant. Subsequent analysis of these polymorphisms in French (Hagar et al., 1993) and Pima native American populations (Celi et al., 1994) did not detect any association with NIDDM, although another group found an increased incidence of the 972 substitution in NIDDM patients (Imai et al., 1994) as well as detecting two other polymorphisms at residues 819 and 1221 in diabetics. In a group of Mexican-Americans, sib-pair analysis for linkage of IRS-1 to NIDDM was also negative (Shipman et al., 1994). Linkage analysis of IRS-1 in MODY did not reveal any association (Vaxillaire et al., 1994). Despite these equivocal results, the candidate gene approach will certainly be used to examine IRS-1 and other signaling molecules as our knowledge of the signaling pathway increases.

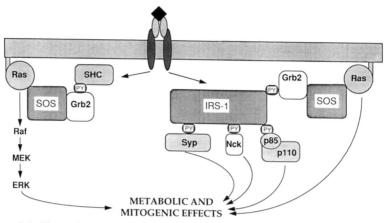

**Figure 3.8.** The insulin receptor and its signaling pathways. Intracellular signaling pathways activated by the occupied insulin receptor (shown here spanning the cell membrane). Phosphotyrosine-SH2 domain interactions are indicated by "PY." p85 and p110 are the subunits of phosphatidyl inositol 3-kinase. See text for details.

## D. Glucose transporter genes

Facilitated glucose transport in tissue is mediated by a family of membrane-bound glycoproteins (Figure 3.9). At least five distinct glucose transporter species have been identified and each is the product of a separate gene (Bell *et al.*, 1993). These glucose transporters are structurally related and share a great deal of sequence identity. Each is characterized by tissue-specific expression. For example, GLUT4 (also termed the insulin-regulatable glucose transporter) is only expressed in the classical insulin target tissues of skeletal muscle, cardiac muscle, and adipose tissue. In these tissues, glucose transport is rate limited for overall glucose metabolism. Skeletal muscle is the major site for postprandial glucose uptake and is the major site of insulin resistance in NIDDM. GLUT2 is expressed predominantly in liver and pancreatic β-cells and GLUT1 is expressed in most tissues including brain and erythrocytes.

The GLUT1 gene was the first glucose transporter gene to be cloned. Shortly thereafter a number of studies examined the relationship of GLUT1 polymorphisms with NIDDM. Linkage analysis indicated association of the GLUT1/*Xbal*/6.5 kb allele with diabetes in three populations examined (Li *et*

**Figure 3.9.** The insulin-sensitive glucose transporter. Model of the insulin-sensitive glucose transport (GLUT4) with exon–function relationships. Individual amino acids are indicated by circles. There are 12 putative membrane-spanning domains numbered M1–M12 and these are shown as rectangles. Adapted with permission from Kusari *et al.* (1991).

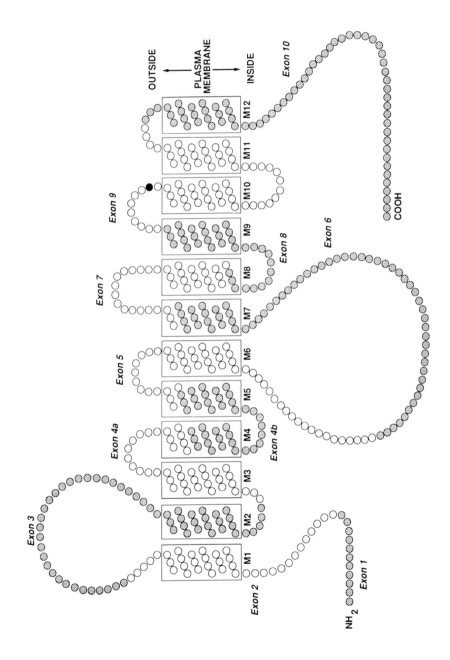

*al.*, 1991) but further studies of several ethnic groups failed to demonstrate any association of GLUT1 RFLPs with NIDDM (Li *et al.*, 1990; Cox *et al.*, 1988). Family studies in MODY and NIDDM pedigrees did not demonstrate any co-segregation (Vinik *et al.*, 1988).

GLUT2 has been considered as a candidate gene since glucose transport may be a rate-limiting step in glucose-induced insulin synthesis and secretion in β-cells and/or may be essential for glucose metabolism in the liver. Examination of a variety of GLUT2 RFLPs has failed to find any association with NIDDM (Matsutani *et al.*, 1990). Very recently, however, a mutation in GLUT2 has been linked to diabetes in a single patient (Mueckler *et al.*, 1994). The contribution of such mutations to the overall incidence of NIDDM remains to be determined.

GLUT4 protein and mRNA content are normal in skeletal muscle of NIDDM subjects, indicating that the decreased skeletal muscle glucose transport is due to impaired translocation of GLUT4 transporters to the cell surface, a decrease in GLUT4 intrinsic activity, or both. These findings and the genetic basis for NIDDM sparked a search for genetic defects in the GLUT4 gene in NIDDM. Sequencing of the GLUT4 coding regions from DNA of six NIDDM patients revealed that one patient was heterozygous for an Ile383 → Val mutation, whereas the coding sequence was normal in the remaining patients (Kusari *et al.*, 1991b). This raised the possibility that GLUT4 mutations may exist in a small subgroup of NIDDM patients, although no single mutation will be common. GLUT4 *Kpn*I RFLPs have not been linked to NIDDM in several ethnic groups (Bell, 1991). Further studies of the *Kpn*I polymorphism of the GLUT4 gene in British Caucasian and South Indian subjects did not demonstrate any association (Oelbaum, 1992; Baroni *et al.*, 1992b).

## E. Maternally inherited diabetes and deafness (MIDD)

For many years, a maternal effect in the transmission of diabetes has been consistently observed (Alcolado and Alcolado, 1991). Mitochondrial genes are maternal in origin and, hence, this suggested a role for a gene encoded by mitochondrial DNA (Wallace, 1992). Mitochondrial DNA is circular and is 16,569 bp in length (Figure 3.10). This DNA codes for 13 enzymes involved in oxidative phosphorylation as well as 2 ribosomal RNAs and 22 transfer RNAs (tRNA) needed for mitochondrial protein synthesis. Oxidative phosphorylation plays a crucial role in insulin secretion and mutations in mitochondrial DNA may affect islet cell function in this way (Ballinger *et al.*, 1992).

Pedigrees with maternal transmission of NIDDM have been examined and an A to G substitution at position 3243 in the mitochondrial tRNA[Leu(UUR)] gene has been found to link with the disease in all these pedigrees. Sensorineural deafness, a common feature of mitochondrial disease, was found to accompany

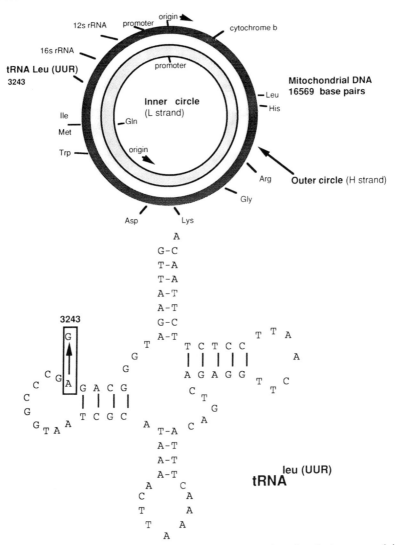

**Figure 3.10.** The mitochondrial genome and the proposed secondary cloverleaf structure of the tRNA<sup>LeuUUR</sup> gene. The tRNA genes are indicated by the three-letter amino acid codes. The location of the tRNA<sup>Leu(UUR)</sup> mutation at 3243 is indicated. The A to G mutation at base 3243 in the dihydrouridine loop is indicated. Adapted with permission from van den Ouweland (1994).

the diabetes (Ballinger *et al.*, 1992; Kadowaki *et al.*, 1994; van den Ouweland *et al.*, 1994). The term maternally inherited diabetes and deafness (MIDD) has been proposed to describe the syndrome (van den Ouweland *et al.*, 1994).

The 3243 mutation disrupts the mitochondrial DNA binding site for a protein factor that terminates transcription between the 16S ribosomal RNA and tRNA$^{Leu(UUR)}$ genes. This affects both the synthesis of tRNA$^{Leu(UUR)}$ and the binding of transcription termination factor resulting in defects in the synthesis of mitochondrial proteins and impairment of insulin secretion. Diabetes associated with this mutation occurs in a significant proportion of NIDDM and IDDM patients presenting as either slowly progressive IDDM or insulin-deficient NIDDM. The mitochondrial tRNA$^{Leu(UUR)}$ accounts for approximately 1% of maternally transmitted diabetes (Kadowaki *et al.*, 1994). The same mutation is associated with the MELAS syndrome (mitochondrial myopathy, encephalopathy, lactic acidosis and stroke-like episodes) although it is not clear how it gives rise to varying phenotypes. There may be additional mutations in mitochondrial DNA of patients with MIDD or variations in the clinical phenotype may arise from differences in the levels of normal and mutant DNA in different cells (heteroplasmy). MELAS and MIDD have been observed in the same pedigree (Remes *et al.*, 1993) and the appearance of either phenotype appeared to correlate with the amount of mutant mitochondrial DNA present in peripheral blood in that the mutant DNA was present in higher proportion in the blood of patients with MELAS than with MIDD.

## F. Islet amyloid polypeptide (IAPP)

Islets from NIDDM subjects contain characteristic interstitial amyloid deposits and it is now known that the protein component of this amyloid material is a 37 amino acid peptide termed IAPP or amylin (Cooper *et al.*, 1989; O'Brien *et al.*, 1993; Kahn *et al.*, 1990; Nishi *et al.*, 1992). *In vitro* and *in vivo* studies have indicated that at pharmacologic concentrations this peptide can produce insulin resistance, and it is also possible that accumulation and deposition of IAPP in islets may adversely affect β-cell function. However, to date, sequence and linkage studies have failed to identify any genetic abnormalities at the level of the IAPP gene in NIDDM (Nishi *et al.*, 1990, 1992). Furthermore, transgenic mice overexpressing human IAPP in pancreatic β-cells have been generated. Despite greatly elevated circulating levels of IAPP, these animals showed no sign of insulin resistance or diabetes (Fox *et al.*, 1993; de Koning *et al.*, 1994). In addition, infusion of IAPP at subpharmacologic levels failed to cause insulin resistance in dogs. Taken together, these studies indicate that IAPP is unlikely to play a significant role in the insulin resistance of NIDDM. Although it remains possible that the islet amyloid deposits commonly found in NIDDM have an adverse effect on islet function, it is interesting to note that the transgenic mice did not exhibit pancreatic amyloid deposition, despite greatly elevated rates of human IAPP secretion (Fox et al., 1993; de Koning et al., 1994). A number of studies have investigated linkage of the IAPP locus and NIDDM. No mutations in the amylin gene have been found in the small number of patients who have been studied (Nishi et al., 1990; Cook et al., 1991; Nishi et al., 1992; Tokuyama et al., 1994).

## G. The human glucagon-like peptide-1 receptor

The mammalian preproglucagon gene has been shown to encode glucagon as well as two glucagon-like peptides, GLP-1 and GLP-2, respectively (Bell et al., 1983). GLP-1 and glucagon are conserved in all mammals examined thus far. In the distal ileal L-cells, preproglucagon is proteolytically processed to form GLP-1-(1–37), GLP-1-(7–36)-amide, and GLP-1-(7–37). They are potent postprandial stimulators of insulin secretion. The GLP-1 receptor is expressed on islet cells and is a 463 amino acid G-protein-coupled receptor (Thorens et al., 1993a,b; Dillon et al., 1993). The GLP-1 receptor facilitates glucose responsiveness in glucose-resistant islet cells. Consequently, the GLP-1 receptor has been considered to be a candidate gene for the development of insulin secretory defects in diabetes. Using simple sequence repeat polymorphisms to examine alleles on chromosome 6p, no linkage with NIDDM was observed in African- and Caucasian-Americans indicating that this is not a major risk factor in NIDDM in these racial groups (Tanizawa et al., 1994) although this does not exclude a potential role in NIDDM in other populations.

## H. Glycogen synthase

In muscle, the insulin-sensitive glucose transporter, GLUT4, is the major route for cellular glucose uptake (Bell et al., 1993). Once inside the cell, a sizeable fraction of this glucose is routed toward glycogen synthesis for storage purposes. Key enzymatic steps which regulate the process involve glycogen synthase. Insulin activates glycogen synthase through a complex phosphorylation cascade (Dent et al., 1990).

In addition to glycogen synthesis, glucose can also be oxidized to carbon dioxide and water via the Krebs cycle or converted to lactate and pyruvate through glycolytic metabolism. In vivo, glycogen synthesis plus glycolysis are classified as nonoxidative glucose metabolism and Krebs cycle activity is termed oxidative metabolism.

Glycogen synthase activity can be measured in muscle biopsy samples. The level of enzyme activity correlates with the magnitude of nonoxidative glucose disposal. In the insulin-stimulated state, 60% of glucose metabolism is nonoxidative. With moderately severe NIDDM, total insulin-stimulated glucose metabolism is reduced by 75% and proportional defects in nonoxidative and oxidative metabolism are observed. In subjects with impaired glucose tolerance, only nonoxidative metabolism is reduced. Thus, the defect in muscle glucose uptake in NIDDM could arise from abnormalities of intracellular glucose processing as well as from defects in glucose transport. Since glycogen synthesis is a major component of nonoxidative metabolism, much attention has been focused on potential abnormalities of this pathway in NIDDM. Numerous studies have shown a variety of defects in insulin-stimulated glycogen synthase activity in NIDDM and impaired glucose tolerance (Shulman *et al.*, 1990; Vestergaard *et al.*, 1991; Thorburn *et al.*, 1991) and, hence, attention has focused on this enzyme as a potential candidate gene. Mutations in the glycogen synthase gene were considered possible based on the observations that glycogen synthesis and glycogen synthase activity was abnormal in muscle-biopsy specimens obtained from normoglycemic relatives of patients with NIDDM (Vaag *et al.*, 1992). Furthermore, the expression of glycogen synthase was also decreased in skeletal muscle from diabetic patients (Vestergaard *et al.*, 1993).

The glycogen synthase gene is located on chromosome 19q13.3 adjacent to the apolipoprotein E and apolipoprotein C-II genes. Two polymorphic alleles ($A_1$ and $A_2$) for glycogen synthase have been identified by *Xbal* digestion (Groop *et al.*). These differ by a single base in an intron. The A1 allele lacks this *Xbal* site. An association between the *Xbal* RFLP ($A_2$) allele of the glycogen synthase gene and NIDDM was observed in Finnish subjects (Groop *et al.*, 1993). Thirty percent of Finnish NIDDM patients demonstrated the $A_2$ allele, while the $A_2$ allele was only found in 8% of control subjects. This polymorphism ($A_2$ allele) was linked to a subtype of NIDDM characterized by a prominent family history, insulin resistance, and hypertension. This allele was also found in a small number of patients who had no family history of NIDDM. Since this polymorphism is located in a noncoding region, it is unlikely that any structural changes result from this, although it could regulate expression of the gene. However, expression of the glycogen synthase gene was found to be normal in these patients. It is possible that the *Xbal* polymorphism is in linkage disequilibrium with another diabetes-susceptibility gene in Finnish patients. The association between the $A_2$ allele and NIDDM was not observed in French or Japanese NIDDM patients (Kuroyama *et al.*, 1994) despite the similar frequency of the $A_2$ allele in the general population.

The identification of a simple tandem repeat DNA polymorphism $(TG)_n$ (Vionnet and Bell, 1993) allowed the study of associations between this polymorphism and NIDDM in Japanese patients (Kuroyama *et al.*, 1994). Nine alleles were identified in the group of subjects studied. One particular allele, the 2G allele, was found to be significantly associated with NIDDM in Japanese diabetic subjects (Kuroyama *et al.*, 1994). Taken together, all these studies implicate the glycogen synthase gene as a potential marker for NIDDM (Groop *et al.*, 1993; Kuroyama *et al.*, 1994) although its precise role as a candidate gene remains to be fully clarified. Given the polygenic etiology of NIDDM, it is possible that defects in the glycogen synthase gene may act in concert with defects in other genes to give rise to the NIDDM phenotype.

## I. Maturity-onset diabetes of the young (MODY)

MODY is a form of non-insulin-dependent diabetes characterized by an early age of onset and autosomal dominant inheritance suggesting that it may result from a single gene defect (Fajans, 1990). Insulin resistance is usually not present and patients typically display a reduced and delayed insulin response to glucose. The diagnosis is often made in the second decade of life and the diabetes is usually mild. MODY is unusual in comparison to classical late-onset NIDDM and probably accounts for <1% of NIDDM cases. Despite the dominant inheritance, MODY does appear to be heterogeneous as demonstrated by the diverse clinical and metabolic profiles observed in different families and populations (Fajans, 1990). Molecular geneticists have paid particular attention to MODY because of its early age of onset (phenotypic expression) and the availability of well-delineated pedigrees.

The relatively early onset of the disease (<25 years) and the high penetrance enabled the assembly of families with multiple generations. In a large well-studied American MODY kindred, the RW family, 75 DNA markers were tested for linkage to MODY in 39 affected individuals. A marker close to the adenosine deaminase gene was found to cosegregate with the disease in this family localizing a putative diabetes-susceptibility gene to chromosome 20q (Bell *et al.*, 1991). Polymorphisms close to the ADA gene are being used as markers to identify this gene by positional cloning. In this kindred, linkage between this putative gene and MODY has been found in some but not all branches indicating the existence of two distinct susceptibility genes for MODY in this family (Bowden *et al.*, 1992). In a French study of patients with MODY, linkage with the ADA region was only found in two cohorts testifying to the genetic heterogeneity of MODY (Froguel *et al.*, 1992). Linkage of the ADA region with MODY was also absent in a small collection of British and Italian families studies (Baroni *et al.*, 1992a).

Because insulin secretory defects exist in MODY, genes for proteins involved in glucose-induced insulin secretion are likely candidate genes. These include GLUT2, glucokinase, and insulin. Some pedigrees of MODY were hypoinsulinemic and insulin gene mutations were considered in the etiology (Fajans, 1990). After several pedigrees were studied, it was apparent that this locus was not involved in MODY (Permutt, 1991). In addition, the GLUT2 locus was excluded by linkage analysis in MODY (Froguel *et al.*, 1991).

Prior to genetic studies of MODY, it had been suggested that glucokinase was the major glucose sensor in the pancreatic β-cell (Meglasson and Matschinsky, 1984). This was based on the premise that hexose metabolism must occur before insulin secretion is stimulated and glucokinase has a $K_m$ for glucose (6.1 to 8.9 mmol/liter) in the physiological range. GLUT2 transports glucose rapidly across the cell membrane over a wide range of glucose concentrations indicating that it is not a rate-limiting step in glucose-induced insulin secretion. Glucokinase phosphorylates glucose producing glucose-6-phosphate and the high $K_m$ enables direct coupling of glucose-6-phosphate production to the extracellular glucose concentration. For example, a 15% decrease in glucokinase would shift the threshold for glucose-stimulated insulin secretion from 5 to 6 mmol/liter (Meglasson and Matschinsky, 1984). This process may be critical in the insulin response to glucose and led to the hypothesis that this gene may be involved in diabetes mellitus.

The rat liver glucokinase gene was the first enzyme to be cloned (Andreone *et al.*, 1989) and was followed by the cloning of a rat islet cell glucokinase (Iynedjian, 1993). The islet cell cDNA was found to differ from the liver enzyme in its 5′-end. Both enzymes are encoded by the same gene and are alternatively spliced variants which are regulated by tissue-specific promoters (Magnuson *et al.*, 1989) (Figure 3.11). Insulin is the key regulator of the liver enzyme, while the islet enzyme is regulated by plasma glucose levels (Magnuson *et al.*, 1989; Permutt *et al.*, 1992). The cloning of the cDNA for human liver and islet glucokinases (Tanizawa *et al.*, 1991; Koranyi *et al.*, 1992) provided the basis for examining the role of genetic defects in glucokinase. The human glucokinase gene is located on chromosome 7p and contains 12 exons. The liver and islet enzymes differ in the first exon and share exons 2–10 (Iynedjian, 1993) (Figure 3.11).

Three DNA polymorphisms located near the glucokinase gene on Chr 7p were found to be tightly linked to MODY with a LOD score of 25 (Froguel *et al.*, 1992). This score implied that the chance of random association between the glucokinase gene and MODY was less than 1 in $10^{25}$. This linkage was not established in all the MODY families studied and, therefore, suggested the presence of other diabetes-susceptibility genes. In the French study, the glucokinase gene was linked in approximately 60% of families (Froguel *et al.*, 1993a,b). Similar results were found in a study of British families (Hattersley *et al.*, 1992).

Molecular scanning of the glucokinase gene using SSCP analysis identified variant exons and direct sequencing of these exons revealed a nonsense mutation in exon 7 (codon 279) in one kindred (Vionnet *et al.*, 1992). This mutation produced a truncated enzyme devoid of activity. Further analysis revealed two missense mutations in exon 7 in other kindreds (Froguel *et al.*, 1993a,b). Altogether, 32 different mutations in the glucokinase gene have been identified (Froguel *et al.*, 1993a,b). These were found to cosegregate with early

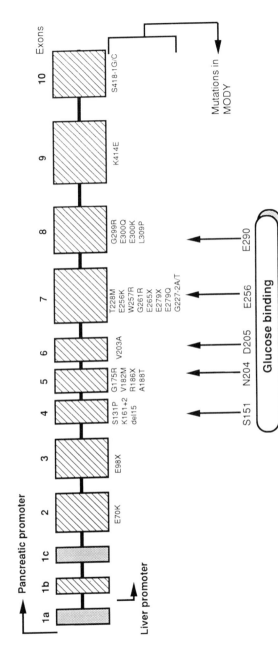

**Figure 3.11.** The glucokinase gene. Organization of the glucokinase gene with structure–function relationships and the locations of some the mutations identified. The exons transcribed in pancreatic β-cells are shown. Adapted with permission from Iynedjian (1993) and Takeda *et al.* (1993). Single-letter amino acid codes are used in this figure.

onset NIDDM and most likely to be the cause of hyperglycemia (Vionnet *et al.*, 1992; Stoffel *et al.*, 1992; Froguel *et al.*, 1993a,b). The mutations are largely heterogeneous and only 3 have been found to occur in more than one kindred. About half of the mutations occur in a CpG dinucleotide "hotspot." The mutations described thus far include nonsense, missense, and deletions and result in decreased glucokinase activity in the pancreas (Gidh-Jain *et al.*, 1993). As expected, the threshold for glucose-stimulated insulin secretion is raised. This manifests as a normal first-phase insulin secretion with a blunted second phase in response to glucose (Velho *et al.*, 1992). Most of the mutations occur in exons 5, 7, and 8 and can be sorted into two groups. One group affects the active site cleft or the surface loops that lead into the cleft. These mutations result in dramatic reductions of enzyme activity. The second group includes residues located distantly from the active site and have smaller effects on the $V_{max}$ and $K_m$ (Gidh-Jain *et al.*, 1993).

The mutations identified thus far have only been found in the heterozygous state. A single abnormal allele would only decrease function by about 50%. Hence, such patients would present with mild NIDDM. It is also apparent that there is incomplete penetrance of the phenotype and consequently only 50% of the heterozygous patients develop diabetes which does not appear to worsen with age unless there is some other superimposed factor which modulates insulin sensitivity.

Mutations in glucokinase domains which are important for enzyme activity (i.e., ATP or glucose-binding sites) may impair glucokinase activity to an extent that manifests as impaired glucose tolerance or diabetes early in life, presenting as MODY. Less severe mutations, which cause only small changes in enzyme activity (in the $V_{max}$ or $K_m$ for glucose), may only cause diabetes when superimposed upon other factors which induce an element of insulin resistance (e.g., obesity, aging, or other genetic factors). The presentation of diabetes in this instance may be delayed until a later age and, hence, it may be important to distinguish cryptic MODY from late-onset NIDDM. Some GCK alleles are associated with late-onset NIDDM in African-Americans and Creole Mauritians (Chiu *et al.*, 1992a,b). However, linkage analysis of larger kindreds and molecular scanning by SSCP analysis (Zouali *et al.*, 1993; Chiu *et al.*, 1993) indicate that glucokinase is a rare major susceptibility gene in late-onset NIDDM (Cook *et al.*, 1992).

The role of nine other candidate genes has been evaluated in MODY (Vaxillaire *et al.*, 1994). These included hexokinase II, IRS-1, fatty acid-binding protein 2, glucagon-like peptide-1 receptor, apolipoprotein C-II, glycogen synthase, adenosine deaminase, and phosphoenolpyruvate carboxykinase. None of these genes showed evidence of linkage with MODY and, hence, it was concluded that these loci are not major susceptibility genes in MODY.

## J. Glucokinase regulatory protein

In approximately 50% of patients with MODY, there is no evidence of linkage with the glucokinase gene implying that other genes may be involved (Permutt *et al.*, 1992). One such protein is the glucokinase regulatory protein (Van Schaftingen *et al.*, 1994). The glucokinase regulatory protein is composed of 628 amino acids with no significant identity with any known protein (Detheux *et al.*, 1994). The gene for the glucokinase regulatory protein has been localized to chromosome 2p22.3 (Vaxillaire *et al.*, 1994). An *Xbal* RFLP has been identified in the gene (Vaxillaire *et al.*, 1994). The glucokinase regulatory protein functions to inhibit glucokinase competitively with respect to glucose. Although its role in the pancreatic β-cell is unclear it has been postulated that mutations in the protein may affect its response to disinhibitors causing decreased glucokinase activity.

## K. The fatty acid-binding protein 2 (FABP2) locus

In the Pima population, insulin resistance is a familial characteristic which precedes the development of NIDDM (Lillioja *et al.*, 1987). The magnitude of insulin resistance shows a large degree of variance which is not accounted for by differences in obesity, fitness, sex, or age. The variance coaggregates in families, suggesting a role for genetic determinants (Lillioja *et al.*, 1987). Maximal insulin action, as defined by glucose clamp studies, is trimodally distributed (Bogardus *et al.*, 1989) and is related to obesity. Taken together, these data indicate that insulin action is determined by an autosomal codominant gene with two alleles. The expression of a gene which influences insulin action would be a strong candidate in the etiology of NIDDM in this population.

Red cell antigen markers were used to investigate linkage with insulin action because the chromosomal locations for most of these are known and these are fairly polymorphic among the Pima population. Sib-pair analysis indicated linkage between the red cell glycophorin A/B locus on chromosome 4q. These findings were extended using additional DNA markers for chromosome 4q and linkage was confirmed in 123 siblings (Prochazka *et al.*, 1993). Two of the markers tested were derived from FABP2 and the annexin V (ANX5) gene which are separated by 1 centimorgan. Annexin is an intracellular calcium-dependent phospholipid-binding protein of unclear physiological function. FABP2 functions in the uptake and transport of long-chain fatty acids in enterocytes. Changes in the FABP2 gene may conceivably alter the amount and types of fatty acids absorbed in the diet. Muscle from diabetic subjects displays changes in fatty acid composition (Borkman *et al.*, 1993) and competition between fatty acids and glucose may contribute to insulin resistance (Randle *et al.*, 1963). A specific point mutation in the FABP2 gene has been identified which is particularly common in the Pimas (Prochazka *et al.*, unpublished). Its precise relationship to the development of NIDDM remains to be determined, however, since studies in European populations (Humphreys *et al.*, 1994) have not demonstrated any genetic defects at this locus, implying that this may be specific to the Pima population. It is possible that linkage disequilibrium has been maintained in the Pimas but lost in heterogeneous European populations.

## VI. THE "THRIFTY" GENE AND SILENT POLYMORPHISM HYPOTHESES

NIDDM may be particularly prevalent in modern society if genes which turn out to be diabetogenic underwent positive evolutionary selection because they provided a survival advantage during periods of starvation (Neel, 1962). A thrifty genotype would allow efficient utilization of fuels and prevent weight loss. With an abundance of food, this genotype would lead to weight gain, obesity, and diabetes. The high prevalence of NIDDM in ethnic groups which have adopted a more westernized lifestyle lends support to this hypothesis. In the past, such groups may have faced starvation or near starvation and, hence, the thrifty genotype would have been advantageous (Dowse and Zimmet, 1993). The Pima native Americans in Arizona and the Nauruans in the South Pacific exemplify the changes in lifestyle over the past century with an increase in the incidence of NIDDM (Zimmet, 1992).

A number of genes may predispose toward the development of a thrifty phenotype: these include those involved in postprandial thermogenesis and fat deposition. Thrifty genes may induce selective insulin resistance. Resistance to the suppression of gluconeogenesis by insulin may result in increased lipogenesis and fat storage (Turner *et al.*, 1993).

Some mutations may have been silent during the periods of a food-deprived lifestyle and may not have conferred any selective advantage (Turner *et al.*, 1993). With the abundance of food, such mutations would lead to disease. Genetic determinants of impaired β-cell function may have existed as previously neutral polymorphisms. With the superimposition of obesity and insulin resistance overt hyperglycemia and NIDDM would develop (Turner *et al.*, 1993).

## VII. CONCLUSION

Considerable effort has been devoted to elucidating the genetic factors underlying NIDDM. A number of genetic defects, each affecting small subgroups of patients, have been clearly identified in specific genes: glucokinase, insulin, the insulin receptor, and the mitochondrial genome.

Despite these advances, however, the genetic basis for NIDDM in the vast majority of patients remains to be determined. This is undoubtedly confounded by the heterogeneous and polygenic nature of NIDDM.

The precise molecular mechanism of insulin action remains to be clearly elucidated although substantial advances have been made in recent years (White and Kahn, 1994). Since, insulin resistance is a primary feature of NIDDM, it is likely that diabetes-susceptibility genes will involve proteins in the insulin-signaling pathways. As novel proteins in this cascade are identified, it will be important to characterize their primary sequences and their regulatory control in large numbers of patients with NIDDM.

Genetic defects are generally regarded as affecting coding regions of gene where a defective protein is produced and is causally linked to the disease. Examples include cystic fibrosis, sickle cell anemia, and Duchenne muscular dystrophy. However, mutations in regions of a gene which regulate transcription may also result in disease. This is particularly evident with the thalassemias (Weatherall, 1991). It is quite possible that defects in the regulatory regions of the candidate genes described may be identified in the future. Defects in insulin action or insulin secretion could potentially arise from such mutations which would affect gene expression of candidate proteins.

It is anticipated that rapid advances in molecular genetics in this, the last decade of the 20th century, will bring forth a speedy resolution of these fundamental issues in our understanding of non-insulin-dependent diabetes.

## Acknowledgments

T.S.P. was supported by a Juvenile Diabetes Foundation International fellowship. W.J.L. was supported by a fellowship from the Medical Research Council of Canada.

## References

Alcolado, J. C., and Alcolado, R. (1991). Importance of maternal history of non-insulin dependent diabetic patients. Br. Med. J. **302**:1178–1180.

Almino, K., Bjorbaek, C., Vestergaard, H., Hanson, T., Echwald, S., and Pedersen. O. (1993). Aminoacid polymorphisms of insulin related substrate-1 in non-insulin dependent diabetes mellitus. Lancet **342**:828–832.

Andreone, T. L., Printz, R. L., Pilkis, S. J., Magnuson, M. A., and Granner, D. K. (1989). The amino acid sequence of rat liver glucokinase deduced from cloned cDNA. J. Biol. Chem. **264**:363–369.

Bahary, N., Leibel, R. L., Joseph, L., and Friedman, J. M. (1994). Molecular mapping of the mouse db mutation. Proc. Natl. Acad. Sci. USA **87**:8642–8646.

Ballinger, S. W., Shoffner, J. M., Hedaya, E. V., Trounce, I., Polak, M. A., Koontz, D. A., and

Wallace, D. C. (1992). Maternally transmitted diabetes and deafness associated with a 10.4 kb mitochondrial DNA deletion. *Nature Genet.* **1:**11–15.

Barnett, A. H., Eff, C., Leslie, R. D., and Pyke, D. A. (1981). Diabetes in identical twins: A study of 200 pairs. *Diabetologia* **20:**87–93.

Baron, V., Kaliman, P., Gautier, N., and Van Obberghen, E. (1992). The insulin receptor activation process involves localized conformational changes. *J. Biol. Chem.* **267:**23290–23294.

Baroni, M. G., Alcolado, J. C., Needham, E. W., Pozzilli, P., Stocks, J., and Galton, D. J. (1992a). Sib-pair analysis of adenosine deaminase locus in NIDDM. *Diabetes* **41:**1640–1643.

Baroni, M. G., Oelbaum, R. S., Pozzilli, P., Stocks, J., Li, S. R., Fiore, V., and Galton, D. J. (1992b). Polymorphisms at the GLUT1 (HepG2) and GLUT4 (muscle/adipocyte) glucose transport genes and non-insulin-dependent diabetes mellitus (NIDDM). *Hum. Genet.* **88:**557–561.

Bell, G. I. (1991). Lilly lecture 1990. Molecular defects in diabetes mellitus. *Diabetes* **40:**413–422.

Bell, G. I., Burant, C. F., Takeda, J., and Gould, G. W. (1993). Structure and function of mammalian facilitative sugar transporters. *J. Biol. Chem.* **268:**19161–19164.

Bell, G. I., Xiang, K. S., Newman, M. V., Wu, S. H., Wright, L. G., Fajans, S. S., Spielman, R. S., and Cox, N. J. (1991). Gene for non-insulin-dependent diabetes mellitus (maturity-onset diabetes of the young subtype) is linked to DNA polymorphism on human chromosome 20q. *Proc. Natl. Acad. Sci. USA* **88:**1484–1488.

Bell, G. I., Santerre, R. F., and Mullenbach, G. T. (1983). Hamster preproglucagon contains the sequence of glucagon and two related peptides. *Nature* **302:**716–718.

Bell, G. I., Selby, M. J., and Rutter, W. J. (1982). The highly polymorphic region of the human insulin gene is composed of simple tandemly repeating sequences. *Nature* **295:**31–33.

Bell, G. I., Karam, J. H., and Rutter, W. J. (1981). Polymorphic DNA region adjacent to the 5' end of the human insulin gene. *Proc. Natl. Acad. Sci. USA* **78:**5759–5761.

Bellanné-Chantelot, C., Lacroix, B., and Ougen, P. (1992). Mapping the whole human genome by fingerprinting yeast artificial chromosomes. *Cell* **70:**1059–1068.

Benecke, H., Flier, J. S., and Moller, D. E. (1993). Alternatively spliced variants of the insulin receptor protein. *J. Clin. Invest.* **89:**2066–2072.

Bogardus, C., Lillioja, S., Nyomba, B. L., Zurlo, F., Swinburn, B., Esposito-Del Puente, A., Knowler, W. C., Ravussin, E., Mott, D. M., and Bennett, P. H. (1989). Distribution of in vivo insulin action in Pima Indians as mixture of three normal distributions. *Diabetes* **38:**1423–1432.

Borkman, M., Storlien, L. H., Pan, D. A., Jenkins, A. B., Chisholm, D. J., and Campbell, L. V. (1993). The relationship between insulin sensitivity and the fatty acid composition of skeletal muscle phospholipids. *N. Engl. J. Med.* **328:**238–244.

Botstein, D., White, R. L., Skolnick, M., and Davis, R. W. (1980). Construction of a genetic linkage map in man using restriction fragment length polymorphisms. *Am. J. Hum. Genet.* **32:**314–331.

Bowden, D. W., Gravius, T. C., Akots, G., and Fajans, S. S. (1992). Identification of genetic markers flanking the locus for maturity-onset diabetes of the young on human chromosome 20. *Diabetes* **41:**88–92.

Brown, V. I., and Greene, M. I. (1991). Molecular and cellular mechanisms of receptor-mediated endocytosis. *DNA Cell Biol.* **10:**399–409.

Cama, A., Patterson, A. P., Kadowaki, T., Kadowaki, H., Siegel, G., D'Ambrosio, D., and Lillioja, S. (1990). The amino acid sequence of the insulin receptor is normal in an insulin-resistant Pima Indian. *J. Clin. Endocrinol. Metab.* **70:**1155–1161.

Celi, F. S., Walston, J., Silver, K., Austin, S., and Shuldiner, A. R. (1994). Evidence against insulin receptor substrate-1 (IRS-1) polymorphisms in Pima native Americans. *Diabetes* **43**(Suppl. 1):224A.

Charles, M. A., Fontbonne, A., Thibult, N., Warnet, J. M., Rosselin, G. E., and Eschwege, E.

(1991). Risk factors for NIDDM in white population. Paris prospective study. *Diabetes* **40**:796–799.

Chiu, K. C., Province, M. A., Dowse, G. K., Zimmet, P. Z., Wagner, G., Serjeantson, S., and Permutt, M. A. (1992a). A genetic market at the glucokinase gene locus for type 2 (non-insulin-dependent) diabetes mellitus in Mauritian Creoles. *Diabetologia* **35**:632–638.

Chiu, K. C., Province, M. A., and Permutt, M. A. (1992b). Glucokinase gene is genetic marker for NIDDM in American blacks. *Diabetes* **41**:843–849.

Chiu, K. C., Tanizawa, Y., and Permutt, M. A. (1993). Glucokinase gene variants in the common form of NIDDM. *Diabetes* **42**:579–582.

Cocozza, S., Porcellini, A., Rticcardi, G., Monticelli, A., Condorelli, G., Ferrara, A., and Pianese, L. (1992). NIDDM associated with mutation in tyrosine kinase domain of insulin receptor gene. *Diabetes* **41**:521–526.

Coghlan, M. P., Pillay, T. S., Tavare, J. M., and Siddle, K. (1994). Site-specific insulin receptor phosphoserine/phosphothreonine antibodies: Identification of serine 1327 as a novel site of phorbol ester-induced phosphorylation. *Biochem. J.* **303**:893–899.

Cook, J. T., Hattersley, A. T., Christopher, P., Bown, E., Barrow, B., Patel, P., Shaw, J. A., Cookson, W. O., Permutt, M. A., and Turner, R. C. (1992). Linkage analysis of glucokinase gene with NIDDM in Caucasian pedigrees. *Diabetes* **41**:1496–1500.

Cook, J. T., Patel, P. P., Clark, A., Hoppener, J. W., Lips, C. J., Mosselman, S., O'Rahilly, S., Page, R. C., Wainscoat, J. S., and Turner, R. C. (1991). Non-linkage of the islet amyloid polypeptide gene with type 2 (non-insulin-dependent) diabetes mellitus. *Diabetologia* **34**:103–108.

Cooper, G. J., Day, A. J., Willis, A. C., Roberts, A. N., Reid, K. B., and Leighton, B. (1989). Amylin and the amylin gene: Structure, function and relationship to islet amyloid and to diabetes mellitus. *Biochim. Biophys. Acta* **1014**:247–258.

Cotton, R. H. G. (1989). Detection of single base changes in nucleic acids. *Biochem. J.* **263**:1–10.

Cox, N. J., and Bell, G. I. (1989). Disease associations. Chance, artifact, or susceptibility genes? *Diabetes* **38**:947–950.

Cox, N. J., Xiang, K. S., Bell, G. I., and Karam, J. H. (1988). Glucose transporter gene and non-insulin-dependent diabetes [letter]. *Lancet* **2**:793–794.

Cox, N. J., Xiang, K. S., Fajans, S. S., and Bell, G. I. (1992). Mapping diabetes-susceptibility genes. Lessons learned from search for DNA marker for maturity-onset diabetes of the young. *Diabetes* **41**:401–407.

DeFronzo, R. A., Bonadonna, R. C., and Ferrannini, E. (1992). Pathogenesis of NIDDM: A balanced overview. *Diabetes Care* **15**:318–368.

DeKoning, E. J. Hoppener, J. W., Verbeek, J. S., Oosterwijk, C., Van Hulst, K. L., Baker, C. A., Lips, C. J., Morris, J. F., and Clark, A. (1994). Human islet amyloid polypeptide accumulates at similar sites in islets of transgenic mice and humans. *Diabetes* **43**:640–644.

Dent, P., Lavoinne, A., Nakielny, S., Caudwell, F. B., Watt, P., and Cohen, P. (1990). The molecular mechanism by which insulin stimulates glycogen synthesis in mammalian skeletal muscle. *Nature* **348**:302–308.

Detheux, M., Vandekerckhove, J., and Van Schaftingen, E. (1994). Cloning and sequencing of rat liver cDNAs encoding the regulatory protein of glucokinase. *FEBS Lett.* **339**:312.

Dillon, J. S., Tanizawa, Y., Wheeler, M. B., Leng, X. H., Ligon, B. B., Rabin, D. U., Yoo-Warren, H., Permutt, M. A., and Boyd, A. E. (1993). Cloning and functional expression of the human glucagon-like peptide-1 (GLP-1) receptor. *Endocrinology* **133**:1907–1910.

Dons, R. F., Ryan, J., Gorden, P., and Wachslicht-Rodbard, H. (1981). Erythrocyte and monocyte insulin binding in man. *Diabetes* **30**:896–902.

Dowse, G., and Zimmet, P. (1993). The thrifty genotype in non-insulin-dependent diabetes. *Br. Med. J.* **306**:532–533.

Dunaif, A. (1993). Insulin resistance and ovarian dysfunction. *In* "Insulin Resistance" (D. Moller, Ed.), pp. 301–326. Wiley, Chichester, U.K.

Ebina, Y., Ellis, L., Jarnagin, K., Edery, M., Graf, L., Clauser, E., and Jing-Hsuing, O. (1985). The human insulin receptor cDNA: The structural basis for hormone-activated transmembrane signalling. *Cell* **40**:747–758.

Egan, S. E., Giddings, B. W., Brooks, M. W., Buday, L., Sizeland, A. M., and Weinberg, R. A. (1993). Association of Sos Ras exchange protein with Grb2 is implicated in tyrosine kinase signal transduction and transformation [see comments]. *Nature* **363**:45–51.

Elbein, S., Rotwein, P., Permutt, M. A., Bell, G. O., Sanz, N., and Karam, J. H. (1985). Lack of association of the polymorphic locus in the 5′-flanking region of the human insulin gene and diabetes in American blacks. *Diabetes* **34**:433–439.

Elbein, S. C., Corsetti, L., Ullrich, A., and Permutt, M. A. (1986). Multiple restriction fragment length polymorphisms at the insulin receptor locus: A highly informative marker for linkage analysis. *Proc. Natl. Acad. Sci. USA* **83**:5223–5227.

Fajans, S. S. (1990). Scope and heterogeneous nature of MODY. *Diabetes Care* **13**:49–64.

Fox, N., Schrementi, J., Nishi, M., Ohagi, S., Chan, S. J., Heisserman, J. A., Westermark, G. T., Leckstrom, A., Westermark, P., and Steiner, D. F. (1993). Human islet amyloid polypeptide transgenic mice as a model of non-insulin-dependent diabetes mellitus (NIDDM). *FEBS Lett.* **323**:40–44.

Frattali, A. L., Treadway, J. L., and Pessin, J. E. (1992). Insulin/IGF-1 hybrid receptors: Implications for the dominant-negative phenotype in syndromes of insulin resistance. *J. Cell. Biochem.* **48**:43–50.

Friedenberg, G. R., Reichart, D., Olefsky, J. M., and Henry, R. R. (1988). Reversibility of defective adipocyte insulin receptor kinase activity in non-insulin dependent diabetes mellitus. *J. Clin. Invest.* **82**:1398–1406.

Froguel, P., Zouali, H., Sun, F., Velho, G., Fukumoto, H., Passa, P., and Cohen, D. (1991). CA repeat polymorphism in the glucose transporter GLUT 2 gene. *Nucleic Acids Res.* **19**:3754.

Froguel, P., Vaxillaire, M., Sun, F., Velho, G., Zouali, H., Butel, M. O., Lesage, S., Vionnet, N., Clement, K., and Fougerousse, F. (1992). Close linkage of glucokinase locus on chromosome 7p to early-onset non-insulin-dependent diabetes mellitus. *Nature* **356**:162–164.

Froguel, P., Velho, G., Passa, P., and Cohen, D. (1993a). Genetic determinants of type 2 diabetes mellitus: Lessons learned from family studies. *Diabetes Metab.* **19**:1–10.

Froguel, P., Zouali, H., Vionnet, N., Velho, G., Vaxillaire, M., Sun, F., Lesage, S., Stoffel, M., Takeda, J., and Passa, P. (1993b). Familial hyperglycemia due to mutations in glucokinase. Definition of a subtype of diabetes mellitus. *N. Engl. J. Med.* **328**:697–702.

Gidh-Jain, M., Takeda, J., Xu, L. Z., Lange, A. J., Vionnet, N., Stoffel, M., Froguel, P., Velho, G., Sun, F., and Cohen, D. (1993). Glucokinase mutations associated with non-insulin-dependent (type 2) diabetes mellitus have decreased enzymatic activity: Implications for structure/function relationships. *Proc. Natl. Acad. Sci. USA* **90**:1932–1936.

Granner, D. K., and O'Brien, R. M. (1992). Molecular physiology and genetics of NIDDM. Importance of metabolic staging. *Diabetes Care* **15**:369–395.

Groop, L. C., Kankuri, M., Schalin-Jantti, C., Ekstrand, A., Nikula-Ijas, P., Widen, E., Kuismanen, E., Eriksson, J., Franssila-Kallunki, A., and Saloranta, C. (1993). Association between polymorphism of the glycogen synthase gene and non-insulin-dependent diabetes mellitus. *N. Engl. J Med.* **328**:10–14.

Haffner, S. M., Stern, M. P., Mitchell, B. D., Hazula, H. P., and Patterson, J. K. (1990). Incidences of Type II diabetes in Mexican Americans predicted by fasting insulin and glucose levels, obesity, and body fat distribution. *Diabetes* **39**:283–288.

Hager, J., Zouali, H., Velho, G., Froguel, P. (1993). Insulin receptor substrate (IRS-1) gene polymorphisms in French NIDDM families. *Lancet* **342**:1430.

Harris, M. I., Hadden, W. C., Knowler, W. C., and Bennett, P. H. (1987). Prevalence of diabetes and impaired glucose tolerance and plasma glucose levels in US populations aged 20–74 yr. *Diabetes* **36**:523–534.

Hattersley, A. T., Turner, R. C., Permutt, M. A., Patel, P., Tanizawa, Y., Chiu, K. C., O'Rahilly, S., Watkins, P. J., and Wainscoat, J. S. (1992). Linkage of type 2 diabetes to the glucokinase gene. *Lancet* **339**:1307–1310.

Hitman, G. A., and McCarthy, M. I. (1991). Genetics of non-insulin dependent diabetes mellitus. *Baillieres. Clin. Endocrinol. Metab.* **5**:455–476.

Humphreys, P., McCarthy, M., Tuomilehto, J., Tuomilehto-Wolf, E., Stratton, I., Morgan, R., Rees, A., Owens, D., Stengard, J., and Nissinen, A. (1994). Chromosome 4q locus associated with insulin resistance in Pima Indians. Studies in three European NIDDM populations. *Diabetes* **43**:800–804.

Imai, Y., Accili, D., Suzuki, Y., Sesti, G., Fusco, A., and Taylor, A. (1994). Variant sequences of IRS-1 in NIDDM. (Suppl. 1): *Diabetes* **43**:168A.

Imano, E., Kadowaki, H., Kadowaki, T., Iwama, N., Watarai, T., Kawamori, R., and Kamada, T. (1991). Two patients with insulin resistance due to decreased levels of insulin receptor mRNA. *Diabetes* **40**:548–557.

Iynedjiam, P. B. (1993). Mammalian glucokinase and its gene. *Biochem. J.* **293**:1–13.

Kadowaki, H., Kadowaki, T., Camo, A., Marcus-Samuels, B., Rovira, A., Bevins, C. L., and Taylor, S. I. (1990a). Mutagenesis of lysine 460 in the human insulin receptor. *J. Biol. Chem.* **265**:21285–21296.

Kadowaki, T., Bevins, C. L., Cama, A., Ojamaa, K., Marcus-Samuels, B., Kadowaki, H., and Beitz, L. (1988). Two mutant alleles of the insulin receptor gene in a patient with extreme insulin resistance. *Science* **240**:787–789.

Kadowaki, T., Kadowaki, H., Accili, D., and Taylor, S. I. (1990b). Substitution of lysine for asparagine at position 15 in the alpha subunit of the human insulin receptor. *J. Biol. Chem.* **265**:19141–19150.

Kadowaki, T., Kadowaki, H., Mori, Y., Tobe, K., Sakuta, R., Suzuki, Y., Tanabe, Y., Sakura, H., Awata, T., and Goto, Y. (1994). A subtype of diabetes mellitus associated with a mutation of mitochondrial DNA. *N. Engl. J. Med.* **330**:962–968.

Kadowaki, T., Kadowaki, H., Rechler, M. M., Serrano-Rios, M., Roth, J., Gordon, P., and Taylor, S. I. (1990c). Five mutant alleles of the insulin receptor gene in patients with genetic forms of insulin resistance. *J. Clin. Invest.* **86**:254–264.

Kadowaki, T., Miyake, Y., Hagura, R., Akanuma, Y., Kajinuma, H., Kuzuya, N., Takaku, F., and Kosaka, K. (1984). Risk factors for worsening to diabetes in subjects with impaired glucose tolerance. *Diabetologia* **26**:44–49.

Kahn, C. R., Flier, J. S., Bar, R. S., Archer, J. A., Gordon, P., Martin, M. M., and Roth, J. (1976). The syndromes of insulin resistance and acanthosis nigricans. *N. Engl. J. Med.* **294**:739–745.

Kahn, S. E., D'Alessio, D. A., Schwartz, M. W., Fujimoto, W. Y., Ensinck, J. W., Taborsky, G. J. J., and Porte, D. J. (1990). Evidence of cosecretion of islet amyloid polypeptide and insulin by beta-cells. *Diabetes* **39**:634–638.

Karasik, A., and Hattori, M. (1994). Animal models of NIDDM. *In* "Joslin's Diabetes Mellitus" (C. R. Kahn and G. C. Weir, Eds.), pp. 317–350. Lea & Febiger, Philadelphia.

Kim, H., Kadowaki, H., Sakura, H., Adawara, M., Momomura, K., Takahashi, Y., and Miyazaki, Y. (1992). Detection of mutations in the insulin receptor gene in patients with insulin resistance by analysis of single-stranded conformational polymorphisms. *Diabetologia* **35**: 261–266.

Knowler, W. C., Bennett, P. H., Pettitt, D., and Savage, P. J. (1981). Diabetes incidence in Pima Indians: Contributions of obesity and parental diabetes. *Am. J. Epidemiol.* **113**:144–156.

Koranyi, L. I., Tanizawa, Y., Welling, C. M., Rabin, D. U., and Permutt, M. A. (1992). Human islet glucokinase gene. Isolation and sequence analysis of full-length cDNA. *Diabetes* **41**:807–811.

Kubar, J., and Van Obberghen, E. (1989). Oligomeric states of the insulin receptor: Binding and autophosphorylation properties. *Biochemistry* **28**: 1086–1093.

Kuroyama, H., Sanke, T., Ohagi, S., Furuta, M., Furuta, H., and Nanjo, K. (1994). Simple tandem repeat DNA polymorphism in the human glycogen synthase gene is associated with NIDDM in Japanese subjects. *Diabetologia* **37**:536–539.

Kusari, J., Olefsky, J. M., Strahl, C., and McClain, D. A. (1991a). Insulin receptor cDNA sequence in NIDDM patients homozygous for insulin receptor gene RFLP. *Diabetes* **40**:249–254.

Kusari, J., Verma, U. S., Buse, J. B., Henry, R. R., and Olefsky, J. M. (1991b). Analysis of the gene sequences of the insulin receptor and the insulin-sensitive glucose transporter (GLUT-4) in patients with common-type non-insulin-dependent diabetes mellitus. *J. Clin. Invest.* **88**:1323–1330.

Kusari, J., Verma, V. S., Buse, J. B., Henry, R. R., and Olefsky, J. M. (1991c). Analysis of the gene sequences of the insulin receptor and the insulin sensitive glucose transporter (GLUT-4) in patients with common-type non-insulin-dependent diabetes mellitus. *J. Clin. Invest.* **89**:1323–1330.

Levy-Toledano, R., Caro, L. H. P., Accili, D., and Taylor, S. I. (1994). Investigation of the mechanism of the dominant negative effect of mutations in the tyrosine kinase domain of the insulin receptor. *EMBO J.* **13**:835–842.

Lewis, R. E., Volle, D. J., Sanderson, S. D. (1994). Phorbol ester stimulates phosphorylation on serine 1327 of human insulin receptor. *J. Biol. Chem.* **42**:26259–26266.

Li, S. R., Alcolado, J. C., Stocks, J., Baroni, M. G., Oelbaum, R. S., and Galton, D. J. (1991). Genetic polymorphisms at the human liver/islet glucose transporter (GLUT2) gene locus in Caucasian and West Indian subjects with type 2 (non-insulin-dependent) diabetes mellitus. *Biochim. Biophys. Acta* **1097**:293–298.

Li, S. R., Oelbaum, R. S., Bouloux, P. M., Stocks, J., Baroni, M. G., and Galton, D. J. (1990). Restriction site polymorphisms at the human HepG2 glucose transporter gene locus in Caucasian and west Indian subjects with non-insulin-dependent diabetes mellitus. *Hum. Hered.* **40**:38–44.

Liang, P., and Pardee, A. B. (1992). Differential display of eukaryotic messenger RNA by means of the polymerase chain reaction. *Science* **257**:967–971.

Lillioja, S., Mott, D. M., Howard, B. V., Bennett, P. H., Yki-Jarvinen, H., Freymond, D., Nyomba, B. L., Zurlo, F., Swinbum, B., and Bogardus, C. (1988). Impaired glucose tolerance as a disorder of insulin action: Longitudinal and cross-sectional studies in Pima Indians. *N. Engl. J. Med.* **318**:1217–1225.

Lillioja, S., Mott, D. M., Zawadzki, J. K., Young, A. A., Abbott, W. G., Knowler, W. C., Bennett, P. H., Moll, P., and Bogardus, C. (1987). In vivo insulin action is familial characteristic in nondiabetic Pima Indians. *Diabetes* **36**:1329–1335.

Lisitsyn, N., and Wigler, M. (1993). Cloning the differences between two complex genomes. *Science* **259**:946–951.

Maegawa, H., McClain, D. A., Friedenberg, G., Olefsky, J. M., Napier, M., Lipari, T., and Dull, T. J. (1988). Properties of a human insulin receptor with a COOH-terminal truncation. *J. Biol. Chem.* **263**:8912–8917.

Magnuson, M. A., Andreone, T. L., Printz, R. L., Koch, S., and Granner, D. K. (1989). Rat glucokinase gene: Structure and regulation by insulin. *Proc. Natl. Acad. Sci. USA* **86**:4838–4842.

Matsutani, A., Koranyi, L., Cox, N., and Permutt, M. A. (1990). Polymorphisms of GLUT2 and GLUT4 genes. Use in evaluation of genetic susceptibility to NIDDM in blacks. *Diabetes* **39**:1534–1542.

McClain, D. A., Henry, R. R., Ullrich, A., and Olefsky, J. M. (1988). Restriction fragment length polymorphism in insulin receptor gene and insulin resistance in NIDDM. *Diabetes* **37**:1071–1075.

Meglasson, M. D., and Matschinsky, F. M. (1984). New perspectives on pancreatic islet glucokinase. *Am. J. Physiol.* **246:**E1–13E.

Moller, D. E., Yokota, A. A., and Flier, J. S. (1989). Normal insulin receptor cDNA sequence in Pima Indians with NIDDM. *Diabetes* **38:**1496–1500.

Moller, D. E., Yokota, A. A., White, M. F., Pazianos, A. G., and Flier, J. S. (1990). A naturally occurring mutation of insulin receptor alanine 1134 impairs tyrosine kinase function and is associated with dominantly inherited insulin resistance. *J. Biol. Chem.* **265:**14979–14985.

Moller, D. E., Benecke, H., and Flier, J. S. (1991). Biologic activities of naturally occurring human insulin receptor mutations. *J. Biol. Chem.* **266:**10995–11001.

Moller, D. E., Cohen, O., Yamaguchi, Y., Assiz, R., Grigorescu, F., Eberle, A., and Morrow, L. A. (1994). Prevalence of mutations in the insulin receptor gene in subjects with features of the Type A syndrome of insulin resistance. *Diabetes* **43:**247–255.

Mueckler, M., Kruse, M., Strube, M., Riggs, A. C., Chiu, K. C., and Permutt, M. A. (1994). A mutation in the GLUT2 glucose transporter gene of a diabetic patient abolishes transport activity. *J. Biol. Chem.* **269:**17765–17767.

Neel, J. V. (1962). Diabetes mellitus: A thrifty genotype rendered detrimental by 'progress'? *Am. J. Hum. Genet.* **14:**353–362.

Nishi, M., Bell, G. I., and Steiner, D. F. (1990). Islet amyloid polypeptide (amylin): No evidence of an abnormal precursor sequence in 25 type 2 (non-insulin-dependent) diabetic patients. *Diabetologia* **33:**628–630.

Nishi, M., Sanke, T., Ohagi, S., Ekawa, K., Wakasaki, H., Nanjo, K., Bell, G. I., and Steiner, D. F. (1992). Molecular biology of islet amyloid polypeptide. *Diabetes Res. Clin. Pract.* **15:**37–44.

Nishio, Y., Aiello, L. P., and King, G. L. (1994). Glucose-induced genes in bovine aortic smooth muscle cells identified by mRNA differential display. *FASEB. J.* **8:**103–106.

O'Brien, T. D., Butler, P. C., Westermark, P., and Johnson, K. H. (1993). Islet amyloid polypeptide: A review of its biology and potential roles in the pathogenesis of diabetes mellitus. *Vet. Pathol.* **30:**317–332.

O'Rahilly, S., Choi, W. H., Patel, P., Turner, R. C., Flier, J. S., and Moller, D. E. (1991). Detection of mutations in insulin receptor gene in NIDDM patients by analysis of single-stranded conformational polymorphisms. *Diabetes* **40:**777–782.

O'Rahilly, S., Trembath, R. C., Patel, P., Galton, D. J., Turner, R. C., and Wainscoat, J. S. (1988). Linkage analysis of the human insulin receptor gene in Type 2 (non-insulin dependent) diabetic families and a family with maturity onset diabetes of the young. *Diabetalogia* **31:**792–797.

Oelbaum, R. S. (1992). Analysis of three glucose transporter genes in a Caucasian population: No associations with non-insulin-dependent diabetes and obesity. *Clin. Genet.* **42:**260–266.

Olansky, L., Welling, C., Giddings, S., Adler, S., Bourey, R., Dowse, G., Serjeantson, S., Zimmet, P., and Permutt, M. A. (1992). A variant insulin promoter in non-insulin-dependent diabetes mellitus. *J. Clin. Invest.* **89:**1596–1602.

Olefsky, J. M. (1989). DeGroot: Endocrinology, 2nd ed. W. B. Saunders, Philadelphia.

Olefsky, J. M. (1990). The insulin receptor. A multifunctional protein. *Diabetes* **39:**1009–1016.

Olefsky, J. M. (1991). Cecil Textbook of Medicine. W. B. Saunders, Philadelphia.

Orita, M., Suzuki, Y., Sekiya, T., and Hayashi, K. (1989). Rapid and sensitive detection of point mutations and DNA polymorphisms using the polymerase chain reaction. *Genomics* **5:**874–879.

Owerbach, D., Bell, G. I., Rutter, W. J., and Shows, T. B. (1980). The insulin gene is located on chromosome 11 in humans. *Nature* **286:**82–84.

Permutt, M. A. (1991). Use of DNA polymorphisms for genetic analysis of non-insulin dependent diabetes mellitus. *Baillieres. Clin. Endocrinol. Metab.* **5:**495–526.

Permutt, M. A., Chiu, K. C., and Tanizawa, Y. (1992). Glucokinase and NIDDM. A candidate gene that paid off. *Diabetes* **41:**1367–1372.

Permutt, M. A., and Elbein, S. C. (1990). Insulin gene in diabetes. Analysis through RFLP. *Diabetes Care* **13**:364–374.

Pillay, T. S., Whittaker, J., Lammers, R., Ullrich, A., and Siddle, K. (1991). Multisite serine phosphorylation of the insulin and IGF-I receptors in transfected cells. *FEBS Lett.* **288**:206–211.

Porte, D., Jr. (1991). Banting lecture 1990: Beta-cells in Type II diabetes mellitus. *Diabetes* **40**:166–180.

Prochazka, M., Lillioja, S., Tait, J. F., Knowler, W. C., Mott, D. M., Spraul, M., Bennett, P. H., and Bogardus, C. (1993). Linkage of chromosomal markers on 4q with a putative gene determining maximal insulin action in Pima Indians. *Diabetes* **42**:514–519.

Rabson, S. M., and Mendenhall, E. N. (1956). Familial hypertrophy of pineal body, hyperplasia of adrenal cortex and diabetes mellitus: Report of three cases. *Am. J. Clin. Pathol.* **26**:283–290.

Rabuodi, S. H., Mitchell, B. D., Stern, M. P., Eifler, C. W., Haffner, S. M., Hazuda, H. P., and Frazier, M. L. (1989). Type 2 diabetes mellitus and polymorphism of insulin receptor gene in Mexican Americans. *Diabetes* **38**:975–980.

Rajagopalan, M., Neidigh, J. L., and McClain, D. A. (1991). Amino acid sequences Gly–Pro–Leu–Tyr and Asn–Pro–Glu–Tyr in the submembranous domain of the insulin receptor are required for normal endocytosis. *J. Biol. Chem.* **266**:23068–23073.

Randle, P. J., Garland, P. B., Hales, C. N., and Newsholmes, E. A. (1963). The glucose fatty-acid cycle: It's role in insulin sensitivity and the metabolic disturbances of diabetes mellitus. *Lancet* **1**:785–789.

Remes, A. M., Majamaa, K., Herva, R., and Hassinen, I. E. (1993). Adult-onset diabetes mellitus and neurosensory hearing loss in maternal relatives of MELAS patients in a family with the tRNA (Leu(UUR)) mutation. *Neurology* **43**:1015–1020.

Reynet, C., and Kahn, C. R. (1993). Rad: A member of the Ras family overexpressed in muscle of Type II diabetic humans. *Science* **262**:1441–1444.

Saad, M. F., Knowler, W. C., Pettitt, D. J., Nelson, R. G., Charles, M. A., and Bennett, P. H. 1991). A two step model for development of non-insulin dependent diabetes mellitus. *Am. J. Physiol.* **90**:229–235.

Schumacher, R., Soos, M. A., Schlessinger, J., Brandenberg, D., Siddle, K., and Ullrich, A. (1993). Signalling-competent receptor chimeras allow mapping of major insulin receptor binding domain determinants. *J. Biol. Chem.* **268**:1087–1094.

Seino, S., Seino, N., Nishi, S., and Bell, G. I. (1989). Structure of the human insulin receptor gene and characterization of its promotor. *Proc. Natl. Acad. Sci. USA* **86**:114–118.

Serjeantson, S., and Zimmet, P. (1987). Diabetes in the Pacific: Evidence for a major gene. *In* "Diabetes Mellitus: Recent Knowledge on Aetiology, Complications and Treatment" (S. Babe, M. Gould, and P. Zimmet, Eds.), pp. 23–40. Academic Press, Sydney.

Serjeantson, S. W., and Zimmet, P. (1991). Genetics of non-insulin dependent diabetes mellitus in 1990. *Baillieres. Clin. Endocrinol. Metab.* **5**:477–493.

Shipman, P., Kammerer, C., O'Connell, P., and Stern, M. (1994). Insulin receptor substrate-1 is not linked to type II diabetes in Mexican Americans. *Diabetes* **43**(Suppl.1):168A.

Shulman, G. I., Rothman, D. L., Jue, T., Stein, P., DeFronzo, R. A., and Shulman, R. G. (1990). Quantitation of muscle glycogen synthesis in normal subjects and subjects with non-insulin-dependent diabetes by 13C nuclear magnetic resonance spectroscopy. *N. Engl. J. Med.* **322**:223–228.

Steiner, D. F., Tager, H. S., Chan, S. J., Nanjo, K., Sanke, T., and Rubenstein, A. H. (1990). Lessons learned from molecular biology of insulin-gene mutations. *Diabetes Care* **13**:600–609.

Stoffel, M., Froguel, P., Takeda, J., Zouali, H., Vionnet, N., Nishi, S., Weber, I. T., Harrison, R. W., Pilkis, S. J., and Lesage, S. (1992). Human glucokinase gene: Isolation, characterization, and identification of two missense mutations linked to early-onset non-insulin-dependent (type 2) diabetes mellitus [published erratum appears in *Proc. Natl. Acad. Sci. USA* 1992 Nov 1, **89**(21):10562]. *Proc. Natl. Acad. Sci. USA* **89**:7698–7702.

Takata, Y., Webster, N. J. G., and Olefsky, J. M. (1991). Mutation of the two carboxyl-terminal tyrosines results in an insulin receptor with normal metabolic signaling but enhanced mitogenic signaling properties. *J. Biol. Chem.* **266:**9135–9139.

Tanizawa, Y., Koranyi, L. I., Welling, C. M., and Permutt, M. A. (1991). Human liver glucokinase gene: Cloning and sequence determination of two alternatively spliced cDNAs. *Proc. Natl. Acad. Sci. USA* **88:**7294–7297.

Tanizawa, Y., Riggs, A. C., Elbein, S. C., Whelan, A., Donis-Keller, H., and Permutt, M. A. (1994). Human glucagon-like peptide-1 receptor gene in NIDDM. Identification and use of simple sequence repeat polymorphisms in genetic analysis. *Diabetes* **43:**752–757.

Taylor, S. I. (1992). Lilly Lecture: Molecular mechanisms of insulin resistance. Lessons from patients with mutations in the insulin receptor gene. *Diabetes* **41:**1473–1490.

Taylor, S. I., Royh, J., Blizzard, R. M., and Elders, M. J. (1981). Qualitative abnormalities in insulin binding in a patient with extreme insulin resistance; Decreased sensitivity to alterations in temperature and pH. *Proc. Natl. Acad. Sci. USA* **78:**7157–7161.

Thorburn, A. W., Gumbiner, B., Bulacan, F., Brechtel, G., and Henry, R. R. (1991). Multiple defects in muscle glycogen synthase activity contribute to reduced glycogen synthesis in non-insulin dependent diabetes mellitus. *J. Clin. Invest.* **87:**489–495.

Thorens, B., Porret, A., Buhler, L., Deng, S. P., Morel, P., and Widmann, C. (1993). Cloning and functional expression of the human islet GLP-1 receptor. Demonstration that exendin-4 is an agonist and exendin-(9-39) an antagonist of the receptor. *Diabetes* **42:**1678–1682.

Thorens, B., and Waeber, G. (1993). Glucagon-like peptide-I and the control of insulin secretion in the normal state and in NIDDM. *Diabetes* **42:**1219–1225.

Tokuyama, Y., Kanatsuka, A., Suzuki, Y., Yamaguchi, T., Taira, M., Makino, H., and Yoshida, S. (1994). Islet amyloid polypeptide gene: No evidence of abnormal promoter region in thirty-five type 2 diabetic patients. *Diabetes Res. Clin. Pract.* **22:**99–105.

Treadway, J. L., Morrison, B. D., Soos, M. A., Siddle, K., Olefsky, J., Ullrich, A., and McClain, D. A. (1991). Transdominant inhibition of tyrosine kinase activity in mutant insulin/insulin-like growth factor 1 hybrid receptors. *Proc. Natl. Acad. Sci. USA* **88:**214–218.

Turner, R. C., Levy, J. C., and Clark, A. (1993). Complex genetics of Type 2 diabetes: Thrifty genes and previously neutral polymorphisms. *Q. J. Med.* **86:**413–417.

Ullrich, A., Bell, J. R., Chen, E. Y., Herrera, R., Petruzelli, I. M., Dull, T. J., and Gray, A. (1985). Human insulin receptor and its relationship to the tyrosine kinase family of oncogenes. *Nature* **313:**756–761.

Ullrich, A., Dull, T. J., and Gray, A. (1980). Genetic variation in the human insulin gene. *Science* **200:**612.

Ullrich, A., and Schlessinger, J. (1990). Signal transduction by receptors with tyrosine kinase activity. *Cell* **61:**203–212.

Vaag, A., Henriksen, J. E., and Beck-Nielsen, H. (1992). Decreased insulin activation of glycogen synthase in skeletal muscles in young nonobese Caucasian first-degree relatives of patients with non-insulin-dependent diabetes mellitus. *J Clin. Invest.* **89:**782–788.

van den Ouweland, J. M., Lemkes, H. H., Trembath, R. C., Ross, R., Velho, G., Cohen, D., Froguel, P., and Maassen, J. A. (1994). Maternally inherited diabetes and deafness is a distinct subtype of diabetes and associated with a single point mutation in the mitochondrial tRNA (Leu(UUR)) gene. *Diabetes* **43:**746–751.

Van Schaftingen, E., Detheux, M., and Veiga Da Cunha, M. (1994). Short-term control of glucokinase activity: Role of a regulatory protein. *FASEB J.* **8:**414–419.

Vaxillaire, M., Vionnet, N., Vigouroux, C., Sun, F., Espinosa, R., Lebeau, M. M., Stoffel, M., Lehto, M., Beckmann, J. S., and Detheux, M. (1994). Search for a third susceptibility gene for maturity-onset diabetes of the young. Studies with eleven candidate genes. *Diabetes* **43:**389–395.

Velho, G., Froguel, P., Clement, K., Pueyo, M. E., Rakotoambinina, B., Zouali, H., Passa, P., Cohen, D., and Robert, J. J. (1992). Primary pancreatic beta-cell secretory defect caused by mutations in glucokinase gene in kindreds of maturity onset diabetes of the young. *Lancet* **340**:444–448.

Vestergaard, H., Bjorbaek, C., Andersen, P. H., Bak, J. F., and Pedersen, O. (1991). Impaired expression of glycogen synthase mRNA in skeletal muscle of NIDDM patients. *Diabetes* **40**:1740–1745.

Vestergaard, H., Lund, S., Larsen, F. S., Bjerrum, O. J., and Pedersen, O. (1993). Glycogen synthase and phosphofructokinase protein and mRNA levels in skeletal muscle from insulin-resistant patients with non-insulin-dependent diabetes mellitus. *J. Clin. Invest.* **91**:2342–2350.

Vinik, A. J., Cox, N. J., Xiang, K., Fajans, S. S., and Bell, G. I. (1988). Linkage studies of maturity onset diabetes of the young—R. W. pedigree [letter]. *Diabetologia* **31**:778.

Vionnet, N., and Bell, G. I. (1993). Identification of a simple tandem repeat DNA polymorphism in the human glycogen synthase gene and linkage to five markers on chromosome 19q. *Diabetes* **42**:930–932.

Vionnet, N., Stoffel, M., Takeda, J., Yasuda, K., Bell, G. I., Zouali, H., Lesage, S., Velho, G., Iris, F., and Passa, P. (1992). Nonsense mutation in the glucokinase gene causes early-onset non-insulin-dependent diabetes mellitus. *Nature* **356**:721–722.

Wallace, D. C. (1992). Diseases of the mitochondrial DNA. *Annu. Rev. Biochem.* **61**:1175–1212.

Warram, J. H., Martin, B. C., Krolewski, A. S., Soeldner, J. S., and Kahn, C. R. (1990). Slow glucose removal rate and hyperinsulinemia precede the development of Type II diabetes in the offspring of diabetic parents. *Ann. Int. Med.* **13**:909–915.

Weatherall, D. J. (1991). The New Genetics and Clinical Practice. Oxford Univ. Press, Oxford.

Weir, G. C., and Leahy, J. L. (1994). Pathogenesis of non-insulin-dependent (type II) diabetes mellitus. *In* "Joslin's Diabetes Mellitus" (C. R. Kahn and G. C. Weir, Eds.), pp. 240–264. Lea & Febiger, Philadelphia.

Weissenbach, J., Gyapay, G., Dib, C., Vignal, A., Morisette, J., Millasseau, P., Vaysseix, G., and Lathrop, M. (1992). A second generation linkage map of the human genome. *Nature* **359**:794–801.

White, M. F., and Kahn, C. R. (1994). The insulin signaling system. *J. Biol. Chem.* **269**:1–4.

White, M. F., Livingston, J. N., Backer, J. M., Lauris, V., Dull, T. J., Ullrich, A., and Kahn, C. R. (1988). Mutation of the insulin receptor at tyrosine 960 inhibits signal transmission but does not affect its tyrosine kinase activity. *Cell* **54**:641–649.

Wicking, C., and Williamson, B. (1994). From linked marker to gene. *Trends Genet.* **7**:288–293.

Wilden, P. A., Siddle, K., Haring, E., Backer, J. M., White, M. F., and Kahn, C. R. (1992). The role of insulin receptor kinase domain autophosphorylation in receptor-mediated activities. *J. Biol. Chem.* **267**:13719–13727.

World Health Organization (WHO). (1985). Study group on Diabetes Mellitus. *In* "Technical Report Series 727." WHO, Geneva.

Yip, C. C. (1992). The insulin-binding domain of the insulin receptor is encoded by exon 2 and exon 3. *J. Cell. Biochem.* **48**:19–25.

Zimmet, P. Z. (1992). Kelly West Lecture 1991: Challenges in diabetes epidemiology from West to East. *Diabetes Care* **15**:232–252.

Zouali, H., Vaxillaire, M., Lesage, S., Sun, F., Velho, G., Vionnet, N., Chiu, K., Passa, P., Permutt, A., and Demenais, F. (1993). Linkage analysis and molecular scanning of glucokinase gene in NIDDM families. *Diabetes* **42**:1238–1245.

# 4

# The Hemophilias

**P. M. Green, J. A. Naylor, and F. Giannelli**
Division of Medical and Molecular Genetics
United Medical and Dental Schools of Guy's and St. Thomas's Hospitals
7/8th floors Guy's Hospital Tower
London Bridge
London SE1 9RT, United Kingdom

## I. TWO DISTINCT FORMS OF HEMOPHILIA

Possibly the first historical reference to hemophilia may be found in the Mish-neh, a second century compilation of Jewish law. In this, Rabbi Judah the Patriarch writes that if a mother has had two sons circumcised and who died as a result, then the third son must not be circumcised. However, it is unclear whether this and later rabbinical writings specifically refer to hemophilia, since deaths following circumcision could be due to many causes. Sometimes, how-ever, specific reference to "loose" blood is made (Rosner, 1969).

An accurate account of hemophilia was published by Otto in 1803 and the X-linked nature of the disease, which affects males and is transmitted by females, was clearly described. This description, of course, precedes Mendel's discovery of the laws of inheritance by half a century, and that of X-linkage (Morgan, 1910) by more than a century. A significant advance in the under-standing of hemophilia was the correct description in 1937 of the function of an antihemophilic factor (Patek and Taylor, 1937), but the existence of two distinct forms of hemophilia was not yet suspected. Possibly the first hint of such hetero-geneity can be found in the investigation of the linkage between hemophilia and color blindness by Haldane and Smith (1947). In this study, 1 of 17 pedigrees was exceptional in showing two recombinations in four scorable meioses, while others showed clear linkage. The authors briefly entertained the possibility of genetic heterogeneity but could not pursue it further because no more informative families were available.

In 1952, the discovery of complementation of the coagulation defect in mixtures of plasma from unrelated patients clearly indicated the existence of two different hemophilias (Aggeler *et al.*, 1952; Biggs *et al.*, 1952). In the United Kingdom the two hemophilias were named classical hemophilia and Christmas disease (the latter from the name of the first patient identified), while in the United States the terms hemophilia A and B were used.

The coagulation factors deficient in hemophilia A and B are now called factors VIII and IX, respectively. The latter is a serine protease, while the former is a cofactor of factor IX; both circulate in the blood in an inactive form. When activated, factors IX and VIII cooperate to cleave and activate factor X, thus together contributing to the proteolytic cascade that controls the conversion of fibrinogen to fibrin.

Modern theory suggests that following tissue damage activated coagulation factor VII (VIIa) and tissue factor form a complex that activates both factor IX and X. The latter then activates prothrombin and both thrombin and factor Xa activate factor VIII. The action of factor VIIa and tissue factor is then inhibited by the tissue factor pathway inhibitor (previously known as LACI—lipoprotein associated coagulation inhibitor) and thus the task of maintaining the coagulation pathway rests with activated factor IX and VIII (Broze, 1992). Factor IX is also activated by a branch of the proteolytic coagulation pathway called the intrinsic pathway, where factor XI is the element immediately preceding factor IX and therefore responsible for the activation of the latter (Bajaj *et al.*, 1983).

Since hemophilia A and B affect the same blood coagulation step, patients with either type of hemophilia have essentially the same features. These include spontaneous bleeding, especially into joints and muscles, prolonged bleeding from minor cuts, and severe bleeding from lacerations. Repeated hemarthroses usually lead to crippling joint deformities. Spontaneous bleeding is usually limited to patients with moderate and severe disease, while mildly affected patients may show symptoms only after clear traumas. The severity of the disease is generally correlated with the degree of specific coagulant deficits. Patients with coagulant levels of $<2$, $2$–$10$, and $>10\%$ of normal are usually severely, moderately, and mildly affected, respectively. Approximately 48% of the patients with hemophilia A have severe disease, while a lower proportion (36%) has severe hemophilia B (Rizza and Spooner, 1983). Differential diagnosis for the two diseases is based on coagulation assays for factors VIII and IX. The management of the diseases, which is now based on the administration of the appropriate coagulation factor, has markedly improved patients' prognosis, but recently infection with the HIV virus has caused premature death, and the risk of other viral infections, such as hepatitis B, C, and δ, has also raised much concern. A serious immunological complication may also arise during treatment, as some patients develop antibodies (inhibitors) against the therapeutic coagulation factor. This complication is more common in hemophilia A than in hemophilia B: 10–20 and 2–5% of patients, respectively (Roberts, 1971; Green *et al.*, 1991b; Kessler, 1991).

## II. GENETICS OF THE HEMOPHILIAS

Both hemophilias are X-linked recessive diseases and the genes for factor VIII and IX are both on the long arm of the X chromosome. The first, in band Xq28, is the most telomeric of the known X-linked genes associated with human disease (Freije and Schlessinger, 1992). Factor IX is more proximal and has been mapped close to the boundary between bands Xq26 and Xq27 (Schwartz *et al.*, 1987).

The population genetics of the two hemophilias is similar. Haldane in 1935 argued that since affected males had a much reduced chance of reproducing, a proportion of the hemophilia genes in the population should be lost at each generation. Taking into account the fact that males carry one-third of the X-chromosomes in the population, he calculated that the loss of hemophilia genes at each generation should be equal to $1/3(1-f)l$, where $f$ is the chance that a patient produces offspring relative to that of a normal male, and $l$ is the incidence of the disease. He was then able to show that if such a loss had not been compensated by new mutations the observed frequency of hemophilia would imply that at the time of the Norman conquest every Englishman would have to have been hemophilic. By this *reductio ad absurdum* he proved the concept that an equilibrium existed between selection and mutation in hemophilia. Recent empiric data have suggested that prior to the introduction of modern replacement therapy the value of $f$ was about 0.5 for both hemophilia A and B (Francis and Kasper, 1983; Ferrari and Rizza, 1986). This figure implies a renewal rate in the hemophilia A and B population gene pool of about $1/6$ per generation, and this in turn suggests that both diseases should be mutationally very heterogeneous.

Unrelated patients, therefore, should usually carry mutations of independent origin. This, of course, is particularly true of mutations causing severe disease as these are subject to the most severe selection. By contrast, mild mutations may persist in the population and even attain relatively high population frequency through genetic drift and founder effects (Bottema *et al.*, 1990; Thompson *et al.*, 1990). In recent times the introduction of therapy has increased the survival and chance of reproduction of hemophilic patients, thus modifying the equilibrium between mutation and selection. This as well as the longer life span presumably explain the marked increase in the number of patients registered with hemophilia centers. For example, in the United Kingdom there were 883 hemophilia B patients in 1983 and 1088 in 1992 (Rizza and Spooner, 1983, and personal communication).

Many attempts have been made to determine whether the mutation rates for either hemophilia differ in males and females. Haldane in 1947 was the first to suggest a 10-fold higher rate in males than in females. However, indirect estimates of the sex-specific mutation rates suffer from the uncertain values of the patients' effective reproduction ($f$); the effect of ascertainment bias on segregation ratios and doubtful carrier diagnoses. Thus, in hemophilia A discordant results have been reported even if the weight of evidence seems to favor a higher mutation rate in the male (Tuddenham and Giannelli, 1994). In hemophilia B indirect estimates of the sex-specific mutation rate lead to inconclusive results. However, the first direct estimate of such mutations suggested a higher mutation rate in males (Montandon *et al.*, 1992). Data in favor of a higher mutation rate in males have also been reported by Ketterling *et al.* (1993b).

## III. FACTOR IX: THE GENE AND THE PROTEIN

The gene coding for factor IX (Figure 4.1) has been cloned by a number of groups using different approaches. The first genomic clones were isolated in 1982 by Choo *et al.* (1982) using a cDNA probe of bovine factor IX which had been isolated with oligonucleotidic probes designed from the protein sequence of bovine factor IX. cDNA clones were isolated by Kurachi and Davie (1982) using a complex strategy that used both baboon cDNA enriched for factor IX and oligonucleotidic probes, and also by Jaye *et al.* (1983) who used a 52-base probe deduced from the amino acid sequence of bovine factor IX. The gene structure was determined by Anson *et al.* in 1984, and the complete genomic sequence was published in 1985 by Yoshitake *et al.* Thus, we now know that the factor IX protein is encoded by a mRNA of 2802 nt that includes a 5' untranslated region of 29–50 nt (depending on which of three methionines is the major start site of translation) plus a 3' untranslated tail of 1390 nt that includes the polyadenylation signal (AATAAA) 15 nt upstream of the start of the poly A tail. The mRNA is transcribed from a genomic region encompassing 34 kb of DNA that includes eight exons which encode the protein domains of factor IX (Figure 4.1). These domains are shared by a number of other proteins but particularly close homology is found between the genes for factor IX, protein C, and factors VII and X (Foster *et al.*, 1985; Leytus *et al.*, 1986; O'Hara *et al.*, 1987). These four genes probably represent duplications of a common ancestral gene.

The first exon of factor IX codes for a prepeptide sequence that is necessary for transport into the endoplasmic reticulum. This is common to all secreted proteins and is characterized by a core of hydrophobic residues flanked

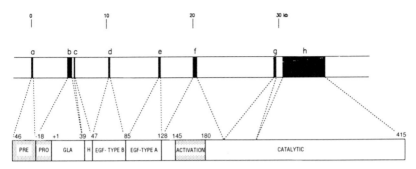

**Figure 4.1.** Factor IX gene (top) and protein (bottom). Exons are indicated by full boxes and dotted lines show segments of protein encoded by each exon. Shaded areas represent parts of the protein that are excised during maturation and activation. Numbers indicate amino acids at domain boundaries.

by a charged residue. Immediately after entering the endoplasmic reticulum, the prepeptide is cleaved by a signal peptidase at the Cys −19 to Thr −18 bond. Exon b codes for the propeptide and gla domain. The former, of 18 residues, includes a recognition site for a vitamin K-dependent γ-carboxylase that adds a second carboxyl group to the first 12 glutamic acid residues of factor IX. The propeptide is cleaved by a specific peptidase to produce the mature protein which begins with Tyr+1 of the gla domain. This contains 11 γ-carboxylated glutamic acid residues, confers affinity for phospholipidic membranes, and its residues 3–11 appear to be necessary for binding to the endothelial cell surface—possibly by way of a receptor protein (Cheung et al., 1992). The gla domain binds $Ca^{2+}$, essential for the correct folding and function of the region. The next factor IX domain is the hydrophobic stack, encoded by exon c. This is a short domain of largely hydrophobic residues that forms an α-helix, together with the carboxyl end of the gla domain. Exons d and e both code for epidermal growth factor (EGF)-like domains. Such domains occur in many proteins and have been subdivided into three types: A, B, and C. The first EGF domain of factor IX is of type B and contains aspartic acid at position 64 which is β-hydroxylated in about one-third of all factor IX molecules (Rees et al., 1988). This residue, together with Asp47, Asp49, and Gln50, forms a high-affinity binding site for $Ca^{2+}$ ions (Handford et al., 1990, 1991). Tyrosine 69 is one of the conserved features of this type of EGF domain and, although this residue is not a ligand for calcium, it is essential for normal clotting activity (Hughes et al., 1993). This EGF domain not only undergoes hydroxylation of Asp64 but also glycosylation of serines 53 and 61 (Hase et al., 1988; Nishimura et al., 1992). The second EGF domain of factor IX is of type A and contains no β-hydroxylated aspartic acid or $Ca^{2+}$ binding site. Its function in factor IX is not understood, but it has been suggested that it contains residues important in binding factor $VIII_a$ or factor X (Nishimura *et al.*, 1993). Exon f codes for the activation peptide and flanking sequences including two glycosylatable asparagines (157 and 167). During the clotting mechanism, factor $XI_a$ and factor $VII_a$ in the presence of tissue factor cleave factor IX after Arg145 and Arg180 to release a 35-aa activation peptide. This transforms factor IX into an active protease (factor $IX_a$) that consists of a light chain (residues 1–145) and a heavy chain (181–415) linked by a single disulfide bridge (Cys132–Cys289). Exons g and h together code for the catalytic domain of this serine protease that includes the obligatory catalytic triad: His221, Asp269, and Ser365 as well as Asp359 at the bottom of the binding pocket which confers an affinity for basic residues in the substrate. Upon cleavage of the activation peptide, it is believed that Val181 interacts with Asp364 in the catalytic domain (Titani and Fujikawa, 1982).

# IV. THE HEMOPHILIA B MUTATIONS

## A. Development of detection methods

The first hemophilia B mutations were detected by Southern blotting as a result of the specific analysis of patients who had developed antibodies (inhibitors) to therapeutic factor IX (Giannelli *et al.*, 1983).

In the same year, protein sequencing identified on Arg145 → His substitution that could be considered the result of a single base substitution. In the following 5 years "point" mutations in the factor IX of hemophilia B patients were characterized by constructing genomic DNA libraries of each patient, cloning their factor IX gene, and sequencing each exon after appropriate subcloning (Rees *et al.*, 1985; Winship, 1986; Bentley *et al.*, 1986; Davis *et al.*, 1987; Ware *et al.*, 1988).

The development of the polymerase chain reaction (PCR) has revolutionized many aspects of molecular biology, not the least of which is the study of mutations. Thus, Tsang *et al.* (1988) were able to confirm by PCR of genomic DNA a mutation found by conventional cloning and sequencing. DNA cloning was then rapidly abandoned in favor of the impressively faster methods of PCR amplification and direct sequencing (Green *et al.*, 1989). A different approach, genomic amplification with transcript sequencing, was also introduced and was used extensively by Dr. Sommer and his colleagues (Stoflet *et al.*, 1988; Koeberl *et al.*, 1989). This uses factor IX specific primers containing in addition a promoter sequence that after PCR amplification is used to transcribe and, hence, further to amplify the region of interest, which is then sequenced. In order to increase the speed of mutation detection even more, methods were developed to screen DNA segments for sequence changes. This then allowed sequencing to be targetted to a particular exon or region containing the mutation. Montandon *et al.* (1989) used a chemical mismatch detection procedure (variously called the "HOT", "AMD," or "CCM" method) to scan all the essential regions of the factor IX gene, amplified by PCR, in such a way as to locate any sequence change within the segments amplified. Using this method a hybrid is formed between the target sequence of the patient and the normal sequence which is then subjected to chemicals that will react with, and cleave at, any mismatched base (Cotton *et al.*, 1988). The resultant products are then visualized and sized on a polyacrylamide gel, where the presence and size of a fast-moving band indicate, respectively, a mutation and its approximate position. Attree *et al.* (1989) used denaturing gradient gel electrophoresis (DGGE) to identify patients with mutations in the region encoding the catalytic domain of factor IX. Later, Fraser *et al.* (1992) used single-strand conformation polymorphism to identify mutation-containing segments. However, this method, which is popular for its simplicity, detects only a proportion of all sequence changes (Sheffield *et al.*, 1993).

## B. Types of mutation causing hemophilia B

Gross rearrangements of the factor IX gene are rare. Only 2–4% of patients carry gross deletions. These may involve the whole gene and additional sequences. At least four of the complete gene deletions have been shown to remove the adjacent *mcf2* transforming gene, and this did not cause any additional clinical feature (Anson *et al.*, 1988). One patient with a large deletion that includes the newly discovered gene Sox-3 has been reported to suffer from hemophilia and mental retardation (Stevanović *et al.*, 1993). Several partial gene deletions have also been identified, and only one has significant factor IX antigen in circulation (Factor IX$_{Strasbourg\ 1}$: Vidaud *et al.*, 1986). This deletion removes exon d and should not alter the reading frame of the mRNA.

Deletion junctions have been examined in six cases: simple deletions were found in two cases (Chen and Scott, 1990; Ludwig *et al.*, 1992), three patients had inversions at the deletion junction (Peake *et al.*, 1989; Solera *et al.*, 1992; Ketterling *et al.*, 1993a), and one patient had a 16-bp insertion at the deletion junction (Green *et al.*, 1988). Additionally, two large insertions have been reported, and one appears to result from a new transposition of an *Alu* repeat (Chen *et al.*, 1988; Vidaud *et al.*, 1989).

A very large number of "point" mutations has been analyzed since 1989 by many laboratories around the world and these are listed annually in a database currently in its fourth edition. This contains 806 mutations, 734 single base changes, 54 small deletions, 15 small insertions, and 3 combined deletion and insertion (Giannelli *et al.*, 1993). Twelve of the 806 patients had two, and one had three mutations within the factor IX gene due to the association of rare neutral changes with the presumptive detrimental mutations. Of the latter, 378 are unique molecular events, the remainder being repeats. Many of these repeats occur at CpG dinucleotides which are well recognized as hotspots for transition to TpG or CpA. However, some repeatedly observed mutations seem to reveal a founder effect: for example, the transition $T \rightarrow C$ at nucleotide 31,311 that has been reported 30 times, mostly in the United States (Ketterling et al., 1991).

A rare form of hemophilia B ("Leyden" type) improves with the age of the patient, especially at and after puberty, and all affected have been found to have mutations in the promoter of the factor IX gene. To date, 14 different mutations with the "Leyden phenotype" have been described, and they are all clustered in the $-21$ to $+13$ nt region (Reitsma et al., 1988, 1989; Crossley et al., 1989, 1990; Gispert et al., 1989; Royle et al., 1991; Freedenburg and Black, 1991; Hall et al., 1992; Picketts et al., 1992; Ghanem et al., 1993; Reijnen et al., 1993; Giannelli et al., 1993). By contrast, a patient with a $G \rightarrow C$ change at nucleotide $-26$ has not experienced the age-related rise in factor IX. Crossley et al. (1992) have in fact delineated a weak androgen-responsive element (ARE) at nt $-36$ to $-22$ overlapping with a liver-enriched transcription factor (LF-A1/HNF4) at nt $-27$ to $-15$, both of which were disrupted by the $-26$ mutation. All the other mutations disrupt only the above LF-A1/HNF4 site, or a C/EBP site ($+1$ to $+18$), or a site for an as yet unidentified factor in the $-6$ region. It therefore seems probable that the ARE of the factor IX gene promoter is of functional importance and relevant to the age-related changes in factor IX expression.

A total of 37 different mutations have been found that disrupt RNA splicing (Giannelli et al., 1993). All but 4 of these are within the consensus sequences for acceptor and donor splice sites. The remaining 4 mutations are tentatively described as "cryptic" splice sites (i.e., mutations creating a potential acceptor or donor splice site) but proof of this is lacking because the patients' factor IX mRNA is not available for study.

An additional point mutation, identified in a patient with <3% factor IX activity, creates a donor splice site consensus sequence in the 3' untranslated region. The effect of such a mutation on RNA processing is speculative but screening of the entire coding region has revealed no other change, and the mutation appears to have occurred independently three times (Vielhaber et al., 1993). Finally, no point mutations have been found in the polyadenylation signal, but a 653-bp deletion in the 3' untranslated region removes the consensus AATAAA in a mildly affected patient. Downstream of the deletion there is another AATAAA, but whether this is used is uncertain because the patient's mRNA could not be tested. Furthermore, the possibility of other point mutations in the patient's gene was not excluded (De la Salle et al., 1993).

Frameshift and nonsense mutations have been identified throughout the coding sequence of factor IX. With one exception, all patients with such mutations have no detectable factor IX protein, suggesting that either such transcripts are not translated or the resultant peptide is unstable. The exception is factor IX$_{Lincoln\ Park}$. This mutation is a combined 2-bp deletion (at position 31,327-8, codon 402) and 8-bp insertion. This creates a frameshift that is predicted to result in the coding of 36 new amino acids which would result in a mature protein of 438 amino acids (instead of the normal 415). This factor IX is expressed and is fairly stable since the patient has 9% factor IX:Ag (Rao *et al.*, 1990).

Missense mutations account for approximately 80% of all point mutations (Giannelli *et al.*, 1993) and comprise 234 different amino acid substitutions. Fifty-three of the factor IX residues show two (or more) different amino acid substitutions, and this very strongly highlights their importance. Although the spectrum of missense mutations causing hemophilia B has not yet been completely defined, most of the residues that could be expected to be important have been found to be altered in hemophiliacs, e.g., two residues of the catalytic triad (Asp269, Ser365), 7 of the 12 gla residues, Asp47 and Asp64 (involved in $Ca^{2+}$ binding), 21 of the 22 cysteines (all forming disulfide bridges), and the residues adjacent to cleavage sites for signal peptidase, propeptidase, and both factor XI$_a$ and factor VII$_a$.

The distribution of amino acid substitutions is illustrated in Figure 4.2, which shows all the amino acid substitutions of factor IX found in the patients examined so far. This figure also illustrates the degree of conservation of the factor IX residues both in its homologues, namely factor VII, factor X, and protein C, and in the factor IX genes of different mammalian species.

Clearly, hemophilia B mutations preferentially affect residues conserved in the factor IX homologues. Careful consideration of Figure 4.2 would demonstrate that 122 different substitutions have been found among the 107 residues that are absolutely conserved, 48 among the 73 showing only 1 conservative substitution, 17 among the 45 with 1 nonconservative change, and 44 among the 229 less-conserved amino acids. Furthermore, the latter mutations usually affect residues conserved in the factor IX of different mammalian species. The fact that many hemophilia B missense mutations cluster at residues conserved in the factor IX homologues suggests that many detrimental mutations affect residues of importance to the overall structure of these complex multidomain serine proteases. Conversely, mutations that affect residues poorly conserved among the factor IX homologues but highly conserved among the factor IX of different mammalian species may indicate regions important to the specificity of the factor IX function. Interesting in this respect is the clustering of mutations observed around arginine 333. Several of these involves residues conserved only among the factor IX of different mammalian species and affect the activity rather than the concentration of circulating factor IX (i.e., at residues 332–334 and 337, 338, and 340). This region may therefore be important to the specificity of the catalytic activity of factor IX rather than to the fundamental structure of serine proteases.

The most obvious clustering if missense mutations can be seen in the amino end half of the first EGF domain, the regions containing the cardinal residues of the active site (His221, Asp269, Ser365), and two further segments of the catalytic domain: the peptide comprising residues 305 to 313 and that comprising residues 330 to 340. Also, the amino end of the heavy chain of activated factor IX shows a cluster of mutations and this is not surprising since Val181 interacts with Asp364 of the catalytic center (Titani and Fujikawa, 1982).

# V. FUNCTIONAL INTERPRETATION OF OBSERVED SEQUENCE CHANGES AND GENOTYPE/PHENOTYPE CORRELATIONS IN HEMOPHILIA B

The functional interpretation of most mutations now found in hemophilia B patients is relatively easy for the following reasons. A considerable body of information has been gathered in recent years on the factor IX mutations present in hemophilia B patients. There is information on the conservation of each one of the factor IX domains and there are also some tridimensional structural data derived either from the study of factor IX (i.e., the first EGF domain (Baron *et al.*, 1992)) or from the study of factor IX homologues (i.e., the gla region of prothrombin (Soriano-Garcia *et al.*, 1992)). The main residual problem is represented by splice site mutations affecting poorly conserved elements of the splice consensuses or generating new sites as there is at present no ready access to factor IX mRNA from hemophilia B patients.

From the hematologic point of view, patients with hemophilia B may be divided into three groups based on factor IX coagulant and antigen measurements: class I has normal levels of antigen but reduced activity, class II has reduced antigen but greater reduction in activity, and class III has equally reduced antigen and activity. Patients of class I are expected to have mutations

---

Figure 4.2. Sequence and missense mutations of factor IX. Offset white residues in black field are hemophilia B mutations and offset green letters are neutral changes. Normal factor IX residues are shown in fields colored to indicate the degree of conservation in the factor IX homologues (factor VII, X, and protein C) and in the factor IX of eight different mammalian species. Yellow, red, blue, green, and white fields respectively indicate residues: absolutely conserved, showing one conservative substitution, showing one nonconservative change, nonconserved in the factor IX homologues but absolutely conserved in the factor IX of mammalian species, not conserved even in the factor IX of mammalian species. The factor IX sequence is shown to start at M-46 but factor IX could begin instead at M-41 or, most probably, at M-39.

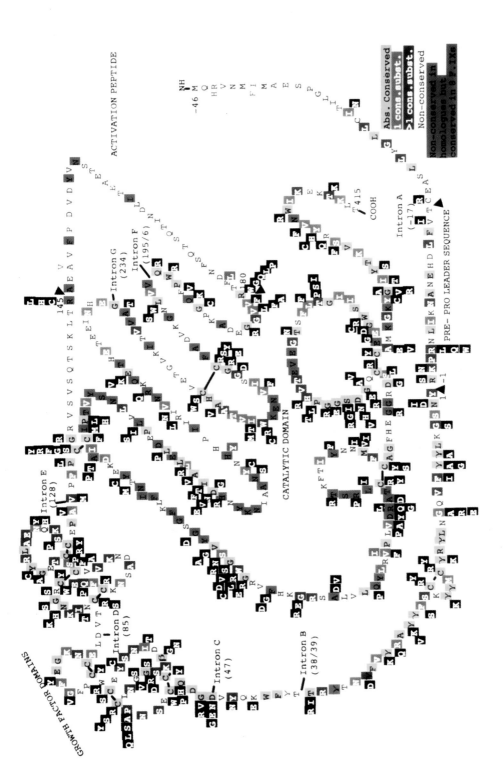

(mostly missense) that compromise (or abolish) the function of factor IX while not affecting its stability or immunologically detectable epitopes. Functional impairment may result either from damage to domains not directly involved in the catalytic activity or to direct damage of the catalytic domain. Examples of the latter are substitutions that affect the cardinal serine of the catalytic center, such as Ser365 → Arg. Examples of the former are mutations of the propeptidase recognition sequence, such as Arg −1 → Ser, which preclude cleavage of the propeptide and full γ-carboxylation. This produces a factor IX of abnormal molecular weight that is grossly dysfunctional.

Class III phenotypes are associated with a very broad spectrum of mutations because they may result from any change that interferes primarily with the synthesis, secretion, or stability of factor IX. Thus, promoter mutations are expected to result in this phenotype, although usually not in its most severe form. The latter is obligatory for patients with deletions removing the whole gene. Experience so far also shows that partial deletions, nonsense mutations, frameshifts, and mutations of the GT and AG oligonucleotides at the start and end of introns usually cause hemophilia where factor IX activity and antigen are too low to be measured. Whether this is due to problems in synthesis (e.g., instability of mRNA) or in protein stability is not known. Milder type III disease is usually observed when the mutations affect the less conserved features of the splice signals. Probably in this case a proportion of the transcript is correctly spliced.

A problem with factor IX transport and secretion may explain the type III phenotype of a severely affected patient with the missense mutation Ile −30 → Asn (Green et al., 1993). In this case the introduction of a hydrophilic residue into the hydrophobic core of the prepeptide creates a block of three hydrophilic amino acids (Thr −31, Asn −30, Cys −29) that grossly modifies the usual structure of prepeptides and is therefore likely to prevent transport of the protein into the endoplasmic reticulum.

Other missense mutations may destabilize factor IX so that it is rapidly degraded or, perhaps less likely, they may alter its tertiary structure so that it becomes both unrecognizable to polyclonal antibodies during factor IX antigen assay and biologically inactive. These are possible scenarios for mutations altering cysteine residues involved in disulfide bridges and, indeed, a severe type III phenotype is observed in the vast majority of these mutations. However, there are exceptions: the Cys18 → Arg mutation causes severe hemophilia with normal levels of factor IX antigen, and both Cys23 → Arg and Cys23 → Tyr cause a class II phenotype with significant levels of antigen (35 and 19%, respectively). Interestingly, Cys18 and Cys23 contribute to the same disulfide bridge.

Patients with reduced antigen levels but greater reduction in activity (class II) usually have missense mutations, and these are expected to reduce mRNA or protein stability or, possibly, to alter the epitope profile of factor IX, but they obviously have an even greater detrimental effect on factor IX function. An example of this is a patient with 1% factor IX activity and 12% factor IX antigen who has an in-frame deletion removing Arg37 from the gla domain of factor IX. This disrupts the amphipathic nature of an $\alpha$-helix predicted on the basis of information on the crystal structure of the gla region of prothrombin (Soriano-Garcia *et al.*, 1992). Such a structural change is entirely consistent with an unstable factor IX of low specific activity.

As mentioned earlier, several hemophilia B mutations have been repeatedly observed in unrelated patients and in many instances such repeats result from independent mutational events. The phenotypic features of patients with these identical mutations appear to be similar enough to suggest that information on the factor IX mutation may serve to predict the severity of the disease. This may be useful in the genetic counseling of families with inadequate clinical information on their affected member(s).

An infrequent but very unfortunate phenotypic feature of hemophilic patients is the development, during the course of treatment, of antibodies against infused factor IX which nullify normal replacement therapy. It was shown for the first time in 1983 that such a complication frequently occurs in patients with gross deletions (Giannelli *et al.*, 1983) and later work has confirmed that patients with gross deletions or functionally equivalent mutations, such as nonsense and frameshift, are predisposed to this life-threatening complication. It has been proposed that this predisposition is a consequence of the failure to produce a factor IX protein capable, during the maturation of the patient's immune system, to induce the development of immune tolerance to normal factor IX (Giannelli *et al.*, 1983; Giannelli and Brownlee, 1986).

## VI. PROGRESS IN CARRIER AND PRENATAL DIAGNOSIS OF HEMOPHILIA B

Any woman related to a hemophilia B patient is at risk of being a carrier of the disease and, since one may expect each patient to have five or six such relatives, the population frequency of women requiring carrier tests for hemophilia B may be as high as 1/5000. Until the cloning of the factor IX gene, carrier diagnoses had to be based on hematological assays but these often provided unhelpful results because the carriers' factor IX values vary from very abnormal to completely normal. This is explained by the random inactivation of one X chromosome early in female embryogenesis and by the ensuing chance that the factor IX-producing cells of individual carriers derive mostly from cells with the normal or, alternatively, the defective allele in the inactive X chromosome. In practice very low titers of factor IX are more diagnostic than intermediate and high values and therefore negative diagnostic errors are more likely than positive ones. In the 1970s, sophisticated statistical procedures were used to provide the best estimates of the risk of heterozygosity by combining pedigree and hematological data, but even so, negative error rates of 15% were expected and absolute answers infrequently obtained (Østavik et al., 1981; Barrow et al., 1982). By contrast, definite prenatal diagnoses were made in expert hands from 1979 by measuring factor IX in fetal blood samples. These samples, however, had to be obtained by fetoscopy at 19 or 20 weeks of gestation (Mibashan et al., 1986).

In centers of excellence this procedure is relatively safe as no significant maternal mortality or morbidity has been reported and the risk to the fetus is small, though real. In a very experienced London center the operative loss and abortion rate was 2.4% and the preterm delivery rate 15% (Mibashan et al., 1986). The acceptance of this procedure by at-risk mothers has been modest because the method is rather invasive and applicable only to midpregnancy.

The cloning of factor IX and the subsequent detection of intragenic polymorphisms (Camerino et al., 1984; Giannelli et al., 1984; Winship et al., 1984) allowed the introduction of DNA-based diagnostic tests in 1984. These tests, in a proportion of families, provide definite carrier and first trimester prenatal diagnoses by determining the intrafamilial segregation of markers associated (in coupling) with the detrimental mutation. The prenatal diagnoses used chorion biopsies taken at 10 weeks of pregnancy and were more readily accepted than fetal blood sampling by mothers at risk. Nevertheless, the indirect carrier and prenatal diagnostic procedures based on the analysis of DNA polymorphisms are still inefficient and rather cumbersome. Each diagnosis requires the analysis of a family group with a clear history of the disease in order to determine the downward transmission of the marker(s) in coupling with the detrimental mutation. Therefore, the group typically comprises the proband, the proband's mother, who must be heterozygous for a marker, and one affected relative (to establish the coupling of polymorphic markers and the disease-causing mutation). In addition, in approximately half of carrier diagnoses, the proband's father must also be examined. Of course, when the father's DNA is examined, definite diagnoses require paternity testing. From the above requirements it follows that diagnoses cannot be made by this procedure when the proband's mother is homozygous for all markers or when essential relatives or positive family history are lacking. In the last situation it is sometimes possible to exclude carrier status or, in a fetus, the disease, but not to arrive at positive diagnoses.

The failure rate for the above indirect DNA diagnostic procedure is inevitably high as the only parameter that can be improved is the number of known gene-specific polymorphic DNA markers. From 1984 to now this has increased and the proportion of Caucasian women heterozygous for at least one factor IX-specific marker has risen from 45 to 89% (Giannelli et al., 1984; Winship et al., 1989). This, however, still leaves 11% of families in which diagnostic tests will fail because the proband's mother is homozygous for all known markers. Furthermore, a recent study of the United Kingdom population has shown that 6% of families have no available affected relative and about 50% have no positive family history because there is a single affected individual (Saad et al., 1994). Finally, it must be noted that the indirect diagnostic procedures provide no information on the molecular biology of the disease.

In order to overcome the above limitations, in 1987 Giannelli advocated the direct detection of gene defects and this was achieved in 1989 by the direct sequencing of all essential regions of the factor IX gene after PCR amplification (Green et al., 1989) or by the detection of mismatches in heteroduplexes formed by amplified test and normal DNA (Montandon et al., 1989). A different amplification and sequencing procedure was also successfully used by Bottema et al. (1989). Other procedures to screen all the factor IX exons for mutations were also introduced later (see for example Fraser et al., 1992; Gahnen et al., 1993).

In 1990 we began to implement a new strategy to maximize the advantages of the direct diagnostic procedures and thus to optimize the provision of both carrier and prenatal diagnoses and genetic counseling for hemophilia B. This strategy, which entails the characterization of the mutation in an index person from each family and the construction of a national confidential register of mutational hematologic and pedigree information to be used for the provision of carrier and prenatal diagnoses, is based on the following considerations. Once the mutation of an index person has been characterized carrier and prenatal diagnoses on any one of his blood relatives for generation after generation can be based solely on the analysis of the region of the gene that is abnormal in the index patient, thus allowing rapid, economic, and yet definite and accurate diagnostic tests. The preliminary study of mutations in the index individuals not only rapidly advances the understanding of the disease but also ensures a sound theoretical basis for the provision of diagnostic services because it thoroughly tests the technical procedures, it yields data to help assess the functional consequences of any sequence change, and shows correlations between mutations and phenotypes that may offer valuable prognostic information. Finally, this work should lead to such interaction between clinicians and mammalian geneticists as to ensure optimal transfer of information and clinicians' access to state-of-the-art technology. After a pilot study on a sample of the Swedish population the strategy was applied to the United Kingdom population (Green et al., 1991b; Giannelli et al., 1992) where it appears to fulfill all of its promises (Saad et al., 1994). National registers of hemophilia B mutations are also being constructed in Sweden (Ljung, personal communication) and New Zealand (van der Water, 1992).

At present, it is therefore possible to offer exact carrier and prenatal diagnoses to almost all families with hemophilia B. The last remaining problem is our ignorance of the incidence of gonadal mosaicism for the hemophilia B mutations. Gonadal mosaicism affects the recurrence risk of the mothers of sporadic patients who test, on somatic cells, as homozygous normal. Since it may be expected that at least a proportion of gonadal mosaics may also show mosaicism in their soma, it is useful to employ carrier detection methods that can detect the mutant gene even when this represents a minor proportion of the factor IX alleles. The mismatch detection method we use (Montandon *et al.*, 1989) seems able to identify mutant alleles representing only 5–10% of the total (Montandon and Green, unpublished observations). However, in order to provide more precise counseling, estimates of the incidence of ovarian mosaicism must be obtained. This requires systematic family studies in large populations, since ovarian mosaicism is revealed every time two children inherit a defective factor IX gene from a mother with homozygous normal somatic cells.

## VII. FACTOR VIII: THE GENE AND THE PROTEIN

The factor VIII gene was cloned in 1984 by two independent groups following somewhat different strategies. One group purified human factor VIII, and after partial protein sequencing synthesized a 36 base oligonucleotide to probe genomic λ libraries and isolate the initial clones containing a 28-kb section of the gene (Gitschier *et al.*, 1984). The remainder of the gene was then isolated by walking from this point. The other group purified and sequenced porcine factor VIII to generate a probe to isolate porcine factor VIII cDNA. This was then used in turn as a probe to isolate the homologous human cDNA (Toole *et al.*, 1984).

In 1984 the factor VIII gene was by far the longest cloned human gene (Figure 4.3), measuring 186 kb in length and containing 26 exons (Gitschier *et al.*, 1984). The promoter of the gene, which has not yet been functionally characterized, contains a GATAAA sequence 30 bp 5' of the presumed transcription start. The 26 exons vary considerably in size from 69 bp (exon 5) to 3106 bp for the exceptionally large exon 14. The exons are separated by introns that also vary considerably in size from 200 bp (intron 17) to 32.4 kb (intron 22). The latter intron is unusual because it contains an HTF island that is associated with two additional transcripts. One, which is of opposite polarity to the factor VIII transcript, is coded entirely by a 1.8-kb segment of intron 22 with an open reading frame uninterrupted by introns (Levinson *et al.*, 1990). The second transcript is of the same polarity as the factor VIII transcript and is processed into a mRNA containing exons 23 to 26 of the factor VIII gene plus a small first exon with eight codons (Levinson *et al.*, 1992). The sequences responsible for these two transcripts have been called genes F8A and F8B, respectively. Interestingly, two additional copies of sequences that contain gene F8A and the first exon of F8B have been found in the human genome, and they are

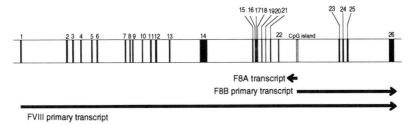

**Figure 4.3.** Factor VIII gene and its transcripts. (Top) Factor VIII gene with exons numbered (black boxes) and CpG island in intron 22. (Bottom) Arrows showing extent and direction of the three transcripts known to arise from the factor VIII gene.

located 500–600 kb telomeric to the factor VIII gene which is so oriented as to have its first exon most telomeric (Frieje and Schlessinger, 1993; Migeon *et al.*, 1993).

The transcript of the factor VIII gene starts 170, or more rarely 172, bases 5' of the translation start. This (the first AUG codon of the message) is followed by an open reading frame comprising 2351 codons and a 3' untranslated tail of 1805 nt that includes a TGA stop codon and an AATAAA polyadenylation signal 19 bases from the start of the poly-A segment. The factor VIII mRNA is therefore 9028 nt long with a coding region of 7053 nt (Gitschier *et al.*, 1984; Toole *et al.*, 1984; Wood *et al.*, 1984).

Factor VIII (Figure 4.4) is synthesized with a signal peptide of 19 aa containing a hydrophobic core of 10 residues flanked by 2 charged residues, as usually found in the propeptide of other secreted proteins. The secreted factor VIII consists of three types of domain called A, B, and C (Vehar *et al.*, 1984). The A domain is present in triplicate and consists of approximately 330 aa. They have a 30% homology with one another and a similar pairwise homology to the domains of ceruloplasmin (Koshinsky *et al.*, 1986). The first two A domains of factor VIII are separated from the third by a B domain of 983 aa which is extremely rich in glycosylatable asparagines and has no homology to other known proteins. Two C type domains are present at the carboxyl end of the protein. These consist of 150 aa and show a 40% homology to each other and 20% homology to the slime mould's discoidin. Recently, a homology has also been found to a milk-fat globule-binding protein (Stubbs *et al.*, 1990). A further feature of the protein is the presence of two small acidic peptides ($a_1$, $a_2$) situated respectively at the amino end of the second and third A domains. Thus, the domain structure of factor VIII can be represented as $A_1a_1A_2Ba_2A_3C_1C_2$.

Coagulation factor V, the cofactor of factor X, has a structure similar to factor VIII but its A domains lack the acidic peptides $a_1$ and $a_2$ and the central large B domain has no homology with that of factor VIII (Jenny *et al.*, 1987).

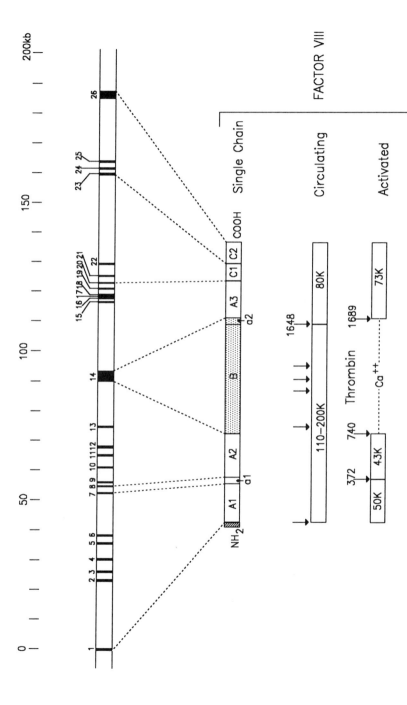

**Figure 4.4.** Factor VIII and its maturation. First bar is factor VIII gene, second bar is primary factor VIII peptide with dotted lines indicating exons coding for different domains. Third and fourth bars show proteolytic cleavages occurring during maturation and activation of factor VIII.

Interestingly, the last three regions are also the least conserved between mouse and human factor VIII (Elder *et al.*, 1993).

The B domain of factor VIII plays no part in blood coagulation and its functional role is unknown. The carboxyl end of the protein ($C_2$ domain) is important for the binding of factor VIII to phospholipidic membranes. The factor VIII protein undergoes extensive post-translational modifications including both glycosylation and sulfation. Glycosylation is both O- and N-linked. Factor VIII contains 25 glycosylatable asparagines, of which 19 are in the B domain (Vehar *et al.*, 1984). A number of tyrosines are sulfated: Tyr346, Tyr718, Tyr719, Tyr723, Tyr1664, Tyr1680 (Pittman *et al.*, 1992). The latter tyrosine is thought to bind von Willebrand factor, a highly polymorphic protein that carries factor VIII in circulation and prevents its premature loss.

Prior to secretion factor VIII is cleaved not only to release the prepeptide but also at the $B/a_2$ boundary (Thr1648–Glu1649) to form a molecule with a light chain consisting of $a_2A_3C_1C_2$ (aa 1649–2232) and a heavy chain consisting of $A_1a_1A_2B$. This chain, however, is of variable length since the B domain is also cleaved at variable internal positions. Conversion of factor VIII to its active form entails cleavage at the $a_1/A_2$ (Arg372–Ser373) and the $a_2/A_3$ (Arg1689–Ser1690) boundaries. The former cleavage transforms factor VIII into a three-chain molecule and the latter releases factor VIII from von Willebrand factor. Activation is also accompanied by cleavage at the $A_2/B$ boundary, so that activated factor VIII is a heterotrimer formed by $Aa_1$, $A_2$, and $A_3C_1C_2$. The first of these chains appears to be bound to the third by salt bridges, while the $A_2$ chain may be held in the complex by other interactions, possibly involving the $a_1$ acidic peptide of the $A_1a_1$ chain. $A_2$ appears to dissociate fairly readily from the other two chains, causing loss of factor VIII activity (Fay *et al.*, 1993).

Activated protein C, an important regulator of coagulation, cleaves factor VIII at the $A_1/a_1$ boundary (Arg336–Met337) and causes its inactivation, but protein C also cleaves at Arg562 and this may represent an additional major target for inactivation (Fay *et al.*, 1991).

## VIII. HEMOPHILIA A MUTATIONS

### A. Development of detection methods

The first hemophilia A mutations were identified in 1985. These were either gross deletions or point mutations that altered the recognition sequence of the *Taq*I restriction enzyme (Gitschier *et al.*, 1985b). *Taq*I restriction proved relatively efficient at detecting some point mutations for two reasons: first, the *Taq*I recognition sequence, TCGA, contains a CpG dinucleotide that often represents a hotspot for transitions (CpG → TpG or CpA); second, the above transitions in the *Taq*I sites present in the coding sequence of the factor VIII gene result in nonsense mutations 4.3 times more frequently than expected by chance and consequently mutations at *Taq*I sites have a relatively high probability of being so clearly detrimental as to cause hemophilia (Green *et al.*, 1991a).

With the introduction of the polymerase chain reaction it became possible to isolate specific regions of the factor VIII gene for the targetted analysis of sites of particular interest, such as codons specifying cleavage activation sites (Gitschier *et al.*, 1988), or sites expected to mutate frequently such as CpG dinucleotides (Pattinson *et al.*, 1990). The direct sequencing of PCR products allowed the analysis of even longer sequences such as individual exons (Higuchi *et al.*, 1990). Further progress came with the adoption of efficient procedures for mutation screening such as DGGE (Kogan and Gitschier, 1990; Traystman *et al.*, 1990). However, the size and complexity of the factor VIII gene continued to be an obstacle to a thorough and complete analysis of all the relevant sequences of the gene. This had two negative consequences: only a proportion of the patients analyzed could be expected to show mutations and even in patients in which a mutation was found it was difficult to be certain that the observed sequence change was the cause of the disease because, often, the possibility could not be excluded that the observed change was neutral while an accompanying detrimental mutation had gone undetected. The presence of neutral and detrimental mutations in the same gene is well documented in hemophilia B (Montandon *et al.*, 1990; Giannelli *et al.*, 1993).

The first mutations detected by a procedure that examined all the essential sequences of the factor VIII gene were published in 1991 (Naylor *et al.*, 1991). This procedure relied in part on the amplification of traces of factor VIII mRNA present in peripheral lymphocytes by means of reverse transcription and PCR. Two important advances stem from the use of mRNA: (1) reduction of labor because this molecule presents the coding sequence as a single stretch of 7053 nt rather than 26 widely spaced exons; and (2) detection of any mutation causing mRNA abnormalities, irrespective of where they are located within the gene. The detection of hemophilia A mutations by the above method entails the PCR amplification of the putative promoter, the 3106-bp exon 14 and the polyadenylation signal region from genomic DNA, and reverse transcription plus PCR amplification of the remainder of the coding sequence with nested primers. A total of eight PCR products are thus obtained: two from exon 14, four from the remainder of the coding sequence, and one from each end of the gene. These segments overlap each other, except for the segment with the polyadenylation signal, and this ensures thorough screening of the whole amplified sequence by the chemical mismatch procedure described earlier. This allows screening of long DNA segments and identifies and locates any mutation, thus reducing the sequencing reactions needed to fully characterize any sequence change to just one or two. Point mutations affecting the coding sequence are readily confirmed by genomic DNA analysis, while gross mRNA changes immediately indicate the segment of genomic DNA to be examined for further characterization of the mutation. The latter was clearly illustrated by one of the first two cases examined, as mismatch screening located a gross change in the cDNA due to the absence of exon 6; sequencing of the splice site at the end of intron 5 then showed an A $\rightarrow$ G transition altering the obligatory AG dinucleotide. This fully accounted for the observed mRNA defect and the patient's phenotype.

A scheme was also developed to achieve complete detection of the hemophilia A mutations by analysis of the genomic DNA with the DGGE screening procedure (Higuchi *et al.*, 1991a). This entailed 47 PCR amplification reactions to amplify all exons, most exon/intron boundaries, and the putative promoter. The PCR products were screened by DGGE and the segment containing a mutation was sequenced fully to define the gene defect. The disadvantage of this procedure is the large number of amplifications required and the inability to detect mutations that could alter mRNA unless they were at the exon/intron boundaries that are examined.

## B. Conventional mutations causing hemophilia A

Southern blotting, focussed sequence analysis, and the more general procedures mentioned above had led to the identification of several factor VIII mutations in patients with hemophilia A (Tuddenham *et al.*, 1991).

Three to 5% of hemophilia A patients has gross deletions of the coding sequence and more than 60 gross deletions have been reported so far. These may involve the whole or a part of the gene (e.g., a single exon) and do not show obvious preferences for any specific region. Usually deletions result in severe disease, but 2 have been found in patients with moderate hemophilia: 1 comprises exon 22 (Youssoufian *et al.*, 1987) and 1 exons 23 and 24 (Lavergne 1992: cited by Tuddenham *et al.*, 1991). The mRNA in these cases is expected to maintain the normal reading frame and presumably the deletions simply lead to the loss of the aa encoded by the missing exons.

Gross insertions and duplications appear to be quite rate but two interesting patients have been reported with insertions of the line repeat sequence in AT-rich regions of exon 14 (Kazazian *et al.*, 1988). These insertions appear to represent recent transpositions and the sequence representing the probable origin of one of these inserted repeats has been located in chromosome 22 (Dombroski *et al.*, 1991).

Two partial duplications have also been reported: one involving 23 kb of intron 22 inserted between exons 23 and 24, and the other exon 13 (Gitschier, 1988; Murru *et al.*, 1990). The former was found in the mother of a deletion patient and it was thought to have predisposed the gene to deletion.

The latter does not alter the reading frame and was found in a mildly affected patient.

The vast majority of conventional hemophilia A mutations are small sequence changes but none so far has been found in the putative promoter of the factor VIII gene (Tuddenham et al., 1991; Diamond et al., 1992; McGinnis et al., 1993; Naylor et al., 1993a). The above mutations therefore are expected to act by affecting RNA processing (splice site mutations) or mRNA translation (nonsense and frameshifts), or the fine structure of factor VIII (missense and amino acid deletions).

At least seven different splice-site mutations have been reported but the mRNA was examined in only one of these (see previous section). Two mutations affected highly conserved residues of the consensuses (GT, AG at ends of introns), while five involved less conserved elements.

Nonsense mutations have been reported at no fewer than 15 different locations and several have been repeatedly observed. At 12 of the 15 locations nonsense codons have arisen by transitions at CpG sites. These mutations are expected to arise frequently and were therefore deliberately sought by targetted screening of CpG sites by TaqI restriction or by oligonucleotide discriminant hybridization (Gitschier et al., 1985b; Pattinson et al., 1990). Nonsense mutations are expected to cause premature termination of translation but recently mRNA analysis has shown that nonsense mutations may sometimes alter RNA processing and cause skipping of the exon containing the nonsense codon (Naylor et al., 1993a): a result that may significantly affect the functional consequences of the mutation. Thus, in one patient with a nonsense mutation at codon 1987, the examined mRNA always lacked exon 19, while another with a nonsense codon at position 2116 showed two mRNAs, one of normal structure and one missing the mutant exon 22.

Frameshift mutations also usually cause premature termination of translation. Frameshifts due to small deletions or insertions have been detected so far at 11 different positions in the factor VIII coding sequence.

More subtle changes to the structure of factor VIII are caused by missense mutations and their analysis should gradually help to define the structural features that are important to the transport, processing, stability, and function of factor VIII. So far 77 different missense mutations have been identified (Table 4.1). Forty-five of these were reported by groups examining the whole coding sequence of factor VIII (Higuchi et al., 1991a,b; McGinnis et al., 1993; Naylor et al., 1991, 1993a), while the others were detected by more limited analysis of the factor VIII gene. Two of the above missenses are due to mutations that alter the normal splice consensus (Gly205 → Trp and Gln565 → Lys; see last two lines in Table 4.1) and could therefore also affect mRNA processing. Some of the missense mutations involve sites of known functional importance such as activation cleavage sites (Arg372 → Cys or Arg372 → His and Arg1689 → Cys or

Table 4.1. Missense Mutations and Single aa Deletions in Factor VIII Gene of Hemophilic Patients[a]

| Line no. | Amino acid change | Severity | Line no. | Amino acid change | Severity | Line no. | Amino acid change | Severity |
|---|---|---|---|---|---|---|---|---|
| 1 | Glu11 → Val | Mild | 29 | Asp542 → Gly | Severe[c] | 57 | Asn1922 → Ser | Moderate[c] |
| 2 | Gly73 → Val | Mild | 30 | Ser558 → Phe | Mild[c] | 58 | Arg1941 → Gln | Moderate/mild[c] |
| 3 | Val85 → Asp | Mild | 31[d] | Ile566 → Thr | Severe | 59 | Arg1941 → Leu | Moderate |
| 4 | Lys89 → Thr | Mild[c] | 32 | Ser577 → Pro | | 60 | Arg1997 → Trp | Moderate[c] |
| 5 | Met91 → Val | Mild[c] | 33[e] | Ser584 → Ile | | 61 | Phe2101 → Leu | Mild[c] |
| 6 | Gly145 → Val | Mild | 34 | Arg593 → Cys | Mild[c] | 62 | Tyr2105 → Cys | Mild[c] |
| 7 | Val162 → Met | Moderate/mild | 35 | Asn612 → Ser | | 63 | Arg2116 → Pro | Severe |
| 8 | Lys166 → Thr | Mild[c] | 36 | Ala644 → Val | Mild[c] | 64 | Ser2119 → Tyr | Mild[c] |
| 9 | Ser170 → Leu | Moderate | 37 | Phe652 deletion | Severe[c] | 65 | Arg2150 → His | Mild[c] |
| 10 | Val266 → Gly | Mild[c] | 38 | Arg698 → Trp | Mild | 66 | Arg2159 → Cys | Mild[c] |
| 11 | Glu272 → Gly | Moderate | 39 | Ala704 → Thr | Mild[c] | 67 | Arg2163 → His | Moderate |
| 12 | Arg282 → His | Severe[c] | 40 | Glu1038 → Lys | Moderate[c] | 68 | Leu2166 → Ser | Severe |
| 13 | Ser289 → Leu | Mild[c] | 41 | Asn1441 → Lys | Moderate[c] | 69 | Pro2205 deletion | Moderate |
| 14 | Phe293 → Ser | Mild[c] | 42 | Val1634 Ala | Moderate[c] | 70 | Arg2209 → Leu | Moderate[c] |
| 15 | Thr295 → Ala | Moderate[c] | 43 | Val1634 → Met | Severe[c] | 71 | Arg2209 → Gln | Severe/mild[c] |
| 16 | Val326 → Leu | Severe/moderate | 44[f] | Tyr1680 → Phe | Mild[c] | 72 | Trp2229 → Cys | Moderate[c] |

| | | | | | | | | | |
|---|---|---|---|---|---|---|---|---|---|
| 17 | Cys329 → Arg | Severe | 45[f] | Tyr1680 → Cys | Severe | 73 | Trp2046 → Arg | Mild |
| 18[b] | Arg372 → Cys | Moderate | 46[b] | Arg1689 → Cys | Moderate/severe | 74 | Pro2300 → Leu | Mild[c] |
| 19[b] | Arg372 → His | Moderate/mild | 47 | Arg1689 → His | Mild | 75 | Arg2304 → Cys | Mild[c] |
| 20 | Leu412 → Phe | Mild[c] | 48 | Tyr1709 → Cys | Moderate | 76 | Arg2307 → Leu | Severe/mild |
| 21 | Lys425 → Arg | Severe[c] | 49[d] | Met1772 → Thr | Severe | 77 | Arg2307 → Gln | Moderate/mild |
| 22 | Tyr473 → His | Mild[c] | 50 | Arg1781 → His | Mild[c] | 78 | Gly205 → Trp | Moderate[c,g] |
| 23 | Tyr473 → Cys | Moderate[c] | 51 | Ser1784 → Tyr | Severe[c] | 79 | Gln565 → Lys | Moderate[g] |
| 24 | Gly479 → Arg | Moderate[c] | 52 | Leu1789 → Phe | Mild | | | |
| 25 | Arg527 → Trp | Mild[c] | 53 | Pro1825 → Ser | Moderate[c] | | | |
| 26 | Arg531 → Cys | Moderat[c] | 54 | Thr1826 → Pro | Mild | | | |
| 27 | Arg531 → Gly | Moderate[c] | 55 | His1848 → Arg | Moderate[c] | | | |
| 28 | Ser535 → Gly | | 56 | Asn1922 → Asp | Severe/moderate[c] | | | |

[a] From Tuddenham e tal. (1991), Diamond et al. (1992), McGinnis et al. (1993), and Naylor et al. (1993a).
[b] Activation cleavage sites.
[c] Reported in papers in which most or all essential factor VIII sequences were screened by fully effective methods.
[d] Generates new glycosylation sites.
[e] Causes loss of glycosylation site.
[f] Sulfated residue involved in von Willebrand factor binding.
[g] Mutations also altering splice site consensuses.

Arg1689 → His) and the sulfated tyrosine at position 1680 that is involved in von Willebrand factor binding (Tyr1680 → Phe and Tyr1680 → Cys). One mutation destroys and 2 create N-glycosylation sites (Ser584 → Ile, Ile566 → Thr and Met1772 → Thr).

## C. Nonconventional mutations causing severe hemophilia A

Unusual hemophilia A mutations that account for 45% of severely affected patients have gone undetected for 8 years since the cloning of the factor VIII gene. The partial, targetted screening of the factor VIII was responsible for this. When in 1991 the putative promoter, all the exons, and most intron/exon boundaries were examined in a substantial group of hemophilia A patients using the genomic DNA screening strategy involving DGGE mentioned above, it became clear that while this procedure detected a mutation in most of the mildly and moderately affected patients, it failed to detect mutations in approximately half the patients with severe disease (Higuchi *et al.*, 1991b). This puzzling result generated hypotheses that postulated the existence of remote factor VIII controlling regions or of genes other than factor VIII responsible for hemophilia A. The puzzle was, however, solved by the mutation detection method that examines factor VIII mRNA. This indicated not only that all patients with hemophilia A had mutations of the factor VIII gene, but also that 45% of the patients with severe disease had an unusual mRNA defect that completely prevented amplification of any segment that crossed the boundary between exons 22 and 23 (Figure 4.5), even though all the coding sequences of the factor VIII gene could be amplified by reactions that avoided the above boundary (Naylor *et al.*, 1992, 1993a). This indicated that the mutations involved intron 22, but since the splice signals and the adjacent regions of intron 22 were normal, the mutations clearly affected internal regions of the intron.

The block to amplification across the exon 22–23 boundary was insurmountable, and it therefore seemed probable that the patients' factor VIII mRNA was interrupted 3' of exon 22 so that exons 23–26 were exclusively expressed as part of the F8B gene controlled by the intron 22 promoter. The successful amplification of the entire F8B message in the patients was in keeping with this idea and therefore the sequences 3' of exon 22 were sought in each of the available patients, using a novel strategy for the isolation of the 3' ends of mRNA (see Figure 4.6 and Naylor *et al.*, 1993b).

The 3' primer for the rapid amplification of cDNA ends (RACE) procedure (Frohman *et al.*, 1988) was used for reverse transcription and the RACE adaptor primer plus a factor VIII-specific primer were used for a first set of 30 PCR cycles. However, due to the low level of factor VIII mRNA in peripheral lymphocytes, this set is not sufficient alone. The PCR products were therefore digested with *Cla*I, which recognizes a sequence in the RACE adaptor primer,

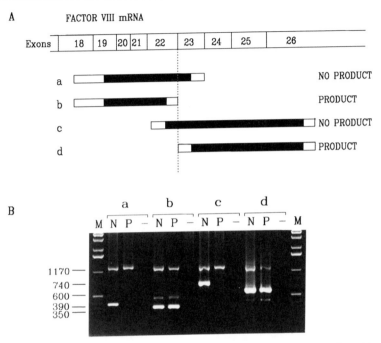

**Figure 4.5.** (A) Exons of the distal segment of the factor VIII mRNA (top) and amplification products of nested PCR reactions a–d. Solid bars are sequences amplified by internal (nested) primer pairs, empty bar extensions are sequences amplified only by external primer pairs, and the vertical dotted line is exon 22/23 boundary. (B) Agarose electrophoresis of the products of reactions a–d and a control reaction that amplifies factor VIII mRNA from exons 8 to 14. Below the positive control bands reactions a and c are negative in the patient (P) but positive in the normal control (N), while reactions b and d are positive in both P and N.

and vectorette cassette (Riley *et al.*, 1990) was ligated to the *Cla*I-digested products. Extension of a second factor VIII-specific primer, situated 3' of the first, then created a template for the vectorette 224 primer, and a second set of 30 PCR cycles was performed using these primers. This resulted in much greater specificity of the amplification reaction and led to the isolation of pure products. Upon sequencing these revealed a set of sequence blocks identifiable as exons on at least three grounds: their precise boundaries, separation by intervening sequences in genomic DNA and, in one case, the observation of a splice site consensus. The novel exon-like sequences identified were variably spliced out of or in to the patients' abnormal factor VIII message (Naylor *et al.*, 1993b) and were not from intron 22. The possibility that the abnormal sequences had been

**A.   RACE**

**B.   Digest with Cla I, ligate vectorette cassette and amplify.**

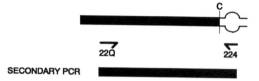

**Figure 4.6.**  New method for isolation of 3′ end of abnormal FVIII mRNA. (A) RACE amplification
of 3′ unknown sequences (Frohman *et al.*, 1988) using factor VIII-specific primer (6A).
(B) Replacement of RACE adaptor sequences with a vectorette cassette at a ClaI site
(C) in the adaptor primer. The noncomplementarity of part of the vectorette sequences
(see bubble on the right-hand side of bar) and the identity of primer 224 and a segment
on the top strand of the bubble make amplification by PCR dependent on factor VIII-
specific primer 22Q lying 3′ of primer 6A.

brought into the patients' intron 22 by insertion or translocation was also excluded because Southern blotting of the intron 22 sequences from the patients'
DNA digested with BamHI, EcoRI, or HindIII did not show gross abnormalities.
It therefore seemed easier to explain our findings by postulating inversions
involving the two telomeric sequences homologous to part of intron 22 as this
would split the factor VIII gene into two fragments but cause minimal disturbance to the restriction map of the intron 22 region (Naylor *et al.*, 1993b). This
hypothesis was confirmed using the novel sequences from the patients' mRNA to
screen three YAC libraries. A contig of YACs was then isolated that mapped the
novel sequences telomeric of the factor VIII gene and possibly in the regions
homologous to part of intron 22 (Figure 4.7). Since the first exon of the normal
factor VIII gene is telomeric to exon 22, the linear arrangement of sequences in
the patients' mRNA clearly indicates inversions. Furthermore, pulse field analysis of the patients' DNA digested with NruI demonstrated that these inversions
involved segments of 500–600 kb (Figure 4.8). These are the sizes expected of
inversions with boundaries in intron 22 and telomeric sequences homologous to
intron 22 (Naylor *et al.*, 1993b). Lakich *et al.* (1993), meanwhile, had thought
that the unusual abnormality we had identified in 45% of severely affected

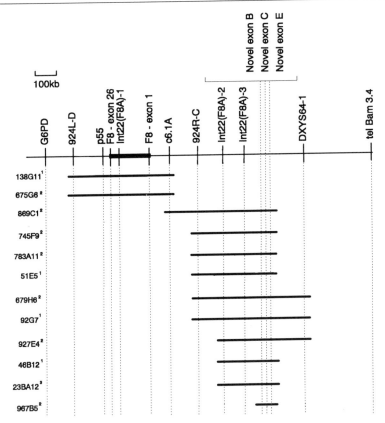

**Figure 4.7.** YAC contig showing location of novel sequences found 3' of exon 22 in patients' mRNA. Top horizontal line is telomeric segment of the long arm of the X chromosome showing factor VIII gene (thicker black line) and landmarks used to order YAC clones isolated with factor VIII and novel sequences (exons B, C, and E) from patients' mRNA. Int22 (F8A) shows the location of three repeat sequences comprising copies of the F8A gene. The landmark content of each YAC is shown by horizontal lines. The novel exons B, C, and E are located in the interval between 924 R-C and DXYS 64-1.

patients could be due to recombination between the homologous sequences in intron 22 and those at the two more telomeric locations. If the latter sequences were in opposite orientation to that in intron 22 homologous intrachromatid recombination would lead to inversions. In the course of testing this hypothesis they found one restriction enzyme (BclI) that resolved the three homologous sequences as a set of three bands of 21.5, 16, and 14 kb that hybridize to the F8A gene sequences. The 21.5-kb band is contributed by intron 22, while those of 14 and 16 kb contain respectively the proximal and distal members of the telomeric pair of repeats. Patients with the intron 22 mutations showed an abnormality of either the 14- or 16-kb band and then invariably of the 21.5-kb band, thus showing that the inversion boundaries on the telomeric side involved either of the two regions containing the telomeric copies of gene F8A (Lakich et al., 1993).

Patients with any of the above inversions show different haplotypes of the factor VIII gene polymorphisms, thus suggesting that the inversions are as independent in origin as other mutations causing hemophilia A (Lakich et al., 1993; Naylor et al., 1993b). We estimate, therefore, that the rate of mutations leading to inversions is of the order of $4 \times 10^{-6}$ per gene per gamete per generation. This rate is equivalent to the mutation rate for the whole spectrum of mutations causing hemophilia B and clearly reveals an unsuspected degree of chromosomal instability (Naylor et al., 1993b).

It is important at this stage to define the structure of the inversion junctions and to gain further insight into the factors that are responsible for the very high observed inversion rate. Furthermore, these results beg the question of whether the chromosomal instability first discovered in hemophilia A may occur in other regions of the genome and thus contribute significantly to man's genetic load.

## IX. FUNCTIONAL INTERPRETATION OF OBSERVED SEQUENCE CHANGES AND GENOTYPE/PHENOTYPE CORRELATIONS IN HEMOPHILIA A

A large proportion of the observed factor VIII mutations can undoubtedly be considered detrimental: that is, gross deletions, insertions, duplications, and the inversions involving intron 22. Together these represent a quarter of all hemophilia A mutations. Further mutations that are clearly detrimental are frameshifts, nonsense mutations, and alterations of the absolutely conserved elements of the splice-site consensuses. A few missense mutations can also readily be recognized as detrimental because they affect regions of known functional importance such as activation cleavage sites and residues important for the factor VIII–von Willebrand binding. The detrimental nature of other missense mutations can be strongly indicated by any one of the following three criteria: (1) the change observed in a patient is the only one affecting his gene, (2) the mutation is of recent origin, or (3) the mutation has occurred independently in two or more families. In order to apply the first of the above criteria it is necessary to use a mutation detection method that screens all the essential sequences of the gene with virtually 100% efficiency such as the method of Naylor et al. (1991). The second criterion relies on family studies and the availability of ascendant relatives, and the third on differences between the patients' factor VIII markers.

Weaker inferences can be based on the evolutionary conservation of the residue involved or on the difference between the mutant and wild-type residues. Factor VIII has few homologues and very little is known about either its tridimensional structure or the functional importance of its individual domains. Site-directed mutagenesis and expression of factor VIII in cultured cells has been used to confirm the importance of factor VIII residues suspected to be functionally important (Eaton *et al.*, 1986; Pittman and Kaufman, 1988, 1989), but this method does not provide a general approach for determining the detrimental nature of factor VIII mutations. In fact, the vast majority of hemophilia A mutations result in deficits of factor VIII protein in circulation and the factors that determine the concentration of factor VIII in circulation cannot be reproduced in cell culture systems. Analysis of the mRNA is important for the functional interpretation of mutations. Of course, any mutation altering splice sites and especially those affecting poorly conserved elements of the splice-site consensus should be examined at the mRNA level. This analysis is also essential for mutations likely to create a new splice site. Gross structural changes, such as partial deletions and duplications or insertions, could also affect splicing in ways that cannot be predicted by DNA analysis and direct examination of the mRNA is appropriate. Furthermore, the results mentioned above indicate that even a single base-pair substitution causing a nonsense codon may result in exon skipping, a phenomenon that may drastically affect the functional consequences of the mutation (Naylor *et al.*, 1993a).

With regard to the correlation between phenotype and genotype, patients with mild and moderate disease usually carry missense mutations but the converse is not necessarily true and a number of missense mutations have been found in patients with severe disease (Higuchi *et al.*, 1991b; Naylor *et al.*, 1993a). Sometimes marked differences in severity have been reported between patients with the same mutation (Tuddenham *et al.*, 1991), but it is difficult at present to assess how much this is due to variations in hematological assays, problems with the characterization of mutations such as failing to detect a second change because of incomplete screening of the essential sequences of the gene, or true biological variation. The latter would seem most probable if differences were observed among relatives examined by the same laboratory, and in this case perhaps variations in the proteins that interact with factor VIII, such as for example von Willebrand factor, could be responsible. Hemophilia A shows a greater proportion of severely affected patients than hemophilia B. The fact that one-fifth of all hemophilia A cases are due to inversions that disrupt the factor VIII gene may, at least in part, explain this difference.

One of the most worrying phenotypic features of hemophilia A patients is the development of antibodies against therapeutic factor VIII (inhibitors). Mutation analysis has clearly indicated that the individual predisposition to develop such a disastrous complication is strongly influenced by the nature of the

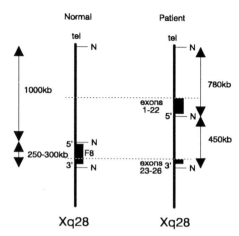

**Figure 4.8.** Distal segment of Xq28 in normal controls and patients as revealed by analysis of pulse field gels. (Left) normal DNA showing *NruI* sites bracketing the factor VIII gene in a 250- to 300-kb interval. (Right) patients' DNA with inversion that splits factor VIII gene in intron 22 so that exons 1–22 and 23–26 are contained within *NruI* fragments of 780 and 450 kb, respectively.

patient's mutation. Thus, 31 of the 111 patients with deletions, nonsense, or frameshift mutations reported so far had inhibitors, while only 3 of the 134 patients with missense or splice-site mutations had this complication.

The correlation between the individual's mutation and his predisposition to the inhibitor complication can best be explained by the hypothesis that failure to produce coagulant factor resembling the normal one does not allow the natural development of immune tolerance to the latter. This hypothesis, proposed first for hemophilia B, where the correlation between type of mutation and presence of inhibitors was first observed, was not readily accepted for hemophilia A (Giannelli *et al.*, 1983; Giannelli and Brownlee, 1986). This is perhaps partly explained by the more complex fate of the inhibitors in hemophilia A, where a higher proportion of such antibodies appears to occur transiently, and partly by the difficulties of mutation analysis that prevented adequate exploration of the hemophilia A mutational spectrum. It is interesting to consider the frequency of inhibitors among groups of patients with different nonsense mutations. Tuddenham *et al.* (1991) have noted that this seems to vary. Thus, none of the six known patients with a nonsense mutation at codon 2116 has inhibitors, while 11 of the 21 with nonsense mutations at codons 1941, 2147, and 2209 have the complication. It is at least conceivable that this difference may be due to the fact, mentioned earlier, that patients with an Arg2116 → Stop mutation produce mRNA where exon skipping eliminates the nonsense codon.

This leads to the synthesis of factor VIII missing only the amino acids encoded by exon 22. Since a patient with a deletion affecting only exon 22 has moderate disease and 2–5% factor VIII antigen in circulation (Youssoufian *et al.*, 1987) it seems that factor VIII missing the amino acids encoded by exon 22 is partially stable and function. Such a factor VIII could allow normal development of immune tolerance toward factor VIII during the maturation of the patient's immune system.

## X. PROGRESS IN CARRIER AND PRENATAL DIAGNOSIS IN HEMOPHILIA A

Delay in the development of rapid procedures for the detection of hemophilia A mutations has meant that carrier tests and prenatal diagnoses have relied so far on indirect DNA-based methods and/or hematological assays (Mibashan *et al.*, 1986; Tuddenham, 1989; Lalloz *et al.*, 1991). The latter have the same limitations described for hemophilia B: X chromosome inactivation which affects carrier tests and the need for late fetoscopy for blood sampling, which causes low acceptance and significant risk in prenatal diagnoses.

Indirect DNA tests based on the intrafamilial segregation of factor VIII DNA polymorphisms became possible with the detection of intragenic markers as early as 1985 (Antonarakis *et al.*, 1985; Gitschier *et al.*, 1985a). However, the small number and linkage disequilibrium of the intragenic markers limited to only 70 the percentage of women heterozygous and therefore informative for any one of these markers (Tuddenham, 1989). Consequently, for many years extragenic markers (DXS15 and DXS52) were also used (Tuddenham, 1989). These showed significant recombination with the factor VIII gene and did not allow definitive tests but simply probabilistic conclusions. In 1991 a polymorphism due to variable number of tandem repeats was found in intron 13 of the factor VIII gene and this greatly increased the informativity of the factor VIII-specific DNA markers (Lalloz *et al.*, 1991). Indirect, polymorphism-based diagnoses are, however, expected to fail in a substantial proportion of families for the reasons already explained in the section on hemophilia B, and therefore the ultimate aim must be to replace indirect tests with the direct detection of the gene defect.

In 1991 the development of a rapid method for the characterization of hemophilia A mutations made this possible for families with conventional mutations (Naylor *et al.*, 1991), but the families with the intron 22 mutation remained a problem because mRNA analysis could not offer adequate carrier diagnosis (Naylor *et al.*, 1992, 1993a). Now, the observation that patients with the intron 22 mutation have inversions that alter two *Bcl*I bands hybridizing to the F8A gene has provided a very rapid direct diagnostic test for families with this unexpected type of mutation (Lakich *et al.*, 1993). As for the hemophilia B mutation, there are no reliable estimates of the incidence of ovarian mosaicism for the hemophilia A mutations.

## XI. CONTRIBUTION OF MOLECULAR BIOLOGY TO TREATMENT

As mentioned earlier, the treatment of both hemophilias is based on the administration of the missing coagulation factor (Rizza and Spooner, 1983). These are generally obtained from blood donations by procedures that have changed over the years and may lead to more or less pure products. Unfortunately, the risk that these may transmit viral diseases has been tragically highlighted by the HIV infections that resulted from the use of inadequately treated products (Pierce *et al.*, 1989). Some risk of viral infection persists even now that virus-inactivation steps are mandatory in the preparation of coagulant factors. Attempts have therefore been made to produce therapeutic coagulation factors *in vitro* from recombinant genes.

Factor VIII has proved easier to produce on an industrial scale and recombinant factor VIII preparations have undergone clinical trials (Lusher *et al.*, 1993). The product appears to have similar pharmacological properties to natural factor VIII. The thorough procedure established to detect inhibitor complications in the course of this trial has detected an incidence of inhibitors higher than expected from previous experience with natural therapeutic products. However, since the latter experience was based on less stringent monitoring procedures, the above finding has been tentatively explained by the detection of low titer and transient immune responses to treatment that may have been missed in the earlier monitoring of natural factor VIII preparations. It is hoped that in the future standardized and stringent criteria will allow the comparison of different therapeutic products in every relevant respect.

The cloning of the factor VIII and factor IX genes has also opened the way to gene therapy. Various experimental approached have been investigated using different gene delivery systems, cells, and test organisms. These experiments have so far yielded results that, though showing some technical success, fall short of what is required for immediate successful translation to man (Hoeben *et al.*, 1993; Gerrard *et al.*, 1993; Kurachi and Yao, 1993).

## XII. CONCLUSION

Over the past 10–12 years the cloning of the factor IX and factor VIII genes has led to clear advances in the understanding of the molecular genetics of hemophilia, important improvements in genetic counseling, and new prospects for therapy.

In particular, work on hemophilia B has led to a fairly detailed definition of the mutational spectrum associated with this disease and to some insight
into the functional consequences of factor IX mutations. Furthermore, methods
for the rapid detection of mutations have permitted carrier and prenatal diagnoses to be based on the direct detection of the gene defect and are also allowing
the establishment of fully efficient strategies for the optimal genetic counseling
and prevention of the disease. In this regard, hemophilia B may represent a
model for other diseases of very high mutational heterogeneity.

Less defined is the mutational spectrum of hemophilia A, but quicker
progress is expected now that rapid procedures exist for the analysis of all
essential sequences of the factor VIII gene. Recent work has unexpectedly
shown that almost half the patients with severe hemophilia A—one-fifth of the
total—carry inversions that split the factor VIII gene at intron 22 and involve
long (500–600 kb) segments of DNA apparently flanked by homologous sequences. It is now important to understand more fully the factors responsible for
such unprecedented chromosomal instability.

Since Southern blotting can be used to detect the above inversions,
and rapid screening of patients' mRNA provides an efficient way to characterize
the other hemophilia A mutations, carrier and prenatal diagnoses based on the
direct detection of the gene defect are now also possible in hemophilia A.

The expression of cloned genes has led to the introduction of recombinant factor VIII in the treatment of hemophilia A, and the genes for both
factors VIII and IX are being used to develop successful approaches toward gene
therapy.

## Acknowledgments

We are very grateful to our colleagues, A. Brinke, S. Hassock, G. Rowley, S. Saad, and L.
Tagliavacca for help with the work presented in this review. We are also grateful to Adrienne Knight
for secretarial help. This work was supported by Action Research and the Medical Research Council
and also by the Spastics Society and the Generation Trust.

## References

Aggeler, P. M., White, S. G., Glendening, M. B., Page, E. W., Leake, T. B., and Bates, G.
(1952). Plasma thromboplastin component (PTC) deficiency: A new disease resembling hemophilia. *Proc. Natl. Acad. Sci. USA* **79**:692–694.
Anson, D. S., Blake, D. J., Winship, P. R., Birnbaum, D., and Brownlee, G. G. (1988). Nulli-
somic deletion of the mcf.2 transforming gene in two haemophilia B patients. *EMBO J.* **7**:2795–
2799.
Anson, D. S., Choo, K. H., Rees, D. J., Giannelli, F., Gould, K., Huddleston, J. A., and
Brownlee, G. G. (1984). The gene structure of human anti-haemophilic factor IX. *EMBO J.*
**3**:1053–1060.
Antonarakis, S. E., Waber, P. G., Kittur, S. D., Patel, A. G., Kazazian, H. H., Jr., Mellis, M. A.,

Counts, R. B., Stamatoyannopoulos, G., Bowie, E. J. W., Fass, D. N., Pittman, D. D., Wozney, J. M., and Toole, J. J. (1985). Haemophilia A: Detection of molecular defects and of carriers by DNA analysis. *N. Engl. J. Med.* 313:842–848.

Attree, O., Vidaud, D., Vidaud, M., Amselem, S., Lavergne, J. M., and Goossens, M. (1989). Mutations in the catalytic domain of human coagulation factor IX: Rapid characterization by direct genomic sequencing of DNA fragments displaying an altered melting behavior. *Genomics* 4:266–272.

Bajaj, S. P., Rapaport, S. I., and Russel, W. A. (1983). Redetermination of the rate-limiting step in the activation of factor IX by factor IXa and by factor VIIa/tissue factor. Explanation for different electrophoretic radioactivity profiles obtained on activation of $^3$H and $^{125}$I-labelled factor IX. *Biochemistry* 22:4047–4053.

Baron, M., Norman, D. G., Harvey, T. S., Handford, P. A., Mayhew, M., Tse, A. G. D., Brownlee, G. G., and Campbell, I. D. (1992). The three-dimensional structure of the first EGF-like module of human factor IX: comparison with EGF and TGF-$\alpha$. *Protein Sci.* 1:81–90.

Barrow, E. S., Miller, C. H., Reisner, H. M., and Graham, J. B. (1982). Genetic counselling in haemophilia by discriminant analysis 1975–1980. *J. Med. Genet.* 19:26–34.

Bentley, A. K., Rees, D. J., Rizza, C., and Brownlee, G. G. (1986). Defective propeptide processing of blood clotting factor IX caused by mutation of arginine to glutamine at position −4. *Cell* 45:343–348.

Biggs, R., Douglas, A. S., MacFarlane, R. G., Dacie, J. V., Pitney, W. R., Merskey, C., and O'Brien, J. R. (1952). Christmas disease: A condition previously mistaken for haemophilia. *Br. Med. J.* 2:1378–1382.

Bottema, C. D. K., Koeberl, D. D., Ketterling, R. P., Bowie, E. J. W., Taylor, S. A. M., Lillicrap, D., Shapiro, A., Gilchrist, G., and Sommer, S. S. (1990). A past mutation at isoleucine 397 is now a common cause of moderate/mild haemophilia B. *Br. J. Haematol.* 75:g212–216.

Bottema, C. D. K., Koeberl, D. D., and Sommer, S. S. (1989). Direct carrier testing in 14 families with haemophilia B. *Lancet* 2:526–528.

Broze, G. J., Jr. (1992). The role of tissue factor pathway inhibitor in a revised coagulation cascade. *Sem. Hematol.* 29:159–169.

Camerino, G., Grzeschik, K. H., Jaye, M., De la Salle, H., Tolstoshev, P., Lecocq, J. P., Heilig, R., and Mandel, J-L. (1984). Regional localisation on the human X chromosome and polymorphism of the coagulation factor IX gene (haemophilia B locus). *Proc. Natl. Acad. Sci. USA* 81:498–502.

Chen, S. H., and Scott, C. R. (1990). Recombination between two 14-bp homologous sequences as the mechanism for gene deletion in factor IX Seattle 1. *Am. J. Hum. Genet.* 47:1020–1022.

Chen, S. H., Scott, C. R., Edson, J. R., and Kurachi, K. (1988). An insertion within the factor IX gene: Hemophilia BEI Salvador. *Am. J. Hum. Genet.* 42:581–584.

Cheung, W. F., Hamaguchi, N., Smith, K. J., and Stafford, D. W. (1992). The binding of human factor IX to endothelial cells is mediated by residues 3–11. *J. Biol. Chem.* 267:20529–20531.

Choo, K. H., Gould, K. G., Rees, D. J., and Brownlee, G. G. (1982). Molecular cloning of the gene for human anti-haemophilic factor IX. *Nature* 299:178–180.

Cotton, R. G. H., Rodrigues, N. R., and Campbell, R. D. (1988). Reactivity of cytosine and thymine in single base pair mismatches with hydroxylamine and osmium tetroxide and its application to the study of mutations. *Proc. Natl. Acad. Sci. USA* 85:4397–4401.

Crossley, M., Ludwig, M., Stowell, K. M., De Vos, P., Olek, K., and Brownlee, G. G. (1992). Recovery from hemophilia B Leyden: An androgen-responsive element in the factor IX promoter. *Science* 257:377–379.

Crossley, M., and Brownlee, G. G. (1990). Disruption of a C/EBP binding site in the factor IX promoter is associated with haemophilia B. *Nature* 345:444–446.

Crossley, P. M., Winship, P. R., Black, A., Rizza, C. R., and Brownlee, G. G. (1989). An unusual case of severe haemophilia B. *Lancet* 1:960.

Davis, L. M., McGraw, R. A., Ware, J. L., Roberts, H. R., and Stafford, D. W. (1987). Factor IX Alabama: A point mutation in a clotting protein results in hemophilia B. *Blood* **69**:140–143.

De la Salle, C., Charmantier, J. L., Baas, M. J., Schwartz, A., Wiesel, M. O., Grunebaum, L., and Cazenave, J. P. (1993). A deletion located in the 3′ non translated part of the factor IX gene responsible for mild haemophilia B. *Thromb. Haemost.* **70**: 370–371.

Diamond, C., Kogan, S., Levinson, B., and Gitschier, J. (1992). Amino acid substitutions in conserved domains of factor VIII and related proteins: Study of patients with mild and moderately severe hemophilia A. *Hum. Mutat.* **1**:248–257.

Dombroski, B. A., Mathias, S. L., Nanthakumar, E., Scott, A. F., and Kazazian, H. H., Jr. (1991). Isolation of an active human transposable element. *Science* **254**:1805–1808.

Eaton, D., Rodriguez, H., and Vehar, G. A. (1986). Proteolytic processing of factor VIII: Correlation of specific cleavages by thrombin, factor Xa and activated protein C with activation and inactivation of factor VIII:C coagulant activity. *Biochemistry* **25**: 505–512.

Elder, B., Lakich, D., and Gitschier, J. (1993). Sequence of the murine factor VIII cDNA. *Genomics* **16**:374–379.

Fay, P. J., Haidaris, P. J., and Huggins, C. F. (1993). Role of the COOH-terminal acidic region of A1 subunit in A2 subunit retention in human factor VIIIa. *J. Biol. Chem.* **268**:17861–17866.

Fay, P. J., Smudzin, T. M., and Walker, F. J. (1991). Activated protein C - catalysed inactivation of human factor VIII and factor VIIIa: Identification of cleavage sites and correlation of proteolysis with cofactor activity. *J. Biol. Chem.* **266**:20139–20145.

Ferrari, N., and Rizza, C. A. (1986). Estimation of genetic risks of carriership for possible carriers of Christmas disease (haemophilia B). *Braz. J. Genet.* **9**:87–99.

Foster, D. C., Yoshitake, S., and Davie, E. W. (1985). The nucleotide sequence of the gene for human protein C. *Proc. Natl. Acad. Sci. USA* **82**:4673–4677.

Francis, R. B., and Kasper, C. K. (1983). Reproduction in haemophilia. *J. Am. Med. Assoc.* **250**:3192–3195.

Fraser, B. M., Poon, M. C., and Hoar, D. I. (1992). Identification of factor IX mutations in haemophilia B: Application of polymerase chain reaction and single strand conformation analysis. *Hum. Genet.* **88**:426–430.

Freedenburg, D. L., and Black, B. (1991). Altered developmental control of the factor IX gene: a new T to A mutation at position +6 of the FIX gene resulting in haemophilia B Leyden. *Thromb. Haemost.* **65**:964.

Freije, D., and Schlessinger, D. (1992). A 1.6 Mb contig of yeast artificial chromosomes around the human factor VIII gene reveals three regions homologous to probes for the DXS115 locus and two from the DXYS64 locus. *Am. J. Hum. Genet.* **51**:66–80.

Frohman, M. A., Dush, M. K., and Martin, G. (1988). Rapid production of full-length cDNA from rare transcripts: Amplification using a single gene-specific oligonucleotide primer. *Proc. Natl. Acad. Sci. USA* **85**:8998–9002.

Gerrard, A., Hudson, D. L., Brownlee, G. G., and Watt, F. M. (1993). Towards gene therapy for haemophilia B using primary human keratinocytes. *Nature Genet.* **3**:180–183.

Ghanem, N., Costes, B., Martin, J., Vidaud, M., Rothschild, C., Foyer-Gazengel, C., and Goossens, M. (1993). Twenty-four novel hemophilia B mutations revealed by rapid scanning of the whole factor IX gene in a French population sample. *Eur. J. Hum. Genet.* **1**:144–155.

Giannelli, F. (1987). The identification of haemophilia B mutations. *In* "Protides of the Biological Fluids, Vol. 25" (H. Peeters, Ed.), pp. 29–32. Pergamon Press, Oxford.

Giannelli, F., and Brownlee, G. G. (1986). Cause of the 'inhibitor' phenotype in the haemophilias. *Nature* **320**:196.

Giannelli, F., Choo, K. H., Rees, D. J., Boyd, Y., Rizza, C. R., and Brownlee, G. G. (1983). Gene deletions in patients with haemophilia B and anti-factor IX antibodies. *Nature* **303**:181–182.

Giannelli, F., Anson, D. S., Choo, K. H., Rees, D. J. G., Winship, P. R., Ferrari, N., Rizza, C. R., and Brownlee, G. G. (1984). Characterisation and use of an intragenic polymorphic marker for detection of carriers of haemophilia B (factor IX deficiency). *Lancet* **1**: 239–241.

Giannelli, F., Saad, S., Montandon, A. J., Bentley, D. R., and Green, P. M. (1992). A new strategy for the genetic counselling of diseases of marked mutational heterogeneity: Haemophilia B as a model. *J. Med. Genet.* **29**:602–607.

Giannelli, F., Green, P. M., High, K. A., Sommer, S., Poon, M. C., Ludwig, M., Schwaab, R., Reitsma, P. H., Goossens, M., Yoshioka, A., and Brownlee, G. G. (1993). Haemophilia B: Database of point mutations and short additions and deletions—fourth edition, 1993. *Nucleic Acids Res.* **21**:3075–3087.

Gispert, S., Vidaud, M., Vidaud, D., Gazengel, C., Boneu, B., and Goossens, M., (1989). A promoter defect correlates with an abnormal coagulation factor IX gene expression in a french family (haemophilia B Leyden). *Am. J. Hum. Genet.* **45**:A189.

Gitschier, J. (1988). Maternal duplication associated with gene deletion in sporadic haemophilia. *Am. J. Hum. Genet.* **43**:274–279.

Gitschier, J., Kogan, S., Levinson, B., and Tuddenham, E. G. D. (1988). Mutations of the factor VIII cleavage sites in haemophilia A. *Blood* **72**:1022–1028.

Gitschier, J., Drayna, D., Tuddenham, E. G. D., White, R. L., and Lawn, R. M. (1985a). Genetic mapping and diagnosis of haemophilia A achieved through BclI polymorphism in the factor VIII gene. *Nature* **314**:738–740.

Gitschier, J., Wood, W. I., Tuddenham, E. G. D., Shuman, M. A., Goralka, T. M., Chen, E. Y., and Lawn, R. M. (1985b). Detection and sequence of mutations in the factor VIII gene of haemophiliacs. *Nature* **315**:427–430.

Gitschier, J., Wood, W. I., Goralka, T. M., Wion, K. L., Chen, E. Y., Eaton, D. H., Vehar, G. A., and Capon, D. J. (1984). Characterization of the human factor VIII gene. *Nature* **312**:326–330.

Green, P. M., Bentley, D. R., Mibashan, R. S., and Giannelli, F. (1988). Partial deletion by illegitimate recombination of the factor IX gene in a haemophilia B family with two inhibitor patients. *Mol. Biol. Med.* **5**:95–106.

Green, P. M., Bentley, D. R., Mibashan, R. S., Nilsson, I. M., and Giannelli, F. (1989). Molecular pathology of haemophilia B. *EMBO J.* **8**:1067–1072.

Green, P. M., Montandon, A. J., Bentley, D. R., and Giannelli, F. (1991a). Genetics and molecular biology of haemophilia A and B. *Blood Coag. Fibrin.* **2**:539–565.

Green, P. M., Montandon, A. J., Ljung, R., Bentley, D. R., Kling, S., Nilsson, I. M., and Giannelli, F. (1991b). Haemophilia B mutations in a complete Swedish population sample. A test of new strategy for the genetic counselling of diseases with high mutational heterogeneity. *Br. J. Haematol.* **78**:390–397.

Green, P. M., Mitchell, V. E., McGraw, A., Goldman, E., and Giannelli, F. (1993). Haemophilia B caused by a missense mutation in the prepeptide sequence of factor IX. *Hum. Mutat.* **2**:103–107.

Haldane, J. B. S. (1935). The rate of spontaneous mutation of a human gene. *J. Genet.* **31**:317–326.

Haldane, J. B. S. (1947). The mutation of the gene for haemophilia and its segregation ratios in males and females. *Ann. Eugen.* **13**:262–271.

Haldane, J. B. S., and Smith, C. A. B. (1947). A new estimate of the linkage between the genes for colour-blindness and haemophilia in man. *Ann. Eugen.* **14**:10–31.

Hall, A., Chansumrit, A., Peake, I. R., and Winship, P. R. (1992). A single base pair deletion in the promoter region of the factor IX gene is associated with haemophilia B. *Br. J. Haematol.* **80**:16.

Handford, P. A., Baron, M., Mayhew, M., Willis, A., Beesley, T., Brownlee, G. G., and Campbell, I. D. (1990). The first EGF-like domain from human factor IX contains a high-affinity calcium binding site. *EMBO J.* **9**:475–480.

Handford, P. A., Mayhew, M., Baron, M., Winship, P. R., Campbell, I. D., and Brownlee, G. G. (1991). Key residues involved in calcium-binding motifs in EGF-like domains. *Nature* 351:164–167.

Hase, S., Kawabata, S., Nishimura, H., Takeya, H., Sueyoshi, T., Miyata, T., Iwanaga, S., Takao, T., Shimonishi, Y., and Ikenaka, T. (1988). A new trisaccharide sugar chain linked to a serine residue in bovine blood coagulation factors VII and IX. *J. Biochem.* 104:867–868.

Higuchi, M., Antonarakis, S. E., Kasch, L., Oldenburg, J., Petersen, E. E., Olek, K., Inaba, H., and Kazazian, H. H. (1991a). Towards complete characterization of mild-to-moderate hemophilia A: Detection of the molecular defect in 25 of 29 patients by denaturing gradient gel electrophoresis. *Proc. Natl. Acad. Sci. USA* 88:8307–8311.

Higuchi, M., Kazazian, H. H., Kasch, L., Warren, T. C., McGinnis, M. J., Phillips, J. A., III, Kasper, C., Janco, R., and Antonarakis, S. E. (1991b). Molecular characterisation of some hemophilia A suggests that about half the mutations are not within the coding regions and splice junctions of the factor VIII gene. *Proc. Natl. Acad. Sci. USA* 88:7405–7409.

Higuchi, M., Wong, C., Kochlan, L., Olek, K., Aronis, S., Kasper, C. K., Kazazian, H. H., Jr., and Antonarakis, S. E. (1990). Characterisation of mutations in the factor VIII gene by direct sequencing of amplified genomic DNA. *Genomics* 6:65–71.

Hoeben, R. C., Fallauk, F. J., van Tilburg, N. H., Cramer, S. J., van Ormondt, H., Briët, E., and van der Eb, A. J. (1993). Toward gene therapy for haemophilia A. Long term persistence of factor VIII-secreting fibroblasts after transplantation into immunodeficient mice. *Hum. Gene Ther.* 4:179–186.

Hughes, P. E., Morgan, G., Rooney, E. K., Brownlee, G. G., and Handford, P. (1993). Tyrosine 69 of the first epidermal growth factor-like domain of human factor IX is essential for clotting activity. *J. Biol. Chem.* 268:17727–17733.

Jaye, M., de la Salle, H., Schamber, F., Balland, A., Kohli, V., Findeli, A., Tolstoshev, P., and Lecocq, J. P. (1983). Isolation of a human anti-haemophilic factor IX cDNA clone using a unique 52-base synthetic oligonucleotide probe deduced from the amino acid sequence of bovine factor IX. *Nucleic Acids Res.* 11:2325–2335.

Jenny, R. J., Pittman, D. D., Toole, J. J., Kriz, R. W., Aldape, R. A., Hewick, R. M., Kaufman, R. J., and Mann, K. G. (1987). Complete cDNA and derived amino acid sequence of human factor V. *Proc. Natl. Acad. Sci. USA* 84:4846–4850.

Kazazian, H. H., Wong, C., Youssoufian, H., Scott, A. F., Phillips, D. G., and Antonarakis, S. E. (1988). Haemophilia A resulting from de novo insertion of L1 sequences represents a novel mechanism for mutation in man. *Nature* 332:164–166.

Kessler, C. M. (1991). An introduction to factor VIII inhibitors: The detection and quantitation. *Am. J. Med.* 9 (Suppl 5A):1–5.

Ketterling, R. P., Bottema, C. D., Phillips, J. A., and Sommer, S. S. (1991). Evidence that descendants of three founders constitute about 25% of hemophilia B in the United States. *Genomics* 10:1093–1096.

Ketterling, R. P., Ricke, D. O., Wurster, M. W., and Sommer, S. S. (1993a). Deletions with inversions: Report of a mutation and review of the literature. *Hum. Mutat.* 2:53–57.

Ketterling, R. P., Vielhaber, E., Bottema, C. D. K., Schaid, D. J., Cohen, M. P., Sexauer, C. L., and Sommer, S. S. (1993b). Germ-line origin of mutation in families with hemophilia B: The sex ratio varies with the type of mutation. *Am. J. Hum. Genet.* 52:156–166.

Koeberl, D. D., Bottema, C. D., Buerstedde, J. M., and Sommer, S. S. (1989). Functionally important regions of the factor IX gene have a low rate of polymorphism and a high rate of mutation in the dinucleotide CpG. *Am. J. Hum. Genet.* 45:448–457.

Kogan, S., and Gitschier, J. (1990). Mutations and a polymorphism in the factor VIII gene discovered by denaturing gradient gel electrophoresis. *Proc. Natl. Acad. Sci. USA* 87:2092–2096.

Koshinsky, M. L., Funk, W. D., van Oost, B. A., and MacGillivray, R. T. A. (1986). Complete cDNA sequence of human preceruloplasmin. *Proc. Natl. Acad. Sci. USA* 83:5056–5090.

Kurachi, K., and Davie, E. W. (1982). Isolation and characterization of a cDNA coding for human factor IX. *Proc. Natl. Acad. Sci. USA* **79**:6461–6464.

Kurachi, K., and Yao, S-N. (1993). Gene therapy in hemophilia B. *Thromb. Haemost.* **70**:193–197.

Lakich, D., Kazazian, H. H., Antonarakis, S. E., and Gitschier, J. (1993). Inversions disrupting the factor VIII gene as a common cause of severe hemophilia A. *Nature Genet.* **5**:236–241.

Lalloz, M. R., McVey, J. H., Pattinson, J. K., and Tuddenham, E. G. D. (1991). Haemophilia A diagnosis by analysis of a hypervariable dinucleotide repeat within the factor VIII gene. *Lancet* **338**:207–211.

Levinson, B., Kenwrick, S., Gamel, P., Fisher, K., and Gitschier, J. (1992). Evidence for a third transcript from the human factor VIII gene. *Genomics* **14**:585–589.

Levinson, B., Kenwrick, S., Lakish, D., Hammonds, G., Jr., and Gitschier, J. (1990). A transcribed gene in an intron of the human factor VIII gene. *Genomics* **7**:1–11.

Leytus, S. P., Foster, D. C., Kurachi, K., and Davie, E. W. (1986). Gene for human factor X: A blood coagulation factor whose gene organization is essentially identical with that of factor IX and protein C. *Biochemistry* **25**:5098–5102.

Ludwig, M., Grimm, T., Brackmann, H. H., and Olek, K. (1992). Parental origin of factor IX gene mutations, and their distribution in the gene. *Am. J. Hum. Genet.* **50**:164–173.

Lusher, J. M., Arkins, S., Abildgaard, C. F., Schwartz, R. S., and the Kogenate Previously Untreated Patient Study Group (1993). Recombinant factor VIII for the treatment of previously untreated patients with haemophilia A. *N. Engl. J. Med.* **328**:453–459.

McGinniss, M. J., Kazazian, H. H., Hoyer, L. W., Bi, L., Inaba, H., and Antonarakis, S. E. (1993). Spectrum of mutations in crm-positive and crm-reduced hemophilia A. *Genomics* **15**:392–398.

Mibashan, R. S., Giannelli, F., Pembrey, M. E., and Rodeck, C. H. (1986). The antenatal diagnosis of clotting disorders. *In* "Advanced Medicine 21" (M. J. Brown, Ed.), pp. 415–438. Royal College of Physicians of London, Churchill Livingstone, Edinburgh.

Migeon, B. R., McGinnis, M. J., Antonarakis, S. E., Axelman, J., Stascowski, B. A., Youssoufian, H., Kearns, W. G., Chung, A., Pearson, P. L., Kazazian, H. H., Jr., and Muneer, R. S. (1993). Severe haemophilia A in a female by cryptic translocation: Order and orientation of factor VIII within Xq28. *Genomics* **16**:20–25.

Montandon, A. J., Green, P. M., Bentley, D. R., Ljung, R., Kling, S., Nilsson, I. M., and Giannelli, F. (1992). Direct estimate of the haemophilia B (factor IX deficiency) mutation rate and of the ratio of the sex-specific mutation rates in Sweden. *Hum. Genet.* **89**:319–322.

Montandon, A. J., Green, P. M., Bentley, D. R., Ljung, R., Nilsson, I. M., and Giannelli, F. (1990). Two factor IX mutations in the family of an isolated haemophilia B patient. Direct carrier diagnosis by amplification mismatch detection (AMD). *Hum. Genet.* **85**:200–204.

Montandon, A. J., Green, P. M., Giannelli, F., and Bentley, D. R. (1989). Direct detection of point mutations by mismatch analysis: Application to haemophilia B. *Nucleic Acids Res.* **17**:3347–3358.

Morgan, T. H. (1910). Sex-linked inheritance in *Drosophila. Science* **32**:120–122.

Murru, S., Casula, L., Pecorara, M., Mori, P., Cao, A., and Pirastu, M. (1990). Illegitimate recombination produced by a duplication within the factor VIII gene in a patient with mild haemophilia A. *Genomics* **7**:115–118.

Naylor, J. A., Green, P. M., Montandon, J. A., Rizza, C. R., and Giannelli, F. (1991). Detection of three novel mutations in two haemophilia A patients by rapidly screening whole essential regions of the factor VIII gene. *Lancet* **337**:635–639.

Naylor, J. A., Green, P. M., Rizza, C. R., and Giannelli, F. (1992). Factor VIII gene explains all cases of haemophilia A. *Lancet* **340**:1066–1067.

Naylor, J. A., Green, P. M., Rizza, C. R., and Giannelli, F. (1993a). Analysis of factor VIII mRNA reveals defects in every one of 28 haemophilia A patients. *Hum. Mol. Genet.* **2**:11–17.

Naylor, J., Brinke, A., Hassock, S., Green, P. M., and Giannelli, F. (1993b). Characteristic mRNA abnormality found in half the patients with severe haemophilia A is due to large DNA inversions. *Hum. Mol. Genet.* **2**:1773–1778.

Nishimura, H., Takao, T., Hase, S., Shimonishi, Y., and Iwanaga, S. (1992). Human factor IX has a tetrasaccharide O-glycosidically linked to serine 61 through the fucose residue. *J. Biol. Chem.* **267**:17520–17525.

Nishimura, H., Takeya, H., Miyata, T., Suehiro, K., Okamura, T., Niho, Y., and Iwanaga, S. (1993). Factor IX Fukuoka: Substitution of Asn[92] by His in the second epidermal growth factor-like domain results in defective interaction with factors VIIIa/X. *J. Biol. Chem.* **268**:24041–24046.

O'Hara, P. J., Grant, F. J., Haldeman, B. A., Gray, C. L., Insley, M. Y., Hagen, F. S., and Murray, M. J. (1987). Nucleotide sequence of the gene coding for human factor VII, a vitamin K-dependent protein participating in blood coagulation. *Proc. Natl. Acad. Sci. USA* **84**:5158–5162.

Ørstavik, K. H., Veltkamp, J. J., Bertina, R. M., and Hermans, J. (1981). Detection of haemophilia B carriers. *In* "Haemophilia" (U. Seligsohn, A. Rimon, and H. Horoszowski, Eds.), pp. 29–38. Castle House Publishing, Tunbridge Wells.

Otto, J. C. (1803). An account of haemorrhagic disposition existing in certain families. *Med. Repos.* **6**:1–4.

Patek, A. J., Jr., and Taylor, F. H. L. (1937). Hemophilia. II. Some properties of a substance obtained from normal human plasma effective in accelerating the coagulation of hemophilic blood. *J. Clin. Invest.* **16**:113–124.

Pattinson, J. K., Millar, J. H., McVey, J. H., Grundy, C. B., Wieland, K., Mibashan, R. S., Martinowitz, U., Tan-un, K., Vidaud, M., Goossens, M., Sampietro, M., Mannucci, P. M., Krawczak, M., Reiss, J., Zoll, B., Whitmore, D., Bowcock, S., Wensley, R., Ajani, A., Mitchell, V., Rizza, C., Maia, R., Winter, P., Mayne, E. E., Schwartz, M., Green, P. J., Kakkar, V. V., Tuddenham, E. G. D., and Cooper, D. N. (1990). The molecular genetic analysis of hemophilia A: A directed search strategy for the detection of point mutations in the human factor VIII gene. *Blood* **76**:2242–2248.

Peake, I. R., Mathews, R. J., and Bloom, A. L. (1989). Haemophilia B Chicago: Severe haemophilia B caused by two deletions and an inversion within the factor IX gene. *Br. J. Haematol.* **71**(suppl):1.

Picketts, D. J., D'Souza, C., Bridge, P. J., and Lillicrap, D. (1992). An A to T transversion at position −5 of the factor IX promoter results in hemophilia B. *Genomics* **12**:161–163.

Pierce, G. F., Lusher, J. M., Brownstein, S. P., Goldsmith, J. C., and Kessler, C. M. (1989). The use of purified clotting factor concentrates in hemophilia: Influence of viral safety, cost and supply on therapy. *JAMA* **261**:3434–3438.

Pittman, D. D., and Kaufman, R. J. (1988). Proteolytic requirements for thrombin activation of anti-hemophilic factor (factor VIII). *Proc. Natl. Acad. Sci. USA* **85**:2429–2433.

Pittman, D. D., and Kaufman, R. J. (1989). Structure–function relationships of factor VIII elucidated through recombinant DNA technology. *Thromb. Haemost.* **61**:161–165.

Pittman, D. D., Wang, J. H., and Kaufman, R. J. (1992). Identification and functional importance of tyrosine sulfate residues within recombinant factor VIII. *Biochemistry* **31**:3315–3325.

Rao, K. J., Lyman, G., Hamsabhushanam, K., Scott, J. P., and Jagadeeswaran, P. (1990). Human factor IX Lincoln Park: A molecular characterization. *Mol. Cell. Probes* **4**:335–340.

Rees, D. J., Jones, I. M., Handford, P. A., Walter, S. J., Esnouf, M. P., Smith, K. J., and Brownlee, G. G. (1988). The role of beta-hydroxyaspartate and adjacent carboxylate residues in the first EGF domain of human factor IX. *EMBO J.* **7**:2053–2061.

Rees, D. J., Rizza, C. R., and Brownlee, G. G. (1985). Haemophilia B caused by a point mutation in a donor splice junction of the human factor IX gene. *Nature* **316**:643–645.

Reijnen, M. J., Peerlinck, K., Maasdam, D., Bertina, R. M., and Reitsma, P. H. (1993). Hemophilia B Leyden: Substitution of thymine for guanine at position −21 results in a disruption of a hepatocyte nuclear factor 4 binding site in the factor IX promoter. *Blood* **82**:151–158.

Reitsma, P. H., Bertina, R. M., Ploos van Amstel, J. K., Riemens, A., and Briet, E. (1988). The putative factor IX gene promoter in hemophilia B Leyden. *Blood* **72**:1074–1076.

Reitsma, P. H., Mandalaki, T., Kasper, C. K., Bertina, R. M., and Briet, E. (1989). Two novel point mutations correlate with an altered developmental expression of blood coagulation factor IX (hemophilia B Leyden phenotype). *Blood* **73**:743–746.

Riley, J., Butler, R., Ogilvie, D., Finniear, R., Jenner, D., Powell, S., Anand, R., Smith, J. C., and Markham, A. F. (1990). A novel, rapid method for the isolation of terminal sequences from yeast artificial chromosome (YAC) clones. *Nucleic Acids Res.* **18**:2887–2890.

Rizza, C. R., and Spooner, R. J. D. (1983). Treatment of haemophilia and related disorders in Britain and Northern Ireland during 1976–1980: Report on behalf of the directors of haemophilia centres in the United Kingdom. *Br. Med. J.* **286**:929–933.

Roberts, H. R. (1971). Acquired inhibitors in haemophilia B. *Thromb. Diath. Harmorrh.* **Suppl 45**:217–225.

Rosner, F. (1969). Hemophilia in the Talmud and Rabbinic writings. *Ann. Intern. Med.* **70**:833–837.

Royle, G., Van de Water, N. S., Betty, E., Ockelford, P. A., and Browett, P. J. (1991). Haemophilia B Leyden arising de novo by point mutation in the putative factor IX promoter region. *Br. J. Haematol.* **77**:191–194.

Saad, S., Rowley, G., Tagliavacca, L., Green, P. M., Giannelli, F., and UK Haemophilia Centres (1994). First report on UK database of haemophilia B mutations and pedigrees. *Thromb. Haemost.* **71**:563–570.

Schwartz, C., Fitch, N., Phelan, M. C., Richer, C. L., and Stevenson, R. (1987). Two sisters with a distal deletion at the Xq26/Xq27 interface: DNA studies indicate that the gene locus for factor IX is present. *Hum. Genet.* **76**:64–67.

Sheffield, V. C., Beck, J. S., Kwitek, A. E., Sandstrom, D. W., and Stone, E. M. (1993). The sensitivity of single-strand conformation polymorphism analysis for the detection of single base substitutions. *Genomics* **16**:325–332.

Solera, J., Magallon, M., Martin-Villar, J., and Coloma, A. (1992). Factor IX Madrid 2: A deletion/insertion in factor IX gene which abolishes the sequence of the donor junction at the exon IV-intron d splice site. *Am. J. Hum. Genet.* **50**:434–437.

Soriano-Garcia, M., Padmanabhan, K., de Vos, A. M., and Tulinsky, A. (1992). The $Ca^{2+}$ ion and membrane binding structure of the Gla domain of the Ca-prothrombin fragment 1. *Biochemistry* **31**:2554–2566.

Stevanović, M., Lovell-Badge, R., Collignon, J., and Goodfellow, P. (1993). *SOX3* is an X-linked gene related to *SRY. Hum. Mol. Genet.* **2**:2013–2018.

Stoflet, E. S., Koeberl, D. D., Sarkar, G., and Sommer, S. S. (1988). Genomic amplification with transcript sequencing. *Science* **239**:491–494.

Stubbs, J. D., Lekutis, C., Singer, K. L., Bui, A., Yuzuki, D., Srinivasan, U., and Parry, G. (1990). cDNA cloning of a mouse mammary epithelial cell surface protein reveals the existence of epidermal growth factor-like domains linked to factor VIII-like sequences. *Proc. Natl. Acad. Sci. USA* **87**:8417–8421.

Thompson, A. R., Bajaj, S. P., Chen, S. H., and McGillivray, R. T. A. (1990). "Founder" effect in different families with haemophilia B mutations. *Lancet* **1**:418.

Titani, K., and Fujikawa, K. (1982). The structural aspects of vitamin K-dependent blood coagulation factors. *Acta Haematol. Japon.* **45**:807–816.

Toole, J. J., Knopf, J. L., Wozney, J. M., Sultzman, L. A., Buecker, J. L., Pittman, D. D., and Kaufman, R. J. (1984). Molecular cloning of a cDNA encoding human antihaemophilic factor. *Nature* **312**:342–347.

Traystman, M. A., Higuchi, M., Kasper, C. K., Antonarakis, S. E., and Kazazian, H. H., Jr. (1990). Use of denaturing gradient gel electrophoresis to detect point mutations in the factor VIII gene. *Genomics* **6**:293–301.

Tsang, T. C., Bentley, D. R., Mibashan, R. S., and Giannelli, F. (1988). A factor IX mutation, verified by direct genomic sequencing, causes haemophilia B by a novel mechanism. *EMBO J.* **7**:3009–3015.

Tuddenham, E. G. D. (1989). Factor VIII and haemophilia A. *In* "The Molecular Biology of Coagulation, Bailliere's Clinical Haematology, Volume 2" (E. G. D. Tuddenham, Ed.), pp. 849–877. Baillière Tindall, London.

Tuddenham, E. G. D., Cooper, D. N., Gitscher, J., Higuchi, M., Hoyer, L. W., Yoshioka, A., Peake, I. R., Schwaab, R., Olek, K., Kazazian, H. H., Lavergne, J-M., Giannelli, F., and Antonarakis, S. E. (1991). Haemophilia A: Database of nucleotide substitutions, deletions and rearrangements of the factor VIII gene. *Nucleic Acids Res.* **19**:4821–4833.

Tuddenham, E. G. D., and Giannelli, F. (1994). Molecular genetics of haemophilia A and B. *In* "Haemostasis and Thrombosis, Third Edition" (A. L. Bloom, C. D. Forbes, D. P. Thomas, and E. G. D. Tuddenham, Eds.), pp. 859–886. Churchill Livingstone, Edinburgh.

van de Water, N. S., Berry, E. W., Ockelford, P. A., and Browett, P. J. (1992). Molecular analysis of the factor IX gene in haemophilia B. *In* "ISH 24th Congress Abstracts," p. 132. Blackwell Scientific, Oxford.

Vehar, G. A., Keyt, B., Eaton, D., Rodriguez, H., O'Brien, D. P., Rutblat, F., Oppermann, H., Keck, R., Wood, W. I., Harkins, R. N., Tuddenham, E. G. D., Lawn, R. M., and Capon, D. J. (1984). Structure of human factor VIII. *Nature* **312**:337–347.

Vidaud, M., Chabret, C., Gazengel, C., Grunebaum, L., Cazenave, J. P., and Goossens, M. (1986). A de novo intragenic deletion of the potential EGF domain of the factor IX gene in a family with severe hemophilia B. *Blood* **68**:961–963.

Vidaud, M., Vidaud, D., Siguret, V., Lavergne, J. M., and Goossens, M. (1989). Mutational insertion of an Alu sequence causes haemophilia B. *Am. J. Hum. Genet.* **45**:A226.

Vielhaber, E., Jacobson, D. P., Ketterling, R. P., Liu, J. Z., and Sommer, S. S. (1993). A mutation in the 3′ untranslated region of the factor IX gene in four families with hemophilia B. *Hum. Mol. Genet.* **2**:1309–1310.

Ware, J., Davis, L., Frazier, D., Bajaj, S. P., and Stafford, D. W. (1988). Genetic defect responsible for the dysfunctional protein factor IX Long Beach. *Blood* **72**:820–822.

Winship, P. R. (1986). D.Phil thesis, Oxford University.

Winship, P. R., Anson, D. S., Rizza, C. R., and Brownlee, G. G. (1984). Carrier detection in haemophilia B using two further intragenic restriction fragment length polymorphisms. *Nucleic Acids Res.* **12**:8861–8872.

Winship, P. R., Rees, D. J. G., and Alkan, M. (1989). Detection of polymorphisms at cytosine phosphoguanidine dinucleotides and diagnosis of haemophilia B carriers. *Lancet* **1**:631–634.

Wood, W. I., Capon, D. J., Simonsen, C. C., Eaton, D. L., Gitschier, J., Keyt, B., Seeburg, P. H., and Smith, D. H. (1984). Expression of active human factor VIII from recombinant DNA clones. *Nature* **312**:330–337.

Yoshitake, S., Schach, B. G., Foster, D. C., Davie, E. W., and Kurachi, K. (1985). Nucleotide sequence of the gene for human factor IX (antihemophilic factor B). *Biochemistry* **24**:3736–3750.

Youssoufian, H., Antonarakis, S. E., Aronis, S., Tsiftis, G., Phillips, D. G., and Kazazian, H. H., Jr. (1987). Characterization of five partial deletions of the factor VIII gene. *Proc. Natl. Acad. Sci. USA* **84**:3772–3776.

# 5

# The Influence of Molecular Biology on Our Understanding of Lipoprotein Metabolism and the Pathobiology of Atherosclerosis

**Thomas P. Knecht and Christopher K. Glass[1]**
Division of Cellular and Molecular Medicine
and Division of Endocrinology/Metabolism
University of California, San Diego
La Jolla, California 92093-0656

## I. INTRODUCTION

In his treatise on the structure of scientific revolutions, Kuhn described a revolution in science as a fundamental change in the understanding of a particular field of science leading to a new set of paradigms and an attendant change in the experimental approaches used to address the questions that these paradigms raise (Kuhn, 1962). Over the past decade, the concepts and techniques of molecular biology have had a revolutionary impact on the course of research in the fields of lipid metabolism and atherosclerosis. During this period, the genes encoding the majority of the structural and enzymatic proteins involved in lipid biosynthesis and metabolism have been cloned, permitting molecular approaches to a broad set of questions of both fundamental and clinical significance. These approaches have provided novel insights into many general aspects of cellular function, such as the molecular mechanisms underlying receptor-mediated endocytosis and how cells control cholesterol. Applications of gene-targeting approaches to complex problems concerning the mechanisms by which lipoproteins are metabolized *in vivo* have permitted direct assessments of the roles of specific proteins involved in these processes. In addition, gene-targeting approaches have been utilized to create mice with defects in lipid metabolism that result in atherosclerosis. These mouse models have already had a dramatic impact on the

[1]To whom correspondence should be addressed.

*Advances in Genetics, Vol. 32*

141

study of how specific genes might promote or inhibit the atherogenic process. Finally, genetic disorders in lipid metabolism represent significant targets for gene therapy and have served as important models for the development of this new therapeutic strategy. The focus of this review will be to describe how molecular approaches have influenced the field of lipoprotein metabolism and our understanding of how aberrant lipid metabolism contributes to the development of atherosclerosis. We will begin with a brief historical perspective and then address the impact of molecular biology on our understanding of the LDL receptor pathway, the biosynthesis and catabolism of triglyceride rich lipoproteins, HDL metabolism, and modifications of LDL that are believed to contribute to the development of atherosclerosis *in vivo.* Our intent in this review is not to be comprehensive, but rather to provide selected examples of in-depth studies demonstrating the major influence of molecular biology on the field.

## II. BACKGROUND AND HISTORICAL PERSPECTIVE

Atherosclerosis is a progressive disease of large arteries that begins as a collection of cholesterol-laden foam cells beneath the endothelium (Figure 5.1). These early lesions may progress in size and complexity to a point at which encroachment on the artery lumen leads to a clinically significant diminution of blood flow to vital organs or limbs. The predilection of complex lesions to lead to thrombus formation results in the acute and often fatal clinical syndromes of myocardial infarction and stroke. In Western societies, complications of atherosclerotic cardiovascular disease are the leading cause of death.

Many risk factors for the development of atherosclerosis have been identified, including a variety of lipid disorders, increasing age, male gender, the postmenopausal state in women, a "western type" diet, tobacco smoking, high blood pressure, diabetes, and a family history of premature atherosclerosis. The relationship between high serum cholesterol levels and premature atherosclerosis was noted by Muller, who first described the syndrome of familial hypercholesterolemia (FH) in the 1930s. Although the Mendelian (single gene) inheritance pattern of the disease was also described at that time, it was not until the 1940s that the "one gene, one protein" hypothesis arose based on work in Nuerospora and *Escherichia coli* with its sweeping implications for the biochemical basis of Mendelian genetics. More complete clinical descriptions of FH were reported in the 1960s, with the characterization of distinct phenotypes associated with heterozygous and homozygous genotypes; the heterozygous form is intermediate in severity between those without the disease and those with the homozygous (most severe) form, in terms of both serum cholesterol level and risk of atherosclerosis. These observations provided strong evidence that elevated

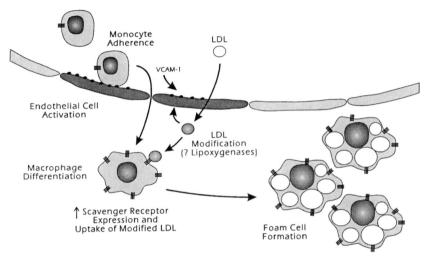

**Figure 5.1.** Model for the development of early atherosclerotic lesions. Hypercholesterolemia promotes the expression of cell adhesion molecules, such as VCAM-1, on the endothelium of large vessels. These adhesion molecules participate in the recruitment of monocytes that subsequently migrate into the subendothelial space. Once located within the artery wall, monocytes undergo a program of macrophage differentiation that includes upregulation of scavenger receptors. Macrophages promote LDL modification by the elaboration of a variety of prooxidant molecules. The specific enzyme systems involved in generating these oxidant molecules have not been clearly defined, but may include 15 lipoxygenase. Modified LDL is subsequently taken up by macrophages via scavenger receptors. Because these receptors are not downregulated by intracellular cholesterol levels, massive levels of cholesterol accumulation can occur, resulting in foam cell formation. Oxidized LDL appears to influence the expression of inflammatory response genes in macrophages, endothelial cells, and smooth muscle cells within the artery wall. Once initialized, oxidation of LDL may promote further recruitment of monocytes, ultimately leading to the formation of a fatty streak.

cholesterol levels could, independent of other factors, cause premature atherosclerosis.

Epidemiologic studies, beginning with the Framingham study that commenced in the 1950s and confirmed by many other studies conducted in the ensuing three decades, indicated that elevated serum cholesterol was a significant risk factor for atherosclerosis in the general population. Further studies of relationships between the various lipoprotein subfractions and cardiovascular disease led to the realization that, while cholesterol contained in the low-density lipoprotein (LDL) fraction was associated with increased risk of atherosclerosis, cholesterol associated with the high-density lipoprotein (HDL) fraction appeared to be protective.

These observations provided the rationale for an intensive and systematic program of research focused on lipid metabolism. Familial hypercholesterolemia provided a particularly important model for working out many critical aspects of cholesterol homeostasis. In the 1970s, Brown, Goldstein, and their colleagues began research at the cellular level to elucidate the mechanisms responsible for elevated cholesterol levels in this syndrome (Goldstein *et al.,* 1977). Using cells from normal individuals and from individuals with homozygous familial hypercholesterolemia, they identified the LDL receptor and characterized its role in maintaining cholesterol homeostasis. LDL receptors were shown, on one hand, to provide a source of extracellular cholesterol for such processes as membrane biogenesis, bile acid synthesis, and steroidogenesis, and on the other hand, to determine circulatory LDL levels. Patients with homozygous familial hypercholesterolemia were shown to lack functional LDL receptors and to develop massively elevated serum cholesterol levels due to the resulting defect in LDL catabolism. The ability of LDL receptors to provide an extracellular source of cholesterol led to the characterization of a finely tuned set of regulatory pathways that control both *de novo* biosynthesis of cholesterol as well as the expression of LDL receptors. Increases in cellular cholesterol content lead to downregulation of HMG-CoA reductase, the enzyme mediating the rate-limiting step in cholesterol biosynthesis, as well as the LDL receptor. Indeed, the approaches that Goldstein and Brown applied to the study of the LDL receptor pathway have served as paradigms for many other receptor-dependent processes. However, the application of molecular biological techniques to the study of the LDL receptor pathway in particular, and to the study of the pathogenesis of atherosclerosis in general, would await the decade of the 1980s and the developing application of molecular biological theory and techniques to eukaryotic systems.

The mechanistic basis for the inverse relationship between HDL cholesterol levels and risk of atherosclerosis has been less well understood than the relationship between LDL and atherosclerosis. Multiple epidemiologic reports have described the relationship between low HDL cholesterol levels and risk of heart attack. It was not until 1968, however, that a physiological mechanism was proposed that may help explain the protective effect of HDL. Glomset proposed that HDL may function to return excess cholesterol from peripheral tissues to the liver (Glomset, 1968), a process described as "reverse cholesterol transport." Despite a formidable body of literature regarding HDL metabolism, including many data *in vivo* and *in vitro* in support of the reverse cholesterol transport hypothesis, there remains no commonly accepted unifying hypothesis at the cellular and molecular level to explain the epidemiologic data. The recent application of molecular biological approaches to questions concerning HDL metabolism has had a profound influence on our understanding of the relationship of HDL to the atherosclerotic process, and many more contributions may be expected in the future.

One of the most significant new concepts to be developed in the past decade concerning the role of lipoproteins in atherogenesis concerns the role of oxidation in altering the atherogenicity of the LDL particle. Recognition that LDL could be modified in ways that altered its metabolism was first described by Goldstein, Brown, and colleagues in the 1970s (Goldstein *et al.*, 1979) when they described that chemical modification of LDL by acetylation or maleylation, which abrogates its recognition by the LDL receptor, markedly enhanced its binding, uptake, and degradation by macrophages, one of the predominant cell types in the atherosclerotic plaque and the cellular precursor of the "foam cell" in the atherosclerotic lesion. It was found that macrophages express negligible amounts of the LDL receptor on their surface and that macrophages incubated with native LDL *in vitro* do not become cholesterol loaded. Instead, the chemically modified LDL was taken up by a separate receptor termed the "scavenger receptor." Unlike the LDL receptor, scavenger receptors were not downregulated by high levels of intracellular cholesterol. As a result, uptake of modified LDL via this pathway permitted massive cellular accumulation of cholesterol. These observations provided a potential solution to the paradoxical observation that patients lacking LDL receptors nevertheless developed foam cells at an accelerated rate. However, the chemical modifications used to render LDL recognizable by the scavenger receptor were not known to occur *in vivo*. Modification of LDL by cells was subsequently described by Henriksen *et al.* in 1981 [see Steinberg *et al.* (1989) for review] who demonstrated that incubation of LDL with endothelial cells resulted in its uptake by macrophages via scavenger receptors, leading to foam cell development. Subsequent studies have revealed that the cellular modification of relevance results from oxidation of LDL lipids. Lipid peroxidation products are subsequently able to react with primary amines in apolipoprotein $B_{100}$, (apoB), the LDL apolipoprotein that serves as the ligand for the LDL receptor. Lipid peroxidation adducts on apoB prevent its recognition by the LDL receptor and in turn renders it a ligand of the scavenger receptor. Oxidation of LDL could also be performed in cell-free systems catalyzed by copper ions or lipoxygenases with similar metabolic consequences. Studies *in vivo* in the Watanabe heritable hyperlipidemic (WHHL) rabbit (an animal model of familial hypercholesterolemia resulting from a mutation in the LDL receptor and characterized by markedly elevated cholesterol levels and atherosclerosis) demonstrated the efficacy of antioxidant treatment in reducing the extent of atherosclerotic lesions independent of overall changes in cholesterol level. A wealth of data has accumulated regarding oxidized LDL metabolism and its relation to atherogenesis [see Witztum *et al.* (1991) for review].

The final common pathway of all the factors influencing the development of atherosclerosis (i.e., the integrated effects of lipoprotein metabolism and the other risk factors alluded to above) must be manifest as alterations in the biology of the artery wall that lead to the development of the atherosclerotic lesion. While a detailed consideration of the cellular and molecular events that contribute to formation of the atherosclerotic lesion is beyond the scope of this review, a brief summary of this extensive and complex field will be an aid to our discussion. The term *atheroma* was first used by 18th century anatomists to describe the gross changes in the artery wall characteristic of atherosclerosis. *Atherosclerosis* is derived from the Greek words *athero,* meaning gruel, and *sclerosis,* meaning hard, accounting for the lay term for the disease, "hardening of the arteries." In 1859, Virchow described atherosclerotic lesions with encroachment on the artery lumen, ulcerative plaques, calcification, and (from the description) necrotic debris and cholesterol crystals (Virchow, 1940). The earliest lesion of the artery wall attributed to the atherosclerotic process is the accumulation of lipid-loaded macrophages, referred to as foam cells, beneath the intact endothelial cells of the intima. The lesion at this stage is referred to as a *fatty streak.* The fatty streak, which in Western societies commonly appears in the first or second decade of life is flat and does not encroach on the artery lumen. The fatty streak may progress to the *fibrous plaque* lesion, with varying degrees of complexity. Fibrous plaques are generally characterized by further accumulation of macrophage foam cells, migration and proliferation of smooth muscle cells, and synthesis of extracellular matrix proteins. These progressive changes lead to a deterioration of the normal anatomy of the intimal and medial layers and encroachment of the lesion into the artery lumen. As lesions become increasingly advanced, foam cells may die, releasing cholesterol and cholesterol esters into the extracellular milieu. Fissuring of the lesions, or loss of endothelial cells from the lesion surface, may result in thrombosis formation and its ensuing complications.

Studies over the past decade have indicated that the initiation and progression of the atherosclerotic process bear many features characteristic of inflammatory responses and aberrant wound repair (Figure 5.1). Accumulation of macrophages within the artery wall requires that circulating monocytes become adherent to the overlaying endothelium. Studies of atherosclerosis in cholesterol-fed rabbits suggest that the cell adhesion molecule, VCAM-1, is upregulated on the endothelium of large arteries in response to hypercholesterolemia and participates in the recruitment of monocytes (Cybulsky *et al.*, 1991). Following migration into the artery wall and differentiation into macrophages, these cells promote the oxidative modification of LDL. Oxidized LDL has been shown to have a number of biological properties beyond its ability to be recognized as a ligand for the scavenger receptor. Oxidized LDL is chemotactic for circulating monocytes, the cellular precursors of the artery wall macrophage, and may thus stimulate the accumulation of macrophages associated with the atherosclerotic lesion. Oxidized LDL has also been shown to inhibit migration of macrophages, thus potentially trapping them in the lesion. These activities may in part reflect the ability of oxidized LDL to regulate a number of genes that control the responses of macrophages to inflammatory stimuli, including monocyte chemotactic protein-1, a monocyte-specific chemotactic factor. These observations suggest that once initi-ated, the process of fatty streak formation could become self-perpetuating, leading to the progressive accumulation of foam cells. The evolution of the fatty streak to more complex lesions, involving the proliferation and migration of smooth muscle cells and the elaboration of extracellular matrix proteins, appears to be controlled by a complex set of growth factors, cytokines, and other hormone-like molecules that are elaborated by macrophages and other cells in the developing lesions (Ross, 1993). While these factors normally play critical roles in develop-ment, tissue remodeling, and wound repair, their actions within the artery wall may ultimately lead to a complex, obstructive lesion that is at high risk of rupture and initiating thrombosis, with the resulting clinical sequalae discussed above.

## III. ATHEROSCLEROSIS RESEARCH IN THE MOLECULAR BIOLOGY ERA

### A. The LDL receptor pathway

The field of atherosclerosis research was ushered into the molecular biology era by Russell, Goldstein, Brown, and their colleagues with their publication in Decem-ber, 1983, describing the cloning of the LDL receptor from a bovine adrenal cDNA library (Russell et al., 1983). [The progression of our understanding of biology of the LDL receptor pathway has been presented in a number of excellent and historical reviews; see Brown et al., 1986; Goldstein et al., 1989; Hobbs et al., 1990).] In the work of Russell et al. (1983), bovine adrenal mRNA was enriched for LDL receptor mRNA by isolation of polysomes associated with nascent LDL receptors via affinity purification, and the mRNA was used to construct a cDNA library. Degenerate oligonucleotide probes, based on partial amino acid sequences of the receptor, were used to isolate partial clones encoding portions of the LDL receptor. Using these partial cDNAs as probes, Northern blotting experiments were performed that demonstrated that the 5.5-kb LDL receptor mRNA was regulated in response to perturbations in cellular cholesterol. Maneuvers that decreased cell cholesterol increased LDL receptor mRNA. These observations implied that cholesterol could regulate gene expression at either the level of transcription or mRNA stability or both.

The following year, in 1984, Yamamoto et al. described the cloning of a full-length human LDL receptor cDNA (Yamamoto et al., 1984). With the complete coding sequence of the gene and their earlier work on the protein chemistry of the LDL receptor, they were able to determine the complete primary amino acid sequence and to identify functional domains of the protein (Figure 5.2). Translation yields a protein of 860 amino acids; proteolytic cleavage of a 21 amino acid signal sequence yields the mature protein sequence of 839 amino

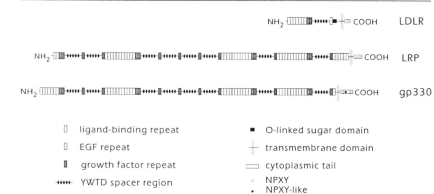

**Figure 5.2.** Predicted primary structures of the LDL receptor, LRP, and GP330. All three proteins contain a C-terminal cytoplasmic tail with NPXY motifs required for clustering in coated pits and internalization. The ligand-binding domains are made up of repeating motifs that occur once in the LDL receptor and four times in LRP and GP330. The ligand-binding repeat also shares sequence homology with complement factors. The EGF precursor homology corresponds to the EGF repeat plus the YWTD spacer region.

acids. The protein is post-translationally modified by the addition of N-linked and O-linked carbohydrate chains. From the extracellular N-terminus of the protein to the cytoplasmic C-terminus of the protein, there are five domains of the mature LDL receptor. Numbering from the N-terminus of the receptor, domains 1, 2, and 3 are extracellular, domain 4 spans the plasma membrane, and domain 5 is cytoplasmic. Analysis of naturally occurring mutations and site-directed *in vitro* mutagenesis with subsequent expression in transfected cells have been used to illuminate the function of these domains. Domain 1 is composed of the N-terminal 292 amino acids. It is a cysteine-rich domain in which all cysteines are involved in disulfide bonds. Domain 1 is organized into seven repeats of 40 amino acids, each of which shares a high degree of sequence homology. There are six cysteine residues per repeat, and there is a clustering of negatively charged amino acids (aspartate and glutamate) toward the C-terminal end of each repeat. Domain 1 was proposed to be the ligand-binding domain based on the known presence and importance of positive charges in the binding regions of the two principle ligands (apolipoproteins $B_{100}$ and E) and on the loss of the ability of the receptor to bind ligands if pretreated with disulfide-reducing agents. *In vitro* mutagenesis experiments revealed differential involvement of these repeats in binding of apoB$_{100}$ versus apoE, thus helping to elucidate the nature of the receptor-ligand interaction. Domain 2 was described as the next roughly 400 amino acids in the C-terminal direction. Domain 2 has 35% sequence homology with the EGF precursor molecule (but this homology is not in the EGF portion of the EGF precursor molecule). This region of the LDL receptor was shown by deletion analysis to be important in LDL binding via $apoB_{100}$ (but not in ligand binding mediated via apoE) and in acid-dependent ligand–receptor dissociation, such as that which occurs in endosomes during the endocytosis/receptor recycling cycle. Domain 3 is a 58-amino-acid sequence to the C-terminal side of domain 2 which is rich in serine and threonine residues and is the region of post-translational O-linked carbohydrate addition. Domain 4, with 22 hydrophobic amino acids, is the membrane-spanning domain. The LDL receptor spans the plasma membrane only once. The last and C-terminal domain, domain 5, consists of 50 amino acids cytoplasmic to the plasma membrane. Mutation analysis of this region revealed a 4-amino acid region of consensus NPXY (where N is asparagine, P is proline, X is any amino acid and Y is tyrosine) to be involved in the clustering of receptors in coated pits on the cell surface and in receptor endocytosis (Figure 5.2).

In 1985, Sudhof et al. described the cloning of the genomic DNA encoding the human LDL receptor (Sudhof et al., 1985). With the cloning of the genomic DNA, it was possible to explore the arrangements of introns and exons and their relation to the domains of the protein as described above. The gene spans 45 kb and is organized into 18 exons interrupted by 17 introns, with a correlation between exon organization and the functional domains of the protein. Comparisons of exon sequence data for the LDL receptor with sequence data from the genes for other proteins provided insights into how two of the functional components of the LDL receptor were derived from two distinct supergene families, one encoding the complement-like ligand-binding repeat and the second the EGF precursor homology. These findings were consistent with the idea, first proposed by Gilbert (1978), that the exon–intron arrangement in the gene facilitated the shuffling of the exons among various genes in the genome, thus increasing the potential for functional molecular diversity, and therefore, evolutionary diversity.

The subsequent contributions of molecular approaches to analysis of the LDL receptor and its role in disease have been rapid and extensive. Analysis of restriction fragment length polymorphisms (RFLPs) has been used as an aid in haplotype identification and to follow segregation of the haplotype within families [see Hobbs et al. (1990) for review] without the need for extensive sequencing of each individual allele. RFLPs spanning the entire LDL receptor gene have been identified. By analysis of RFLPs, it has been possible to demonstrate that a large fraction of patients who are phenotypically homozygous for familial hyper-cholesterolemia are in fact compound heterozygotes, with two different mutant alleles for the LDL receptor.

Studies of the LDL receptor promoter have also established the molecular mechanisms by which cholesterol and oxygenated derivatives, such as 25-hydroxy cholesterol, regulate transcription of the LDL receptor gene. Mutational analysis was used by Goldstein and Brown's group to identify sequences in the 5′

flanking region of the LDL receptor gene required for control of transcription. Two sequences which confer constitutively positive effects via transcription factor Sp1 were found to be required but not sufficient for either maximal transcription or regulation in response to oxysterols. An enhancer sequence termed sterol regulatory element-1 (SRE-1) was likewise identified between the two Sp1 sites that was essential for both maximal transcription and regulation in response to oxysterols [see Goldstein et al. (1990) for review]. Intriguingly, two copies of the SRE-1 sequence were found in the promoter of the HMG-CoA sythase gene, which encodes an enzyme involved in the synthesis of cholesterol and is regulated by cell cholesterol status in a manner analogous to the LDL receptor. A similar, but distinct, sterol regulatory element is present in the HMG-CoA reductase gene promoter. HMG-CoA reductase is the rate-limiting enzyme in the synthesis of cholesterol, dolichols, and the precursors required for prenylation of proteins. Based on the difference in the sequences of the sterol regulatory elements, control of HMG-CoA reductase transcription appears to be mediated by transcription factors that are distinct from those regulating transcription of the LDL receptor and HMG-CoA synthase genes in response to the cholesterol status of the cell (Osborne, 1991).

Using standard biochemical separation techniques and sequence-specific DNA affinity chromatography, a 59- to 68-kDa nuclear protein, termed sterol regulatory element binding protein (SREBP), was purified that bound to the LDL receptor SRE with high affinity and appropriate sequence specificity. Site-directed mutagenesis was employed to demonstrate that base changes which abolished the ability of the SRE-1 sequence to regulate transcription in a sterol-dependent manner also abolished the ability of SREBP to bind to SRE-1. This correlation between DNA binding and function provided convincing evidence that SREBP is, in fact, involved in sterol-regulated transcription of the LDL receptor through its interaction with SRE-1 in the LDL receptor promoter (Briggs et al., 1993; Wang et al., 1993). Partial amino acid sequences were used to isolate two closely related cDNA clones encoding proteins termed SREBP-1 (Yokoyama et al., 1993) and SREBP-2 (Hua et al., 1993). The cDNA encoding SREBP-1 predicts a 1147-amino acid protein of about 121.7-kDa $M_r$ while the SREBP-2 cDNA is predicted to encode a 1141-amino acid protein of similar molecular weight. SREBP-2 shares 47% sequence homology with SREBP-1. The predicted sequences of the N-termini of both SREBPs revealed a basic-helix–loop-helix–leucine zipper domain common to a family of DNA-binding proteins which function as transcription factors. The SREBPs were distinct from other members of this family, however, in their larger size and the specific nucleotide sequence which they recognize (Yokoyama et al., 1993; Hua et al., 1993). The sizes of SREBP-1 and SREBP-2 predicted from the respective cDNAs also suggested that the SREBP first isolated via DNA affinity purification with 59- to 68 kDa $M_r$ was a proteolytic product (or set of proteolytic products) derived from the primary protein; the purified proteins corresponded to the N-terminal half of the primary SREBP-1 or SREBP-2 proteins. When cells were cotransfected with a reporter plasmid containing an SRE-dependent promoter and varying amounts of plasmids constitutively expressing the truncated forms of SREBP-1 or SREBP-2 that corresponded to the size of the proteins purified from the nuclear extracts, SRE-dependent transcription was markedly enhanced, but negative regulation by sterols was abolished. Thus, the N-termini of the SREBPs function as strong, constitutive activators of transcription that are dominant to the regulated function of endogenous SREBP.

The observation that the truncated form of SREBP could activate transcription through the SRE, but did not mediate repression in the presence of sterols, suggested that the entire SREBP protein was required for cholesterol regulation. Indeed, investigation of the cellular localization of nascent SREBP and its subsequent processing has revealed a remarkable mechanism for regulation of cholesterol homeostasis (Figure 5.3) (Wang *et al.*, 1994). When the full-length SREBP-1 cDNA is expressed in cells, a 125-kDa precursor is synthesized that

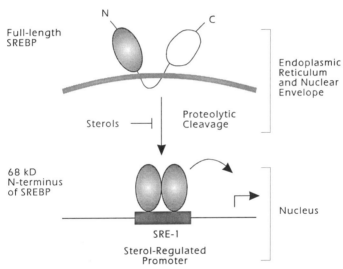

**Figure 5.3.** Mechanisms of sterol-mediated repression of SRE-1 containing promoter. The expression of sterol-responsive genes, such as the LDL-receptor gene, is dependent on the interaction of sterol regulatory element binding proteins (SREBPs) with a sterol regulatory element located upstream of the transcriptional start site. SREBPs are initially synthesized as inactive 125-kDa precursors associated with the endoplasmic reticulum and nuclear membrane. Proteolytic processing of the precursor liberates a 68-kDa N-terminal fragment containing basic helix–loop–helix and leucine zipper motifs found in other classes of transcription factors. This N-terminal fragment translocates to the nucleus where it interacts with the SRE and activates transcription. Because this N-terminal fragment of SREBP turns over rapidly, continuous processing of the 125-kDa precursor is required to maintain transcriptional activity of SRE-containing promoters. Sterols inhibit this processing step, thereby preventing the formation of active SREBP and inhibiting the expression of SRE-containing promoters.

is localized to the endoplasmic reticulum and nuclear envelope. This membrane localization appears to be mediated by two putative transmembrane domains located in the middle of full-length SREBP-1. In contrast, when the truncated forms of SREBP-1 are expressed that lack these transmembrane domains, they are localized to the nucleus regardless of cellular sterol status. Surprisingly, the step regulated by sterols is the proteolysis of the membrane-bound inactive precursor. In sterol-depleted cells, the 125-kDa precursor is proteolytically cleaved, releasing the N-terminal 68-kDa SREBP-1, which translocates to the nucleus and stimulates transcription of SRE-containing genes. Conversely, cholesterol loading inhibits proteolysis, maintaining the active N-terminus in association with the endoplasmic reticulum. Thus, unlike other basic-helix–loop-helix–leucine zipper transcription factors, SREBP is synthesized as an extranuclear integral membrane protein and requires proteolytic processing to release the functional transcription factor. Presumably, the C-terminal portion of the precursor protein remains membrane associated in its extranuclear locus. The function of the C-terminal portion of this molecule is not yet known; it may play a role in sterol recognition and/or proteolysis of the N-terminal (SREBP) portion of the molecule.

## 1. Insights from transgenic and knockout mice

To complete our discussion of the LDL receptor pathway as influenced by advances in molecular biology, we will turn to what has become one of the most powerful tools of the molecular biology revolution, i.e., the new mouse genetics [see Breslow (1993) for review]. In the 1980s, it became possible to introduce foreign DNA into fertilized mouse eggs such that the DNA becomes randomly incorporated into the genome. The mice arising from the implanted eggs subsequently express the foreign gene, and when bred, the foreign genes segregate in a Mendelian fashion. Such mice are referred to as "transgenic." By the end of the 1980s, it became apparent that the ability to manipulate the mouse genome could be carried a step further in that native genes could be disrupted, or "knocked out." Using this technology, animals can be assessed for the effect that loss of the gene has on phenotype. The process involves the use of homologous recombination to introduce mutations into specific genomic loci of pluripotent embryonic stem cells (Capecchi, 1989). In brief, embryonic stem cells are transfected with a DNA construct consisting of a marker gene that permits positive selection flanked by DNA sequences that are homologous to the genomic sequences to be knocked out. A separate marker gene permitting negative selection is frequently incorporated distal to the homologous flanking sequences to help assess the likelihood that incorporation has, in fact, occurred by homologous recombination rather than via random insertion; incorporation via homologous recombination should exclude the terminal marker gene, whereas random incorporation would not be expected to exclude the gene. Thus, homologous recombination inserts the marker gene and disrupts or knocks out the native gene. Cells positive for incorporation of the gene in the appropriate (homologous) genomic location are injected into mouse blastocysts. Some of the resulting chimeric mice that arise from these blastocysts carry the knocked out allele in their germ line. These mice can then be bred to produce homozygotes for the knocked out gene.

Both transgenic and knockout technologies have been applied to the LDL receptor pathway. In 1988, Hofmann *et al.* described the production of transgenic mice harboring the human LDL receptor gene driven by the cadmium-inducible metallothionein promoter (Hofmann *et al.*, 1988). The LDL receptor transgene was expressed in all the tissues examined, with greatest expression in the liver. Expression was further induced in these tissues by treatment with cadmium. The receptors expressed were functional, and the transgenic mice had a greater plasma fractional catabolic rate for LDL, with the greatest tissue uptake by the liver. Subsequently, mice were made transgenic for the human LDL receptor driven by the transferrin promoter; this allowed long-term studies to be carried out without the need for chronic cadmium treatment, which is toxic. With these mice, it was shown that chronic overexpression of the LDL receptor prevented hypercholesterolemia induced by an atherogenic diet (Yokode *et al.*, 1990).

Recently, mice have been produced via homologous recombination in which the native LDL receptor gene has been knocked out (Ishibashi *et al.*, 1993). Mice homozygous for LDL receptor deficiency have elevated intermediate density lipoprotein (IDL) and LDL on standard mouse chow. Plasma fractional catabolic rates for very low-density lipoprotein (VLDL) and LDL were slowed in homozygous receptor-negative mice compared to those in wild-type mice. When fed an atherogenic diet containing a high percentage of fat and cholesterol, the knockout mice develop even more striking hyperlipidemia and exhibit many of the characteristic features of homozygous familial hypercholesterolemia; namely, extensive xanthomata and extraordinary degrees of atherosclerosis, comparable to the extent observed in WHHL rabbits (Ishibashi *et al.*, 1994a). These LDL receptor-negative mice can be expected to become an invaluable model for the study of atherosclerosis and its treatment.

## 2. Toward gene therapy of defects in the LDL receptor pathway

With the genetic basis of certain human diseases well established, the idea of gene therapy, *i.e.*, treatment of genetic diseases by providing a normally functioning gene to correct the underlying gene defect, began to surface in the late 1960s and early 1970s. When first proposed, eukaryotic molecular biological theory and techniques were in their infancy and the prospects of actually accomplishing this goal seemed almost unimaginable. In recent years, however, with the rapidly expanding body of knowledge and techniques in the field of eukaryotic molecular biology, many approaches to gene therapy have been developed, and clinical trials examining some of these approaches have begun (Friedmann, 1992). Familial hypercholesterolemia has provided an important model for evaluation of gene therapy approaches. Mutations in the LDL receptor gene that result in hypercholesterolemia are well characterized, and excellent animal models of homozygous familial hypercholesterolemia are provided by the WHHL rabbit and LDL receptor-deficient mice. The development of gene therapy approaches to FH would be clinically significant because there is no effective form of pharmacologic treatment for this disease, which often results in fatal myocardial infarction within the first or second decades of life.

A critical requirement for the use of gene therapy to treat familial hypercholesterolemia will be to develop vectors that permit high levels of expression of the LDL receptor in receptor-deficient hepatocytes. An initial approach to this problem has been to use the molecular biology of retroviruses to effect efficient delivery of cDNAs to cells in a manner that results in their stable integration into the genome. Using this approach, retroviral sequences that are not required for directing the expression or replication of the viral genome are replaced with an exogenous gene, such as the cDNA encoding the LDL receptor. This recombinant genome is then introduced into a packaging cell line that has been engineered to express the missing viral genes. This cell line is capable of producing recombinant viral particles that can infect other cells, but cannot replicate. Once the genomic information carried by these recombinant viruses has been integrated into the host cell's genome, it can serve to direct the expression of the LDL receptor. Using this approach, Sharkey and colleagues utilized retroviral vectors to direct the expression of the LDL receptor in fibroblasts, resulting in a correction in the cellular defect in the LDL receptor pathway (Sharkey et al., 1990).

In 1991, Chowdhury et al. (1991) took the next step toward the application of in vivo gene therapy when they supplied the LDL receptor gene to receptor-deficient WHHL rabbits and achieved a lowering of the animals' serum cholesterol. Their approach involved the isolation of hepatocytes from WHHL rabbits via partial hepatectomy and subsequent hepatocyte culture in vitro. Recombinant retroviruses were produced by cloning rabbit LDL receptor cDNA behind a strong constitutive promoter (β-actin) in a retroviral vector. Recombinant retroviruses were used to transfect the WHHL hepatocytes in vitro. Transfected hepatocytes were subsequently injected back into the portal veins of the animals from which they had been originally harvested. Studies showed that most of the hepatocytes injected took up residence in the liver. Quantitative analysis of LDL receptor mRNA from liver estimated that the treated animals expressed about 2% of endogenous hepatic LDL receptor transcript. Rabbits so injected with recombinant LDL receptor-expressing hepatocytes exhibited a 30% reduction in serum cholesterol compared to pretreatment levels and continued to express the LDL receptor gene for the duration of the experiment (6.5 months).

More recently, *in vivo* gene replacement of LDL receptor deficiency has been reported in knockout mice. Homozygous LDL receptor-negative mice were injected with a replication-defective adenovirus harboring the human LDL receptor driven by the cytomegalovirus promoter (Ishibashi *et al.*, 1993). Four days following transfection, the receptor was shown to be expressed in liver, and the fractional catabolic rate for VLDL was similar to that for the wild-type mice. Therefore, following LDL receptor knockout, restoration of LDL receptor synthesis could be accomplished via somatic cell infection *in vivo*, correcting the metabolic defect imposed by the gene knockout. The major shortcoming of this approach for gene therapy is the transient period of expression of the exogenous LDL receptor, presumably due to immunologic clearance of adenovirus-infected cells.

Based on the results of *ex vivo* gene replacement experiment in rabbits, a similar approach has recently been used to treat a patient with homozygous familial hypercholesterolemia (Grossman *et al.*, 1994). The patient was 28 years old at the time of treatment and had a history of severe premature coronary artery disease. She was known to be homozygous for a missence mutation that confers a quantitative defect in LDL degradation, resulting in a markedly elevated LDL cholesterol. She had suffered her first heart attack at age 16 and had undergone coronary artery bypass surgery at age 26. This patient underwent partial hepatectomy with recovery and *in vitro* culture of her hepatocytes. The hepatocytes were transfected with a replication-defective retrovirus incorporating the full-length human LDL receptor cDNA driven by a β-actin promoter with an upstream cytomegalovirus enhancer. The patient's LDL receptor-transfected hepatocytes were subsequently reintroduced by injection into her portal vein. Following this procedure, the patient's LDL cholesterol levels fell from a pretreatment value of 482 to 402 mg/dl, and *in situ* hybridization analysis of a liver biopsy indicated the expression of the transfected LDL receptor gene in some cells. However, a rigorous determination of whether the decrease in LDL receptor levels was due to increased LDL clearance based on measurements of the LDL fractional catabolic rate before and after gene therapy was not performed. It is, therefore, unclear at this point whether this particular approach will provide sufficient therapeutic benefit to justify the risks involved in partial hepatectomy and reinfusion of large numbers of genetically modified hepatocytes. Replacement of LDL receptors presents a particularly challenging problem because high levels of hepatic expression will be needed. Clearly, significant improvements in methods of gene delivery will be required in order to obtain stable expression of LDL receptors at levels sufficient to significantly reduce or normalize LDL levels in homozygous familial hypercholesterolemia. The recent development of mice that are homozygous for null mutations of the LDL receptor should facilitate the development and rigorous assessment of new gene therapy approaches. These studies are likely to impact on the therapeutic approaches to other genetic diseases in which relatively high levels of gene expression will be required in a specific organ to achieve therapeutic benefit.

## 3. Molecular biology of apoB and its role in the biosynthesis and clearance of triglyceride-rich lipoproteins

In addition to its role as a ligand for the LDL receptor, apolipoprotein B is required for the assembly and secretion of triglyceride-rich lipoproteins. Triglyceride-rich lipoproteins are generally broken down into two broad categories: VLDL, which are synthesized and secreted by the liver using endogenous lipid (Young, 1990), and chylomicrons, which are synthesized and secreted by the small intestine for the transport and metabolism of dietary lipids.

In the liver, triglycerides and other lipid components of nascent VLDL are assembled with the full-length form of apoB ($apoB_{100}$) in the endoplasmic reticulum. Each VLDL particle consists of a core of neutral lipids, a surface consisting of phospholipid and free cholesterol, and a single molecule of $apoB_{100}$. The neutral lipid core consists primarily of triglycerides, with cholesteryl esters making up a minor fraction. Once formed, these nascent VLDL particles can then associate with additional apolipoproteins, including apoC-I, apoC-II, apoC-III, apoE, and apoA-I. Unlike apoB, these additional apoproteins are not integral components of the VLDL particle and can exchange among different VLDL particles or other classes of lipoproteins, particularly high-density lipoproteins. Once in the circulation, VLDL particles are acted upon by lipoprotein lipase, predominantly in the capillary beds of adipose tissue and muscle, resulting in the extracellular hydrolysis of triglycerides and delivery of the resulting fatty acids to the surrounding cells. With hydrolysis of the core triglycerides, the VLDL particles become relatively enriched in cholesteryl esters and undergo redistribution of phosholipids and exchangeable apolipoprotein components with other lipoproteins. VLDL particles that have undergone this process of partial lipid hydrolysis are referred to as IDL. Further extracellular hydrolysis of triglycerides by lipoprotein lipase and hepatic lipase, exchange of IDL triglycerides for HDL cholesteryl esters via the actions of cholesteryl ester transfer protein (CETP), and loss of the remaining exchangeable surface apoproteins results in the final transformation of IDL to LDL.

The assembly of chylomicrons in the small intestine is also dependent on apoB. Remarkably, the enterocyte synthesizes a truncated form of apoB, termed $apoB_{48}$, that represents the N-terminal 48% of full-length apoB. $ApoB_{48}$ contains the sequences of apoB that are required for the assembly and secretion of triglyceride-rich particles, but lacks the carboxy-terminal sequences present

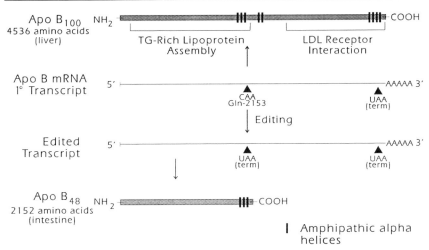

**Figure 5.4.** RNA editing permits two forms of apoB to be expressed from a common primary transcript. In the liver this transcript directs the synthesis of apoB$_{100}$, a 4536 amino acid protein that contains multiple putative lipid-binding domains distributed throughout the primary sequence. Sequences required for interaction with the LDL receptor are located in the C-terminal half of the molecule. In the intestine, the primary apoB transcript is edited at nucleotide 6666 by deamination of cytosine to produce uracil at this position. As a result, the CAA codon that normally encodes glutamine in apoB$_{100}$ is transformed to a UAA termination codon. Translation of this message yields the N-terminal 48% of apoB that retains the structural motifs required for assembly of chylomicrons, but lacks the domain required for interaction with the LDL receptor. The locations of several amphipathic α-helices that are believed to mediate interactions with lipid are shown. In addition, apoB contains many hydrophobic β sheets distributed throughout the molecule that are too numerous to show.

in apoB$_{100}$ that are required for recognition by the LDL receptor (Figure 5.4). Like apoB$_{100}$ present in VLDL particles, apoB$_{48}$ is an integral component of chylomicrons and cannot exchange among other lipoproteins. In addition to apoB$_{48}$, several of the smaller apolipoproteins, including apoC-I, -C-II, -C-III, apoE, apoA-I, apoA-II, and apoA-VI, associate with newly synthesized chylomicrons. Like VLDL particles, the neutral lipid core of chylomicrons consists primarily of triglycerides, with cholesteryl esters being a relatively minor component. These lipids are derived from dietary triglycerides and esterified and unesterified cholesterol that have been hydrolyzed by pancreatic lipases in the intestinal lumen, absorbed by the enterocyte, and reesterified prior to packaging into chylomicron particles. Chylomicrons are secreted into intestinal lymph and subsequently enter the circulation via the thoracic duct. In the circulation, the chylomicrons are rapidly acted on extracellularly by lipoprotein lipase in a manner that is similar to the metabolism of VLDL particles. With hydrolysis of triglycerides, the chylomicrons become smaller and relatively more cholesteryl ester enriched. Concurrent with this is some redistribution of phospholipid and of exchangeable apolipoproteins to other lipoproteins, particularly HDL, resulting in the production of particles that are referred to as chylomicron remnants. Because $apoB_{48}$ lacks the domain of $apoB_{100}$ which mediates binding to the LDL receptor (*i.e.*, the C-terminal half of $apoB_{100}$), chylomicron remnants are cleared from the circulation in a manner that is dependent on apolipoprotein E, as described in detail below.

The dual roles of apoB as structural protein required for the assembly of triglyceride-rich lipoproteins and as the ligand recognized by LDL receptors are particularly intriguing in view of the observation that these functions are dissociated in the enterocyte by its production of a truncated form of apoB. To investigate the structural basis for these two functional roles of apoB and determine the mechanism by which a truncated form of apoB is produced in the intestine, it was necessary to isolate cDNA clones encoding apoB. This was a daunting task because estimates of the molecular weight of $apoB_{100}$ based on polyacrylamide gels were in the neighborhood of 500 kDa, predicting a polypeptide chain of more than 4000 amino acids and a corresponding mRNA of more than 12,000 bases. In fact, when reported in 1986, the $apoB_{100}$ cDNA sequence was 14,121 nucleotides in length and predicted a mature protein of 4,536 amino acids with 512-kDa $M_r$ (Figure 5.4). Sequence analysis revealed a multitude of domains believed to interact with lipid, with nine amphipathic $\alpha$-helices predominantly in the C-terminal half of the molecule and multiple $\beta$-pleated sheets. The presence of multiple lipid-binding domains throughout the molecule was suggested to account for the fact that apoB cannot dissociate from the particle with which it is secreted or exchange between particles. It was shown that there are seven domains of $apoB_{100}$ which bind with high affinity to heparin. Four of these domains are located in the N-terminal half of apoB and are thus present in both $apoB_{100}$ and $apoB_{48}$. These domains may mediate interaction of triglyceride-rich lipoproteins with heparin-like molecules present in capillary endothelium and facilitate the action of lipoprotein lipase. Three of the heparin-binding domains are found in the C-terminal half of the molecule and are characterized by clusters of positively charged amino acids. It was known that heparin could interfere with binding of LDL to the LDL receptor, and that positively charged amino acids are important in the interaction of $apoB_{100}$ with the receptor. It is believed, therefore, that these positively charged groups interact with the negatively charged amino acids in the cysteine-rich repeats in domain 1 of the LDL receptor (above). One of these heparin-binding domains also has sequence homology with the domain of apoE that interacts with the LDL receptor. These domains have been shown to be evolutionarily well conserved.

Also in 1986, the human $apoB_{100}$ gene was cloned from a genomic DNA library [see Young (1990) for review]. The gene had already been mapped to the short arm of human chromosome 2 the year before utilizing both *in situ* analysis of human metaphase chromosomes and analysis of human–mouse somatic cell hybrids. Analysis of the genomic clone revealed an organization of 29 exons and 28 introns spanning 43 kb. Regions in the 5′ end of the gene that dictate liver-specific transcription and binding sites for several transcription factors were mapped. However, there is no evidence that transcriptional control is used to regulate synthesis of apoB$_{100}$ in the liver; the gene appears to be constitutively expressed. In contrast, there is substantial evidence that secretion of apoB is regulated at the level of translation and by proteolysis, and that regulation of plasma levels of apoB$_{100}$ reflect more the rate of clearance of apoB$_{100}$-containing lipoproteins than the rate of secretion.

## 4. ApoB$_{48}$ results from post-transcriptional mRNA editing

In 1987, cDNAs encoding apoB$_{48}$ were cloned from a human intestinal cDNA library (Young, 1990; Hodges *et al.*, 1992; Davidson, 1993). Sequence comparison with apoB$_{100}$ cDNA of hepatic origin revealed sequence identity with the exception of a C to T change at nucleotide 6666 (corresponding to a C to U change at the corresponding position of the mRNA). This base change transforms codon 2153 from a CAA-encoding glutamine in the cDNA from liver to the translation stop codon UAA (TAA in the cDNA) from the intestine (Figure 5.4). This substitution was not present in intestinal genomic DNA, however, indicating that the base change occurred post-transcriptionally. This startling result represented a new and unique mechanism, termed mRNA editing, by which tissue-specific differences in the profile of gene expression can be achieved in eukaryotes and revealed the molecular mechanism by which the intestine synthesizes a truncated form of apoB. Introduction of a translation stop codon at codon 2153 results in the translation of a 241-kDa protein corresponding to the N-terminal 48% of full-length apoB, i.e., apoB$_{48}$.

The details of post-transcriptional apoB mRNA editing have been very actively investigated since discovery of the phenomenon (Davidson, 1993; Hodges *et al.*, 1992; Young, 1990). In 1989, editing of apoB mRNA by a 100,000g supernatant (cytosolic fraction) from rat hepatoma cells was reported; rat liver, unlike human liver, is capable of apoB mRNA editing and synthesizes and secretes both apoB$_{100}$ and apoB$_{48}$. Intestinal editing activity has also been shown in a number of other species that produce apoB$_{48}$, including rat intestine, rabbit intestine, and human intestine. Nuclear extracts have been shown to be active in *in vitro* editing, and editing in the intact cell was shown to be a post-transcriptional nuclear event occurring over the same time course as RNA splicing and polyadenylation. The editing activity was shown to be mediated by a protein or complex of proteins because the activity was abolished by protease treatment, denaturing conditions, disulfide reduction, or chemical modification of specific amino acids. The mechanism of the C to U change appears to be site-specific deamination of cytosine at the 4 position to yield uracil, catalyzed by the editing complex.

In humans, tissue-specific apoB editing activity is developmentally regulated, with fetal intestinal apoB mRNA editing capability being low early in gestation (so that more $apoB_{100}$ than $apoB_{48}$ is produced), increasing throughout gestation, and achieving complete editing by adulthood. Rat intestine and liver both acquire editing capability during development; this developmental regulation of editing accounts for the fact that adult rat liver secretes both $apoB_{100}$ and $apoB_{48}$. Hormonal regulation of editing activity has been reported in rat liver, with thyroid hormone stimulating editing activity. Editing activity has also been reported to respond to dietary perturbations, with fasting decreasing lipogenesis and apoB editing, and refeeding a high carbohydrate diet following fasting (which markedly increases hepatic lipogenesis) stimulating editing.

Analysis of the mRNA sequence requirements for editing of apoB identified an 11-nucleotide sequence beginning 5 nucleotides downstream from the edited C at position 6666 (i.e., positions 6671 to 6681) at which most point mutations destroyed or markedly decreased editing efficiency. This sequence is highly conserved among mammalian species, but not among lower vertebrates which do not edit apoB. The nucleotides between the edited C and the beginning of this sequence, and the nucleotides immediately upstream from the edited C, showed tolerance to base substitutions, demonstrating little or no contribution to the sequence requirements for editing. It has been proposed that this downstream 11-nucleotide sequence is the recognition and/or binding site of the editing complex and that so directed, the editing complex then deaminates a cytosine at a fixed distance upstream.

In 1993, Teng et al. reported the cloning of a cDNA encoding a protein involved in apoB editing [see Teng, et al., (1993); Davidson (1993) for review]. They had previously identified a protein in the cytosolic fraction from chicken intestine, a tissue in which no apoB editing activity is found, which enhances the in vitro editing activity of a rat intestine cytosolic fraction. They then found that a fraction of rat intestine mRNA injected into Xenopus oocytes led to synthesis of a protein which yielded ApoB editing activity when complemented by the chicken intestine cytosolic fraction, but not in its absence. Taking advantage of this observation, they constructed an expression cDNA library from this mRNA fraction. Using the chick intestine complementation assay to screen clones, they were able to isolate a rat cDNA encoding a 229-amino acid protein of 27.3 kDa $M_r$ which in conjunction with the chick intestine cytosolic fraction catalyzed apoB editing. Of note, the rat intestinal cytosolic fraction is also able to complement the cloned editing protein. Expression of this rat cDNA did not confer editing activity in the absence of complementation with either the chick or the rat intestine fraction. In addition, the editing protein, when mixed with a

HepG2 human hepatoma cytosolic extract, was able to effect editing, demonstrating that the complementation by chick intestine cytosol could be produced by human hepatoma cytosol, another cell by itself incapable of apoB editing. Sequence analysis of the editing protein revealed consensus phosphorylation sites for several kinases, raising the possibility of regulation of editing activity by phosphorylation status. The protein also has two leucine zipper motifs. These motifs have been shown to mediate dimerization of DNA-binding proteins, and the possibility that protein dimerization may play a role in the mechanism of RNA sequence recognition and/or editing is raised. Probing rat issues for the presence of mRNA for this editing protein revealed it to be present in the greatest amount in small intestine and liver and absent or present at lower concentrations in other tissues.

At present, therefore, apoB mRNA editing has been shown to be responsible for generating two separate translation products from a single apoB gene by post-transcriptional, site-specific mRNA base modification in a developmentally regulated, tissue-specific manner. This activity is accounted for by at least two proteins, one of which has been cloned, and the other of which can be found in cells that do not edit apoB. The editing complex, therefore, consists of components, some of which have at least some degree of specificity for apoB editing, and other(s) of which presumably serve other functions in the cell unrelated to apoB editing. While many questions remain, rapid advances have been made in our understanding of the mechanism of apoB mRNA editing with the application of molecular techniques, and further advances regarding identification of other proteins involved in the editing complex and regulation of editing may be expected.

## 5. Genetic polymorphisms in apoB

To complete our discussion of the synthesis of triglyceride-rich lipoproteins, we will discuss the genetic variation in the apoB gene and its relationship to risk of atherosclerosis [see Young (1990) for review]. As noted above, there is only one copy of the apoB gene per haploid human genome. Allelic diversity of human apoB was first noted in the 1960s and 1970s with the characterization of antibodies from multiply transfused individuals that recognized specific LDL antigens. Since the cloning of the apoB cDNA and gene in 1986, allelic diversity has been extensively investigated at the genetic level. Multiple RFLPs have been associated with the apoB gene, and a number of these have been correlated with serologically defined apoB diversity. Many studies have examined the relationship between apoB RFLPs and cholesterol levels and/or coronary artery disease. While a number of these studies have noted an association between an RFLP and an outcome (*e.g.*, lipid level or disease), comparison of different studies reveals inconsistencies in these associations. Inconsistencies between an RFLP and an outcome may arise from several variables. Notably, a number of the DNA sequence variations which give rise to apoB RFLPs do not result in altered amino acid sequence of apoB itself; any association of such an RFLP with an outcome would more likely, therefore, be a result of linkage of the RFLP with some other (as yet undefined) gene which may play a role in the outcome measured. Therefore, these RFLP studies must be interpreted with caution.

Direct sequence analysis of the apoB cDNA has revealed a number of mutations which cause predictable lipid abnormalities [see Young (1990) for review]. Most mutations reported in the apoB gene to date have been mutations that result in a truncated variant of apoB or predict a truncated variant which is not detected in plasma. Most of these mutations are single nucleotide transitions or deletions resulting in a premature translation stop codon or a missense protein, but deletions of several nucleotides to entire exons have also been described. These mutations when present in the homozygous or compound heterozygous state produce a clinical syndrome referred to as familial hypobetalipoproteinemia. Homozygous (or compound heterozygous) familial hypobetalipoproteinemia is phenotypically variable, ranging from essentially asymptomatic to severe. In such homozygotes (or compound heterozygotes), plasma levels of VLDL and LDL may be so low as to be clinically indistinguishable from homozygous abetalipoproteinemia. However, abetalipoproteinemia is an autosomal recessive disease in which the apoB gene is normal and hepatic and intestinal apoB is synthesized but in which VLDL and chylomicron synthesis and secretion are defective, causing very low plasma cholesterol and triglyceride levels with no measurable LDL or VLDL, mental retardation, steatorrhea, and deficiency of fat-soluble vitamins (see below). Hypobetalipoproteinemia compound heterozygotes have been reported in patients in whom there is a very low LDL concentration, but normal triglyceride levels and normal dietary fat absorption; these patients were found to harbor apoB mutations which cannot give rise to normal levels of $apoB_{100}$ but which can give rise to normal levels of $apoB_{48}$ (or a truncated variant of very similar size) in addition to production of truncated apoB variants. The ability of the hepatocyte and enterocyte to synthesize a truncated apoB and secrete a triglyceride-rich lipoprotein particle, even at dramatically reduced rates relative to the wild-type apoB gene, appears to be critical in avoiding the clinical phenotype of homozygous abetalipoproteinemia.

Heterozygotes, by virtue of the presence of one normal allele, are characteristically asymptomatic but have low LDL and $apoB_{100}$ concentrations typically one-fourth to one-third of normal. While not conclusively demonstrated, these individuals appear to be at reduced risk of atherosclerosis on the basis of their low LDL concentrations, and anecdotal reports in support of this have been published (Kahn et al., 1978). Clearance of these lipoproteins with truncated apoBs lacking the binding domain for the LDL receptor would be predicted to be mediated by apoE, similar to chylomicrons with their $apoB_{48}$.

Nevertheless, in these patients, the low plasma levels of these lipoproteins appear to be a result of a decreased production rate, reflecting the critical role of apoB in lipoprotein synthesis.

Molecular analysis of the defect responsible for abetalipoproteinemia has led to the identification of mutations in a subunit of the microsomal triglyceride transfer protein (MTP), a lipid transfer protein involved in the intracellular assembly of triglyceride-rich lipoproteins. MTP is a heterodimer of which one subunit is found only in tissues involved in synthesis and secretion of triglyceride-rich lipoproteins and the other is found ubiquitously. The human gene for the tissue-specific subunit of MTP has been cloned from intestine and liver cDNA libraries (Sharp et al., 1993). Patients with abetalipoproteinemia studied to date have all been shown to have mutant alleles for MTP by RFLP analysis or by direct sequence comparison (Sharp et al., 1993; Shoulders et al., 1993). Studies in which an apoB cDNA and the MTP cDNA were transfected into a monkey kidney cell line which normally produces neither apoB nor MTP showed that both genes must be cotransfected in order to get secretion of apoB-containing triglyceride-rich lipoproteins (Leiper et al., 1994). These data demonstrate that apoB and MTP in conjunction appear to be the tissue-specific factors which are both necessary and sufficient to achieve synthesis and secretion of triglyceride-rich lipoproteins, and provide a molecular basis for the clinical similarity between abetalipoproteinemia and severe hypobetalipoproteinemia.

Interestingly, several apoB mutations have been described which result in relatively long truncated variants, which still retain the LDL receptor-binding domain, e.g., apoB$_{87}$. Patients with these mutations appear to have a low LDL level on a basis distinct from that of the aforementioned apoB variants [see Young (1990) for review]. These variants bind with increased affinity to the LDL receptor compared with wild-type apoB$_{100}$. It is, therefore, believed that their low LDL levels may be a result of an increased rate of removal of LDL from plasma, rather than a decreased rate of production.

To conclude our discussion of apoB mutations, we will briefly discuss an ApoB mutation which, contrary to the above mutation, is associated with elevated cholesterol and LDL levels [see Young (1990) for review]. This mutation is a base transition which results in an amino acid substitution of glutamine for arginine at position 3500. Because each LDL particle contains a single molecule of apoB$_{100}$, its ability to bind to the LDL receptor is determined by that apoB. Apolipoprotein B$_{3500}$ has a decreased capacity to bind the LDL receptor and, hence, all LDL particles containing this form of apoB have defective clearance. The lipoprotein disorder associated with this apoB mutation is referred to as familial defective apoB$_{100}$. The clinical manifestations of this syndrome are similar to those of heterozygous familial hypercholesterolemia, with markedly elevated LDL cholesterol levels, xanthomata, and premature atherosclerosis (Rauh et al., 1992). These observations illustrate that elevated LDL levels are sufficient to promote accelerated atherosclerosis regardless of whether the defect in LDL clearance results from the LDL receptor or from the apoB ligand.

## B. Lipoprotein remnant metabolism–The role of apoE

Lipoprotein remnant metabolism is an extremely complex subject for which the reader is referred to several recent reviews (Mahley et al., 1989; Cooper, 1992). We will provide a brief overview of the salient features of lipoprotein remnant metabolism and its relationship to atherogenesis before considering the impact of molecular biology theory and practice on our understanding of this rapidly evolving field. Historically, the field of lipoprotein remnant metabolism evolved with understanding of extracellular cascades of lipoprotein metabolism (Figure 5.5). As noted above, chylomicrons and VLDL particles were shown to be subject to extracellular triglyceride hydrolysis by lipases (lipoprotein lipase and hepatic triglyceride lipase). Chylomicrons and VLDL and their respective remnants share a number of constituent apolipoproteins in addition to the presence of apolipoproteins distinct to each particle type, as discussed above. Some of the associated apolipoproteins are lost and others acquired extracellularly after de novo particle secretion as a consequence of postsecretional extracellular processing pathways. Apolipoproteins associated with remnants serve a variety of functions, including serving as cofactors for enzymes and as ligands for receptors mediating particle uptake. Of greatest import regarding lipoprotein remnant clearance, both VLDL and chylomicron remnants have associated ApoE, which as discussed above, is a ligand for the LDL receptor.

Aberrant remnant metabolism was first described clinically in association with Fredrickson type III hyperlipoproteinemia (dysbetalipoproteinemia) with the recognition that the abnormal lipoproteins which accumulated in this condition were chylomicron and VLDL remnants, referred to collectively as β-VLDL (based on an abnormal electrophoretic migration pattern). It was subsequently shown that a mutant form of apoE was a necessary (but not sufficient) component of the phenotype which resulted from accumulation of these remnant lipoproteins, implicating apoE in the clearance mechanism (Mahley et al., 1989). The type III phenotype characteristically includes elevations of both cholesterol and triglycerides to roughly similar levels associated with β-VLDL, xanthomata of the palmer creases (found in no other hyperlipoproteinemic phenotypes) and tuberous xanthomata, and premature coronary and peripheral atherosclerosis (usually manifest beginning in the fourth to fifth decade).

Classical biochemical studies in the 1970s and early 1980s demonstrated that β-VLDL from patients with type III hyperlipoproteinemia were essentially uniformly associated with a mutant form of apoE. This mutant form was referred to as apoE-2 to distinguish it from the predominant allelic variant,

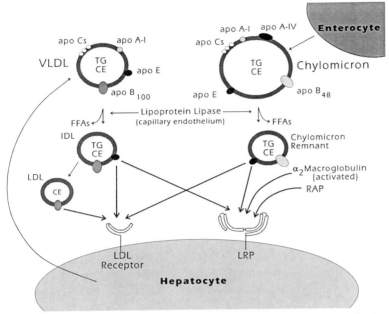

**Figure 5.5.** Metabolism of triglyceride-rich lipoproteins. VLDL particles are synthesized in hepatocytes from endogenous pools of triglyceride and contain $ApoB_{100}$, apoE, apoCs, and
apoA-I. Chylomicrons are synthesized in the enterocyte from triglycerides derived from
dietary fat and contain $apoB_{48}$, apoE, apoCs, and apoA-I, -A-II, and -A-IV.
Chylomicron and VLDL triglycerides are hydrolyzed by lipoprotein lipase in the capillary endothelium, resulting in the formation of chylomicron remnants and intermediate
density lipoproteins (IDL), respectively. Further modifications of IDL lipids by lipoprotein lipase, hepatic lipase, and cholesterol ester transfer protein converts a fraction of
IDL to LDL, which is subsequently taken up through either the LDL receptor or LRP.
Chylomicron remnants are also cleared by the LDL receptor or LRP. Because $apoB_{48}$
that is present on chylomicrons lacks the apoB domain required for recognition by the
LDL receptor, recognition of chylomicrons by the LDL receptor and LRP is mediated by
apoE. LRP can bind ligands other than LRP, including $\alpha_2$ macroglobulin, and is
inhibited by receptor-associated protein (RAP).

apoE-3, and another less frequent variant, apoE-4 (Mahley *et al.*, 1989). The
apoE-2 variant was shown to differ from apoE-3 by a single amino acid substitution of cysteine for arginine in the region of the molecule recognized by the LDL
receptor. This region of apoE was shown to be rich in basic amino acids (arginine and lysine). Chemical modification of these basic amino acids impairs
cellular uptake mediated by apoE by impairing interaction with receptors. The
hypothesis that apoE may mediate lipoprotein remnant uptake and degradation
and that mutant forms of apoE defective in binding prevent the function of this
normal clearance pathway in the liver is further supported by the finding of

patients with apoE deficiency (below) in whom a type III phenotype is present. The complexity of the type III phenotype is exemplified by the fact that while homozygosity for apoE-2 (i.e., apoE-2/2) is required for expression of the type III phenotype (with very rare exceptions, such as absence of apoE production) it is not sufficient, as only a small percentage of people with the apoE-2/2 genotype exhibit the type III phenotype. The remainder, while having detectable β-VLDL, exhibit normal or even relatively low serum lipid and lipoprotein concentrations. Little is known of the myriad potential additional genetic and environmental factors that could play a role in the development of the phenotype. Factors that have been proposed include hypothyroidism, diabetes, obesity, and the genetic copredisposition for familial combined hyperlipidemia.

The human apoE cDNA was cloned and sequenced in 1984, and the genomic sequence was cloned and sequenced the following year (Mahley et al., 1989). Analysis of genomic clones revealed an organization of four exons and three introns within a 3.6-kb fragment. Processing of the primary RNA transcript yields a mature mRNA of approximately 1.2 kb. Translation yields a 317-amino acid protein which includes an 18-amino acid N-terminal signal sequence; proteolytic processing yields the mature protein of 299 amino acids. The LDL receptor-binding domain is localized to the portion of the molecule corresponding to amino acids 140–160, a region rich in basic amino acids. The apoE gene maps to human chromosome 19 and is closely linked to apoC-I. The LDL receptor and apoC-II also map to human chromosome 19, but are not closely linked to the apoE gene.

Transgenic and gene knockout strategies have been applied with remarkable success to the study of the role of apoE in the metabolism of triglyceride-rich lipoproteins [see Breslow (1993) for review]. Transgenic mice that express rat apoE (Shimano et al., 1992a,b) under transcriptional control of the mouse metallothionein promoter exhibit decreased serum levels of VLDL and LDL. The transgenic mice showed a gene dose effect comparing serum rat apoE between heterozygous and homozygous transgenics. Stimulation of serum rat apoE levels with zinc resulted in a further decrease in VLDL and LDL cholesterol levels. In homozygotes, VLDL triglyceride levels were not measurable with or without zinc induction. Consistent with an increased rate of clearance of VLDL and LDL, serum apoB levels were also significantly decreased in transgenic animals. Homozygous transgenic mice were resistant to the hypercholesterolemic effects of cholesterol feeding, and heterozygotes were relatively resistant. In vivo studies revealed that plasma clearance of VLDL, LDL, and chylomicron remnants was markedly accelerated in transgenic animals compared to controls. They also showed that with cholesterol feeding, hepatic LDL receptor expression was decreased similarly in controls and in homozygous transgenic mice; chylomicron remnant clearance, however, was not diminished by cholesterol feeding. These data suggest that increased expression of apoE can stimulate clearance of lipoprotein remnants and can thus lower steady-state plasma concentrations of these lipoproteins in the face of downregulation of the LDL receptor pathway. Implicit is the proposition that lipoprotein remnants may be cleared by a pathway independent of the LDL receptor pathway, as will be discussed in detail below.

As yet, no transgenic mice have been described that overexpress apoE2. However, several reports have been published describing mice transgenic for rare apoE mutations which produce type III hyperlipoproteinemia in the heterozygous state (recall, above, that apoE2 is only associated with the type III phenotype when present in the homozygous state). The dominant inheritance of type III hyperlipoproteinemia with these rare apoE mutations makes their study amenable to transgenic technology. In 1993, mice transgenic for such a human apoE mutation, apoE3-Leiden, were reported that had elevated plasma cholesterol and triglyceride levels (van den Maagdenberg *et al.*, 1993); serum lipids increased dramatically with fat and cholesterol feeding. Similarly, Fazio and colleagues (Fazio *et al.*, 1993) produced mice transgenic for a different human apoE variant which produces type III hyperlipoproteinemia in the heterozygous state. The transgenic mice were shown to have plasma human apoE levels roughly corresponding to transgene copy number, and in turn, to have plasma cholesterol and triglyceride roughly corresponding to human apoE levels. β-VLDL was shown to accumulate in the plasma of the transgenic mice expressing high levels of the variant human apoE. These transgenic mouse β-VLDL were cleared more slowly from the plasma when injected into normal mice than control VLDL from nontransgenic mice, and conversely, were taken up more slowly by the liver than control VLDL. These transgenic mice, therefore, mimic human type III hyperlipoproteinemia in many regards and may serve as a good animal model in the study of this disorder.

In 1992, homologous recombination was used to "knock out" the apoE gene in the mouse (Piedrahita *et al.*, 1992; Plump *et al.*, 1992; Zhang *et al.*, 1992). Zhang and colleagues (Zhang *et al.*, 1992) showed that homozygous apoE knockout mice had marked hypercholesterolemia and modest hypertriglyceridemia. These increases were accounted for in the VLDL/LDL fraction of plasma, with a decreased level of HDL cholesterol. Heterozygous knockout mice had total cholesterol and HDL cholesterol levels similar to controls and only a slight increase in triglyceride. Analysis of the aortas of the homozygous knockout mice fed a normal chow diet revealed atherosclerotic lesions in the proximal aorta by 3 months of age; lesions were absent in heterozygous knockout mice and normal controls. The studies of Plump *et al.* (1992) similarly showed marked hypercholesterolemia and modest hypertriglyceridemia in homozygous knockout mice fed a normal chow diet; feeding a "Western" style high fat/high cholesterol diet produced truly remarkable hyperlipidemia. Increases were in the VLDL + IDL, LDL, and chylomicron remnant fractions of plasma. Heterozygous knockout mice had lipid and lipoprotein profiles similar to controls. Evaluation of arteries in these mice revealed proximal aortic atherosclerotic lesions similar to those observed by Zhang *et al.* (above); Plump *et al.* also described atherosclerotic lesions in the coronary arteries.

These studies provided the first evidence that severe atherosclerosis could be produced in the mouse and raised the possibility that apoE-deficient mice could be used as a background in which to examine the roles of other genes in the atherogenic process. The ultimate utility of this model will require that the basic mechanisms of lesion development in the apoE knockout mouse are similar to the mechanisms involved in human atherosclerosis. From a morphological point of view, lesions in the apoE-deficient mouse progress from fatty streaks to complex lesions with fibrous caps, proliferation of smooth muscle cells, and necrotic cores (Nakashima *et al.*, 1994; Reddick *et al.*, 1994). Furthermore, these lesions contain epitopes that are recognized by antibodies to oxidized LDL (Palinski *et al.*, 1994), suggesting that LDL oxidation may be involved in their generation. Further investigation will be required to characterize the strengths and weaknesses of this model system. It is likely, however, that both the apoE and LDL receptor knockout lines will prove to be invaluable tools in the identification and characterization of other genes that may act to promote or retard the atherogenic process.

## 1. Receptors involved in the clearance of lipoprotein remnants

Cellular uptake and degradation of triglyceride-rich lipoprotein remnants is receptor mediated, resulting in endocytosis and lysosomal degradation, analogous to the LDL receptor pathway described above. While both types of lipoprotein remnants have associated apolipoproteins which may serve as ligands for the LDL (apoB/E) receptor (i.e., apoE $\pm$ apoB$_{100}$), the contributions of cellular receptors mediating hepatic uptake have only very recently been well defined. It was noted that patients with Fredrickson type III hyperlipoproteinemia, characterized by a defect in remnant catabolism due to a point mutation in apoE, have no demonstrable defect in LDL receptor function. In addition, reports of patients lacking serum apoE in whom the typical type III phenotype is also manifest have been published. These patients, homozygous for apoE deficiency, demonstrate no serum apoE by gel electrophoresis or immunodiffusion, and only trace amounts by radioimmunoassay, but have premature atherosclerosis documented as early as the fifth decade of life, again raising a question of the necessity for apoE in macrophage β-VLDL metabolism. Conversely, it was recognized that patients with homozygous familial hypercholesterolemia and, thus, no functional LDL receptors, did not have aberrant postprandial lipoprotein remnant clearance or accumulation of β-VLDL, suggesting that lipoprotein remnants could be cleared by pathways independent of the LDL receptor. An apoE-only receptor (i.e., a receptor which recognized apoE, but not $apoB_{100}$) was hypothesized to account for these (and other consistent) findings.

In 1986, Herz *et al.* reported the cloning of a large liver membrane protein with a high degree of homology to the LDL receptor, which they dubbed the LDL receptor-related protein (LRP) (Herz *et al.*, 1988). From a methodologic point of view, the development of the LRP story is almost opposite that of the LDL receptor; whereas the LDL receptor pathway was first described based on now-classical cell biological and biochemical experiments and subsequently progressed to molecular biological applications, the study of the LRP pathway began with the cloning of the LRP gene. Herz *et al.* initially sought to identify novel members of the complement superfamily by screening cDNA libraries at low stringency for clones containing sequences corresponding to the highly conserved cysteine-rich and acidic domain that is conserved between complement components and the LDL receptor. Using this screening strategy, they were successful in cloning a human cDNA which encoded a large 4544-amino acid membrane protein with 503 kDa $M_r$ (compare with 839 amino acids and 160 kDa $M_r$ of the mature human LDL receptor, above) (Figure 5.2). From sequence analysis, the LRP was shown to consist of multiple repeats each with a high degree of amino acid sequence homology to domains previously described in the LDL receptor (above) including domain 1 (the acidic cysteine-rich ligand-binding domain), domain 2 (the EGF precursor domain, important in $apoB_{100}$ binding and acid-mediated ligand dissociation), and domain 5 (the cytoplasmic tail domain shown to be required for localization to coated pits and for internalization). Rather than representing a complement-like factor, Herz *et al.* speculated that the LRP was, in fact, a lipoprotein receptor based on its sequence similarity to the LDL receptor, its location at the cell surface, and the presence of the conserved sequence in the C-terminal cytoplasmic tail NPXY, which in the LDL receptor is essential for clustering in coated pits and internalization. Furthermore, because of its relatively high levels of expression in the liver, it became a candidate lipoprotein-remnant receptor. With the description of LRP, initiated by cloning the cDNA and sequence analysis with comparison to a well-characterized cell-surface receptor, the way was paved for further molecular and cell biological and biochemical studies. In the following year, Kowal *et al.* (1989) described that LRP and mRNA was present in fibroblasts obtained from patients with homozygous familial hypercholesterolemia. Significantly, they also showed that when β-VLDL from cholesterol-fed rabbits was enriched *in vitro* with apoE, its incubation with familial hypercholesterolemia fibroblasts led to accumulation of esterified cholesterol. This did not occur using β-VLDL that was not enriched in apoE, suggesting it was a critical determinant of cellular recognition. Furthermore, accumulation of cholesterol esters was inhibited by chloroquine, suggesting receptor-mediated endocytotic uptake and delivery of β-VLDL cholesterol. While treatment of normal fibroblasts with 25-hydroxy-cholesterol downregulates the LDL receptor, such treatment was shown to have no effect on LRP expression, suggesting that while the genes for the two receptors are certainly evolutionarily related, they differ in the manner in which they are regulated. Similarly, they showed that polyclonal anti-LRP antibody could block the increase in cholesterol esterification induced by incubation of cells with apoE-enriched rabbit β-VLDL. That same year, Beisiegel et al. (1989) reported that apoE-containing liposomes incubated with human liver cell membranes or the human hepatoma cell line, HepG2, could be cross-linked to LRP via photoaffinity labeling, and that such binding could be competitively inhibited by coincubation with excess β-VLDL (not enriched with additional apoE *in vitro*). Interestingly, they also showed that the allelic variant, apoE2, which as noted above is defective in binding to the LDL receptor, is able to bind to LRP. In this regard, it is noteworthy that the vast majority of patients with apoE2 homozygosity do not develop hyperlipoproteinemia and thus presumably are able to clear lipoprotein remnants at a reasonable rate, despite homozygosity for apoE2. Together, these data lent further credence to the notion that LRP may function as a lipoprotein remnant receptor in the liver.

Since then, the LDL receptor gene family has been studied extensively. Interestingly, a cell-surface glycoprotein, termed gp330, found in renal tubular epithelial cells (Kerjaschki et al., 1982) was partially cloned in 1989 (Raychowdhury et al., 1989). Like LRP, gp330 was shown to have extensive sequence homology with the LDL receptor and the EGF precursor. A complete cDNA sequence has recently been obtained and indicates a domain structure that is very similar to LRP (Figure 5.2). The LDL receptor gene family thus includes at least three members in mammalian cells. In addition, two avian proteins have been proposed as members of the LDL receptor gene family, bringing to five the total number of genes thus far recognized to be in this family.

A further twist in the LRP story came in 1990 when Strickland et al. (1990) reported that the peptide sequence of the $\alpha_2$-macroglobulin receptor correlated perfectly with the amino acid sequence of LRP predicted by Herz et al. (1988), suggesting that the two receptors are the same molecule. The function of the $\alpha_2$-macroglobulin receptor ($\alpha_2$-MR) is to scavenge extracellular proteases and thus regulate their activity. $\alpha_2$-Macroglobulin is a glycoprotein that binds extracellular proteases, and in so doing is "activated" by proteolytic cleavage; cleavage renders the $\alpha_2$-macroglobulin able to inhibit the protease and renders the $\alpha_2$-macroglobulin/protease complex susceptible to binding to the $\alpha_2$-macroglobulin receptor and subsequent receptor-mediated uptake. ($\alpha_2$-Macroglobulin hereafter referred to is activated unless otherwise noted.) It was thus proposed that the $\alpha_2$-MR/LRP is, therefore, a multifunctional receptor with biological roles as diverse as lipoprotein remnant metabolism and regulation of protease activity (Strickland et al., 1990).

LRP has been shown to bind a wide variety of ligands in addition to β-VLDL (via apoE) and $\alpha_2$-macroglobulin/protease complexes. Of most relevance to our discussion, lipoprotein lipase, which as noted above is involved in the pathway of extracellular triglyceride metabolism, has been reported to bind to LRP and to markedly enhance binding and uptake of human chylomicrons and rabbit β-VLDL via LRP (Beisiegel et al., 1991; Chappell et al., 1992, 1993). In their studies, Chappell et al. showed that specific ligands of LRP, such as $\alpha_2$-macroglobulin/protease complex or the 39-kDa receptor-associated protein (RAP, see below), could compete with lipoprotein lipase for binding and uptake via LRP, and that monoclonal antibodies against lipoprotein lipase could block the lipase-mediated binding and uptake. They also provided evidence that cell-surface proteoglycans may be involved in lipoprotein lipase-mediated uptake via LRP. These observations may be of importance in atherosclerotic lesions, based on the ability of macrophages to secrete lipoprotein lipase (Khoo et al.,, 1981). Thus, LPL could bind to triglyceride-rich lipoproteins and subsequently mediate their uptake by macrophages via LRP.

Another ligand for the $\alpha_2$-MR/LRP is the RAP, a 39-kDa $M_r$ protein first described in 1990 as a protein copurifying with $\alpha_2$-MR/LRP (Ashcom et al., 1990). RAP was subsequently shown to bind to LRP with high affinity (Strickland et al., 1990). In 1991, the human RAP cDNA was cloned (Strickland et al., 1994; Herz et al., 1991). Analysis of the amino acid sequence predicted from the cDNA sequence revealed several provocative surprises. RAP corresponds to at least one of the major antigens mediating Heymann nephritis, a murine model of membranous glomerulonephritis in humans. Analysis also demonstrated that RAP has sequence homology with apoE in the region involved in interaction with the LDL receptor (Strickland et al., 1994). Also that year, RAP was reported to modulate the binding of other ligands to LRP (Herz et al., 1991), a finding which has been further developed and has very recently found application in studies using genetically engineered mice (below). In these studies, Herz et al. showed that, when produced as a glutathione S-transferase fusion protein, RAP inhibited LRP-mediated $\alpha_2$-macroglobulin and β-VLDL uptake by LDL receptor-negative human fibroblasts. RAP was shown not to inhibit uptake of ligands via the LDL receptor. Based on the above findings, it perhaps came as no surprise that RAP was subsequently reported to interact with both gp330 and LRP, and that RAP copurifies with gp330 in a similar manner to its copurification with LRP (Kounnas et al., 1992). While the biological role of RAP remains uncertain, these findings have led to the suggestion that RAP may function in vivo in regulating ligand binding to LRP and gp330.

In 1992, the LRP gene was knocked out via homologous recombination by Herz et al. (1992). The effect of the LRP knockout on lipoprotein metabolism or atherosclerosis could not be assessed, however, because homozygotes were embryonic lethals at the embryo implantation stage. Homozygous LRP knockout blastocysts were viable in vitro, however, suggesting that the lethality resulted from inability to implant rather than a lethal defect in cellular metabolism *per se*. They proposed that the defect in implantation resulted from absence of functional LRP needed to serve as a mediator of protease/protease inhibitor complex uptake and degradation by trophoblast cells of the embryo invading the wall of the uterus.

Although insights into the role of LRP in remnant metabolism *in vivo* were not obtained by knockout strategies, several other approaches have been taken. In 1991, Hussain *et al.* (1991) reported that $\alpha_2$-macroglobulin nearly completely inhibited the esterification of cellular cholesterol stimulated by apoE-enriched $\beta$-VLDL incubated with LDL receptor-deficient human fibroblasts *in vitro*. They also reported in *in vivo* studies in mice that $\alpha_2$-macroglobulin delayed plasma clearance and competed for hepatic uptake of chylomicron remnants ($\pm$ apoE enrichment), and conversely, that ApoE-enriched chylomicron remnants delayed plasma clearance of $\alpha_2$-macroglobulin. They concluded that LRP may, therefore, function *in vivo* in chylomicron remnant metabolism. The following year, Szanto *et al.* (1992) reported that treatment of rats with drugs that had one effect on the levels of LDL receptor expression could have opposite effects on LRP expression. The effects that these agents had on plasma chylomicrons and their remnants reflected more the status of LDL receptor expression than that of LRP. Measuring the LDL receptor and LRP via ligand blotting, they showed that compared to controls, rats treated with estradiol or thyroxine had upregulation of LDL receptor expression. In contrast, estradiol treatment downregulated LRP expression, while thyroxine had no effect. They also showed that feeding rats a high cholesterol diet, an intervention known to downregulate the LDL receptor, had no effect on LRP expression. When considering the effect of these treatments on plasma lipoproteins, it was shown that estradiol treatment, which resulted in marked upregulation of the LDL receptor and down regulation of LRP in the liver, also led to marked decreases in total, VLDL, chylomicron, and chylomicron remnant cholesterols. Conversely, cholesterol feeding which resulted in down regulation of the LDL receptor and no change in LRP expression resulted in marked increases in cholesterol in these fractions. They concluded that hepatic expression of these two related receptors was regulated independently (in agreement with the *in vitro* data above), and that the LDL receptor appeared to play a significant role in chylomicron and remnant clearance, whereas LRP did not. The following year, Choi and Cooper (Choi *et al.*, 1993) drew similar conclusions regarding the relative importance of the LDL receptor and LRP in chylomicron remnant clearance. They showed *in vitro* in Chinese hamster ovary cells, which express the LDL receptor, that $\alpha_2$-macroglobulin did not compete with chylomicron remnants for uptake, whereas antibody against the LDL receptor (which does not cross-react with LRP) did block uptake and degradation by roughly 50%. On the other hand, chylomicron remnants inhibited $\alpha_2$-macroglobulin uptake and degradation by at least 50% whereas anti-LDL receptor antibody had no effect. Studies *in vivo* in mice demonstrated that injection of anti-LDL receptor antibody diminished chylomicron remnant plasma clearance more markedly than did $\alpha_2$-macroglobulin, whereas this antibody had no effect on plasma clearance of $\alpha_2$-macroglobulin; they also showed that chylomicron remnants were able to delay plasma clearance of $\alpha_2$-macroglobulin, but to a lesser extent than excess unlabeled $\alpha_2$-macroglobulin. They concluded that while both the LDL receptor and the LRP can mediate remnant uptake, in the mouse (with functioning LDL receptors) the LDL receptor appears to play a fractionally greater role. In concert, these observations suggest that in a normal genetic background (*i.e.*, LDL receptor positive), both the LDL receptor and the LRP may mediate uptake of lipoprotein remnants, and that the majority of particles are cleared by the LDL receptor, while LRP appears to play a relatively minor role in lipoprotein remnant metabolism. In the absence of LDL receptors, such as in familial hypercholesterolemia patients, LRP is likely to be of critical importance in lipoprotein remnant metabolism. With the powerful application of genetically manipulated mice, the hypothesis that lipoprotein remnant metabolism may be divided between two different receptor pathways has been directly tested. In 1994, Ishibashi *et al.* (1994b) reported on mice which had the LDL receptor gene knocked out, the apoE gene knocked out, or both genes knocked out. Mice used in the studies were homozygous for these gene knockouts. Therefore, the double knockout mice lacked the LDL receptor pathway, and lacked the lipoprotein ligand for the LRP pathway, making neither pathway available for lipoprotein metabolism. It was shown that compared to controls fed a normal chow diet, apoE knockout mice were markedly hypercholesterolemic and that the increase was accounted for predominantly by lipoprotein remnants; in agreement with this, apoB$_{48}$ was markedly increased but apoB$_{100}$ was not. On feeding an atherogenic diet, these apoE knockout mice had a remarkable increase in lipoprotein remnants; surprisingly they did not demonstrate a greater increase in apoB$_{48}$ than on the chow diet, but they did demonstrate a marked increase in apoA-IV, which as noted above is normally associated with the triglyceride-rich lipoproteins. The LDL receptor knockout mice had plasma cholesterol intermediate between apoE knockout mice and controls on the chow diet. This increase was accounted for predominantly by an increase in LDL; apolipoprotein analysis revealed a marked increase in apoB$_{100}$ consistent with this and slight increases in apoB$_{48}$ and apoE. On an atherogenic diet, these mice demonstrated dramatic increases in LDL and lipoprotein remnants; apolipoprotein analysis revealed markedly increased apoB$_{48}$ and apoE but only slightly increased apoB$_{100}$ compared to these mice on the chow diet. The overwhelming effect of an atherogenic diet on these mice, therefore, was an accumulation of lipoprotein remnants. The LDL receptor and apoE double knockout mice had plasma cholesterol levels that were not significantly different from the apoE knockout mice on the chow diet. The double knockout mice, like the apoE knockout mice, had markedly elevated lipoprotein remnant levels, and like the LDL knockout mice, had markedly elevated LDL levels. On the atherogenic diet, these double knockout mice developed truly remarkable hyperlipidemia. Apolipoprotein analysis of these mice revealed elevations of both $apoB_{48}$ and $apoB_{100}$ compared to controls; on the atherogenic diet, $apoB_{48}$ was further elevated compared to the chow diet, and as was seen with apoE knockout mice, these mice on the atherogenic diet had a marked increase in apoA-IV. In summary, LDL concentration increased only with LDL receptor deficiency; lipoprotein remnants increased with deficiency of either receptor, but the increase was greater in the absence of apoE (i.e., nonfunctional LRP pathway) than in the absence of the LDL receptor pathway. This clearly demonstrated that the LDL receptor is functional in lipoprotein remnant clearance in addition to LRP and defined the pattern of hyperlipoproteinemia that develops when either or both pathways are interrupted.

Complementary data were reported simultaneously by Willnow and colleagues (Willnow et al., 1994). In these studies, the contribution of LRP to lipoprotein clearance was blocked by overexpression of RAP. This was accomplished by injecting mice with a recombinant adenovirus harboring a RAP cDNA. This approach primarily leads to infection of hepatocytes, which transiently express high levels of RAP. By performing these experiments in normal and LDL receptor knockout mice, they were able to examine the relative contributions of the LDL receptor and LRP systems to lipoprotein clearance. They showed that RAP expression caused a dose-dependent increase in serum cholesterol in LDL receptor-deficient mice, but not in normal mice. The modest increase in cholesterol and triglyceride seen in LDL receptor wild-type mice transiently overexpressing RAP was accountable by the presence of lipoprotein remnants. Apolipoprotein analysis revealed that LDL receptor wild-type mice transiently transgenic for RAP had increased apoE and apoB compared to controls, consistent with some degree of impairment of lipoprotein remnant clearance, even in the presence of a functional LDL receptor pathway. In LDL receptor knockout mice transiently transgenic for RAP, there were slight increases in apoB compared to LDL receptor wild-type mice; apoE levels were markedly elevated compared to those of LDL receptor wild-type mice. Interestingly, regardless of LDL receptor status, overexpression of RAP was associated with decreased apoA-I levels and decreased HDL levels, consistent with a link between HDL metabolism and the LRP pathway of lipoprotein remnant metabolism. To summarize their data, lipoprotein remnants accumulate when either the LDL receptor or the LRP pathway is nonfunctional; combined loss of these pathways leads to massive elevation of plasma lipoprotein remnants. However, in the absence of apoE, further knockout of the LDL receptor makes no difference in total plasma cholesterol levels. Collectively, the studies of Ishibashi et al. and Willnow et al. confirm the two-receptor pathway model of lipoprotein remnant clearance (Figure 5.5). Based on reports presented in the preceding papers, the issue of the relative quantitative contributions of the LDL receptor and LRP pathways to lipoprotein remnant clearance is not yet absolutely resolved. Nevertheless, the field of lipoprotein remnant metabolism has been qualitatively elucidated to a degree unimaginable prior to the application of molecular biology.

## C. HDL metabolism

As noted in the introduction, many epidemiologic studies have established a strong inverse correlation between HDL levels and risk of atherosclerosis. The most widely accepted hypothesis to account for the protective effect of HDL is that it functions to remove excess cholesterol from peripheral tissues and transport it to the liver for excretion in the bile. According to this hypothesis, free cholesterol is initially transferred from cell membranes to the HDL surface. This free cholesterol is then esterified by the enzyme lecithin–cholesterol acyl transferase (LCAT), producing cholesterol esters that translocate to the hydrophobic core of HDL. The LCAT reaction is believed to play an important role in reverse cholesterol transport because by catalyzing the removal of surface cholesterol, the ability of HDL to accept additional cholesterol is extended. Several mechanisms appear to function in the ultimate delivery of HDL cholesterol esters to the liver. First, cholesterol ester-rich HDL may contain apoE on its surface, enabling uptake of intact particles via LDL receptors. Second, HDL cholesterol esters can be transferred to LDL and VLDL through the actions of a plasma CETP, which catalyzes the exchange of neutral lipids. In the case of an exchange involving HDL and VLDL, for example, HDL cholesterol esters would be exchanged for VLDL triglycerides. The net result of this exchange would be for HDL cholesterol esters to take on the metabolic fate of VLDL remnants, which is primarily uptake and degradation in the liver. Finally, HDL particles can deliver cholesterol esters to the liver directly by a mechanism that proceeds independently of the uptake and degradation of HDL particles, a process that has been termed "selective uptake."

The application of molecular techniques to the study of HDL has clarified many details of the roles of specific apolipoproteins and modifying enzymes in its metabolism. In particular, strong, direct evidence of the ability of HDL to retard atherogenesis has been produced *in vivo*, although the precise mechanisms by which this protective effect is achieved remain to be established.

The earliest applications of molecular biology to the HDL field were the cloning and sequencing of genes encoding HDL-associated apolipoproteins in the 1980s. This allowed correlation of mutations of the gene(s) in question with phenotype (usually relating to serum apolipoprotein or cholesterol levels or premature atherosclerosis); the reader is referred to several good reviews regarding mutations in apolipoprotein genes affecting HDL levels and associated clinical manifestations (Breslow, 1985, 1989; Assmann et al., 1993).

The predominant HDL-associated apolipoprotein and the one which defines HDL is apolipoprotein A-I (apoA-I). The human apoA-I gene is located on the long arm of human chromosome 11 and consists of four exons and three introns spanning about 2 kb. The apoA-I gene is physically linked to two other genes encoding HDL-associated apolipoproteins, apoC-III and apoA-IV. It encodes a mature mRNA of 893 nucleotides which includes 5' and 3' untranslated regions; this mRNA is translated to a preproprotein of 267 amino acids. The preproprotein is processed cotranslationally with cleavage of an 18-amino acid N-terminal signal sequence, and subsequently with cleavage of the next 6 N-terminal amino acids postsecretionally to yield the mature protein of 243 amino acids. In addition to its structural and functional role in HDL metabolism, apoA-I is a cofactor required for activity of LCAT, the enzyme in plasma responsible for extracellular esterification of cholesterol.

Limited analysis of correlation between gene exon and protein domain organization and structure/function relationship has yielded some interesting suggestions. Exon 1 encodes the 5' untranslated region. Exon 2 encodes the translation start site and the 18-amino acid signal sequence. Exon 3 encodes the 6-amino propeptide sequence and the N-terminal 42 amino acids and part of the codon for amino acid 43 of the mature protein. Exon 4 encodes the C-terminal 200 amino acids. The region of exon 4 encoding amino acids 99–230 was shown to be composed of six tandem repeats with 64–80% homology, suggesting that this region of the gene arose by intragenic duplications. These tandem repeats each encode an amphipathic α-helix which constitutes the domain of apoA-I which associates with the surface of the lipoprotein. The hydrophobic side of the α-helix associates with the lipid core of the particle, while the hydrophilic side is in contact with the surrounding medium (Breslow, 1985).

Numerous mutations in ApoA-I have been described, some of which result in clinical phenotypes. Homozygosity for deletion of the apoA-I/apoC-III/apoA-IV gene cluster, resulting in the loss of all three of these apolipoproteins and the absence of detectable HDL in plasma, produces a well-defined clinical syndrome including, and most significant for our discussion, premature coronary artery disease, planar xanthomas (cholesterol-laden macrophages in the skin), and corneal opacification. Likewise, homozygosity for an inversion in this gene cluster resulting in fusion gene products of the N-terminus of apoA-I with the C-terminus of apoC-III and of the N-terminus of apoC-III with the C-terminus of apoA-I also produces premature atherosclerosis. Due to the absence of apoA-I, these patients also lack LCAT activity. Taken as a group, patients that lack apoA-I and apoC-III, with or without a lack of apoA-IV, have an absence of normal HDL with severe metabolic consequences (Assmann et al., 1993).

A large number of other genetic defects have been identified affecting only apoA-I. These result in variable clinical findings depending on the nature of the mutation. Deletions toward the 3′ end of the gene that result in small carboxy-terminal deletions of the protein or point mutations throughout the molecule have been described that appear to confer no additional risk of atherosclerosis. The majority of point mutations that have been described that result in single amino acid substitutions are also clinically silent and have been noted only in population screening. Intriguingly, a number of point mutations in apoA-I have been described that are correlated with low HDL levels, altered post-translational processing, abnormal association of the mutant protein with lipid particles, decreased activity in activating LCAT, homo- and heterodimerization (in particular with apoA-II), and in one mutation, familial polyneuropathy. These mutations as a group are notable, however, for absence of increased risk of premature atherosclerosis (Assmann et al., 1993). It must be borne in mind, however, that most of these studies of apoA-I mutations have described patients heterozygous for the genetic defect, and while HDL levels may be affected, the potential clinical consequence(s) of homozygosity for the defect may not be appreciated. For example, a point mutation in the 84th codon that normally encodes glutamine has been described which results in a premature stop codon, and thus functional deletion of the C-terminal 65% of the mature molecule. This region encodes the six amphipathic α-helices that mediate association with the lipoprotein particle. A patient homozygous for this mutation has been reported who had undetectable plasma apoA-I and HDL cholesterol and xanthomatosis and developed premature atherosclerosis in her fifties necessitating coronary artery bypass grafting (Matsunaga et al., 1991).

Fewer frameshift mutations have been described, but there are several instructive examples. A frameshift mutation resulting from a nucleotide deletion in codon 202 and resulting in a wild-type N-terminal 83% of the mature molecule with the remainder C-terminus affected by the frameshift has been described in a male patient homozygous for the mutation. The clinical manifestations of the mutation include very low HDL levels and the corneal opacification characteristic of some of the deletions noted above, as well as decreased LCAT activity, but significantly do not include premature atherosclerosis (Assmann et al., 1993). A report of a frameshift mutation resulting from a nucleotide insertion in the fifth codon, and thus frameshifting essentially the entire molecule, has been described in a review (Schmitz et al., 1990). While no details of this case have been published, this mutation apparently resulted in low HDL levels, xanthomatosis, and premature atherosclerosis. A deletion mutation in which codons 146–160 have been deleted, corresponding to 15 amino acids toward the C-terminus of midline of the primary sequence, has been described in a patient heterozygous for the defect. There was no family history of premature atherosclerosis. Clinically, the patient had very low HDL levels, corneal opacification, and decreased LCAT activity, but notably, no evidence of premature atherosclerosis.

The preceding data are consistent with the proposal that absence of all or a large fraction of the mature apoA-I molecule predisposes to the development of premature atherosclerosis. With some variability, other clinical manifestations commonly encountered include xanthomatosis and corneal opacification. However, deletions or frameshifts affecting a smaller portion of the molecule, while perhaps still associated with corneal opacification, very low HDL levels, and decreased LCAT activity, do not appear to confer any increased risk of premature atherosclerosis. Interestingly, some of the apoA-I mutants that appear to associate poorly with HDL particles are not associated with increased risk of atherosclerosis. These observations raise a number of questions regarding the mechanisms by which apoA-I functions to inhibit the atherogenic process and suggest that some protective functions may be carried out independently of its association with HDL particles.

## 1. Genetically engineered mice and HDL metabolism

Additional insights into the physiologic roles of apoA-I and other HDL-associated apolipoproteins have derived from transgenic animal experiments, which have also provided a powerful tool in the study of the relationship between HDL and atherosclerosis [see Breslow (1993) for review].

In mice, HDL represents the major lipoprotein species. In contrast to human HDL, which in the ultracentrifuge sediments by density as two major subpopulations, $HDL_2$ and $HDL_3$, mouse HDL sediments as a unimodal population that is intermediate in size and density. Transgenic mice have been produced that express high levels of human apoA-I. These mice exhibit correspondingly high HDL cholesterol levels, and unexpectedly, very low levels of mouse apoA-I. The reason for the markedly decreased mouse apoA-I levels could not be accounted for on the basis of downregulation of transcription of the mouse gene, and the mechanism responsible for the decrease remains unknown. It is possible that mouse and human apoA-I compete for a limited number of HDL particles. Due to the relatively low molecular weight of apoA-I, displacement of mouse apoA-I by human apoA-I would be expected to result in rapid clearance in the kidney as a result of glomerular filtration. Interestingly, transgenic mice expressing high levels of human apoA-I were also shown to have an HDL profile similar to that seen in humans, with two populations of HDL particles based on size and density. These observations indicate that the amino acid sequence differences between mouse and human apoA-I account for the different size and density properties of mouse and human HDL. These observations suggested that by introducing human genes involved in lipoprotein metabolism into the mouse, it may be possible to "humanize" the murine lipoprotein profile. Not only would this allow us to gain insights into the mechanistic basis for the differences in lipoprotein metabolism between these two species, but an animal model more similar to human metabolism would be obtained. This idea has been further borne out by studies of cholesterol ester transfer protein as discussed below.

It is extremely difficult to produce atherosclerosis in mouse strains by dietary manipulations alone. However, the mouse strain in which the apoA-I transgene was introduced, the C57BL/6, is unusual because when fed an atherogenic diet high in cholesterol, fat, and bile salts, it develops hypercholesterolemia and fatty streaks near the root of the aorta. Although these fatty streaks do not progress to more complex lesions, they provided the major mouse model for atherosclerosis prior to the advent of the gene knockout strategies that led to the development of the apoE and LDL receptor-deficient mice described above. This C57BL/6 strain, therefore, provided a testing ground for examining the influence of apoA-I levels on fatty streak formation. Nontransgenic control mice showed the predicted fall in HDL cholesterol on feeding a high fat diet accompanied by a marked increase in non-HDL cholesterol (presumably made up of VLDL and LDL). In contrast, human apoA-I transgenic mice exhibited a markedly increased level of apoA-I due to high levels of human apoA-I (with a suppressed level of mouse apoA-I) and correspondingly high HDL cholesterol levels. Histopathologic analysis revealed that the transgenic mice had dramatically less aortic area involved with fatty streaks than controls, consistent with a protective effect mediated by the transgene and the elevated HDL that resulted. A mouse line has also been produced in which the apoA-I gene has been knocked out by homologous recombination. This mouse has no detectable apoA-I in its plasma and very low HDL levels. Intriguingly, these mice do not develop atherosclerosis (Li et al., 1993). This observation may reflect the fact that HDL is the major carrier of cholesterol in the mouse, and that the increased propensity for atherosclerosis associated with low HDL levels may be offset by low plasma cholesterol levels. It should be possible to test this possibility by crossing apoA-I knockout mice with other strains of genetically engineered mice that have high levels of LDL and/or VLDL, such as the LDL receptor or apoE knockout mice.

Apolipoprotein A-II transgenic mice have been produced which overexpress apoA-II using both human and mouse apoA-II. Mice transgenic for human apoA-II produced no consistent change in HDL levels (or LDL or VLDL levels) or suppression of plasma levels of mouse apoA-I or mouse apoA-II. The mice expressing the highest levels of human apoA-II, however, did exhibit a subclass of HDL particles which contained only human apoA-II and which was hypothesized to reflect a high stoichiometric ratio of apoA-II to apoA-I. Furthermore, mice transgenic for both human apoA-I and human apoA-II were produced by crossing the human apoA-I and human apoA-II transgenic mice. Mice transgenic for both human apolipoproteins appeared to exhibit populations of

HDL particles that did not simply represent additive contributions from the populations of particles found in the parent strains. An increase in apoA-II was found across all density fractions, suggesting complex interactions between these apolipoproteins in determining HDL structure. As noted above, expression of human apoA-I decreased the extent of atherosclerosis in cholesterol-fed mice. However, evaluation of the aortas from animals expressing both human apoA-II and apoA-I revealed that the protective effect imparted by expression of the human apoA-I transgene alone was reduced (Schultz *et al.*, 1992). These data suggest that HDL particles containing human apoA-I and apoA-II are not as effective as apoA-I-only particles in protecting against atherosclerosis. Indeed, some investigators have found that the ability of apoA-I particles to remove cholesterol from cells in culture is diminished by addition of apoA-II. Further evidence to support a negative role of apoA-II comes from the study of other transgenic mice that have been produced which express high levels of *mouse* apoA-II (Warden *et al.*, 1993; Hedrick *et al.*, 1993). These mice exhibited a lipoprotein profile different than that seen in the human apoA-II transgenic mice, with elevations of HDL, LDL, and VLDL. The HDL from the transgenic mice showed a shift to larger diameter particles and were relatively depleted in apoE compared to control mice. The transgenic mice also produced a particle found in the LDL density range which has apoA-II and apoE but no apoB and which is present in only trace amounts in control mice; there was no evidence of such a particle in mice transgenic for human apoA-II. Histopathologic analysis revealed that the apoA-II overexpressors had a much greater degree of fatty streak involvement than controls; lesion development was augmented by feeding an atherogenic diet in both groups, but total area involved was greater in the transgenic group. These studies provide insight into possible roles for apoA-II in influencing HDL functional parameters and atherosclerosis, although as noted above, species differences between human and mouse apoA-II warrant exercising caution in interpreting these data and indicate a need for further research.

## 2. Cholesterol ester transfer protein

In performing its task of reverse cholesterol transport, HDL must ultimately deliver cholesterol obtained from peripheral tissues to the liver for excretion. As discussed in our introduction to HDL metabolism, delivery of HDL cholesterol to the liver has been proposed to be achieved through the uptake of intact HDL particles, selective uptake of HDL cholesterol esters independent of particle uptake, or by exchange of cholesterol esters with other lipoprotein species through the actions of cholesterol ester transfer protein. The relative quantitative contributions of these three processes to the metabolism of HDL cholesterol esters have been difficult to determine *in vivo*. Furthermore, it has not been clear, until recently, whether the actions of CETP would be beneficial with respect to risk of atherosclerosis, because it would act to decrease the ratio of HDL cholesterol to LDL and VLDL cholesterol. Insight into the role of CETP in HDL metabolism and atherosclerosis risk came from studies of Japanese families noted to have defects in CETP activity and very high HDL levels [see Inazu *et al.* (1990) for review]. Molecular analysis revealed a point mutation in the CETP gene that prevents normal RNA processing. In the homozygous state, no cholesteryl ester transfer protein is expressed. Heterozygotes have CETP and HDL levels intermediate between homozygous patients and normals. Patients heterozygous or homozygous for CETP deficiency appear to have high HDL cholesterol levels because of decreased transfer of HDL cholesterol esters to other lipoprotein fractions. Although the number of patients with CETP deficiency is small, risk of atherosclerosis appears to be low. On this basis, it was considered that CETP may, in fact, participate in a relatively proatherogenic state, and its absence may impart protection from atherosclerosis.

Transgenic approaches have been quite informative in examining the role of CETP in HDL metabolism because wild-type mice do not express CETP. This negative background therefore permits relatively unambiguous conclusions to be drawn from experiments in which CETP transgenes have been introduced. Transgenic mice expressing high levels of human CETP exhibit markedly reduced levels of HDL. This is complementary to the observations made in human CETP deficiency and consistent with the proposal that CETP plays a role in controlling HDL levels. Furthermore, studies have shown that there is an apparent preference for human apoA-I in mediating this effect, as this effect was shown to be much more marked in mice transgenic for both human CETP and human apoA-I than in mice transgenic for just human CETP and in which the only apoA-I was murine. Transgenic mice have also been produced using another primate CETP source, the cynomolgus monkey. In these mice, expression of the CETP transgene showed a dose-dependent decrease in HDL and apoA-I and a corresponding dose-dependent increase in VLDL and LDL and apoB compared to nontransgenic controls. These mice were fed an atherogenic diet and analyzed for development of aortic fatty streak lesions. It was seen that CETP transgenic mice showed a dose-dependent increase in lesion area which increased over time; lesion area was shown to increase as a function of the ratio of the sum of VLDL and LDL levels to HDL. These data are also complementary to the human data in demonstrating an inverse relationship between CETP activity and HDL levels and suggesting a direct correlation between CETP activity level and risk of atherosclerosis.

Through crossing three independent transgenic mouse lines, it has been possible to develop mice that coexpress three human genes: apoA-I, apoC-III, and CETP. The lipoprotein profile in these mice mimics a lipoprotein profile in humans strongly associated with risk for development of atherosclerosis, with elevated triglycerides in the VLDL fraction, and depressed HDL cholesterol levels (Hayek *et al.*, 1993). Apolipoprotein C-III is an apolipoprotein associated with chylomicrons, VLDL, and HDL. It is encoded by the gene within the human chromosome 11 region that contains the apoA-I/apoC-III/apoA-IV gene cluster. Apolipoprotein C-III levels *in vivo* correlate with triglyceride levels, and a mutation in the human gene identified by restriction fragment length polymorphism has been associated with hypertriglyceridemia. Mice transgenic for human apoC-III alone are hypertriglyceridemic, with a marked increase in VLDL triglyceride and a corresponding decrease in HDL cholesterol. The VLDL in these mice is enriched in apoC-III and relatively depleted in apoE. *In vitro* studies demonstrated a slower rate of uptake of these VLDLs by hepatocytes in culture. Interestingly, VLDLs from these animals were as good a substrate for lipoprotein lipase as normal VLDLs. Furthermore, lipoprotein lipase activity in the transgenic animals was similar to that in controls, lending no support to a hypothesis that apoC-III may contribute to hypertriglyceridemia in part by inhibiting lipolysis. *In vivo* studies showed that the fractional catabolic rate for VLDL triglyceride decreased in a dose-dependent manner in relation to apoC-III expression, and that VLDL levels were inversely correlated with fractional catabolic rate. It was hypothesized that this slower rate of uptake by hepatocytes and slower *in vivo* clearance may be related to a lower association of apoE with these particles, and, thus, a lower rate of uptake via the LDL receptor or LRP. Animals transgenic for both human CETP and apoA-I were described above and were noted to have low HDL levels. To produce the animals transgenic for human apoA-I, apoC-III, and CETP, animals were made transgenic for human apoA-I and apoC-III incorporating both genes on a single DNA fragment and thus in equal copy numbers. These mice were then bred with human CETP transgenic mice to produce mice transgenic for all three human proteins (Hayek *et al.* 1993). The human apoA-I/apoC-III/CETP transgenic mice had markedly decreased HDL levels, below the levels seen in animals transgenic for human apoC-III alone. The size distribution of HDL was also markedly altered compared to controls, even though the tritransgenic animals were hypertriglyceridemic to an extent similar to that seen in animals transgenic for apoC-III alone. The lipoprotein profile seen in the apoA-I/apoC-III/CETP transgenic mice, therefore, is similar to a well-recognized high-risk profile recognized in humans. Furthermore, this profile is recognized with increased frequency in diabetic patients and may contribute to the increased risk of cardiovascular disease in this patient population. Whether or not these mice develop atherosclerosis has not yet been published, but it is predicted that this mouse model may prove very useful in the study of atherogenesis in the setting of this dyslipidemia.

    In summary, these observations demonstrate the utility of transgenic models in the study of complex physiologic problems that are relevant to lipoprotein metabolism in humans. These approaches have had a tremendous impact on our understanding of the physiologic roles of specific apolipoproteins and modifying enzymes in HDL metabolism, particularly apoA-I, apoA-II, and CETP. Although the precise mechanisms remain unclear, firm conclusions can also be drawn regarding a beneficial effect of ApoA-I on the risk of atherosclerosis. Furthermore, the ability to reconstitute in the mouse a particular human lipoprotein phenotype that is multigenic in origin has important implications for the study of many other polygenic disorders. These approaches have begun to allow the study of combinatorial interactions of several specified proteins in the study of atherogenesis and will provide a powerful tool to test hypotheses as well as to test therapies.

## D. Lipoprotein(a) metabolism

Lipoprotein(a) [Lp(a)] was first described in the early 1960s as a lipoprotein-associated antigen variably found in the LDL fraction of human plasma detected with rabbit antibodies raised against human LDL [thus the nomenclature Lp(a) for lipoprotein antigen] [see Utermann (1989), and Scanu et al. (1991) for review]. It was subsequently shown that the antigenic identity resulted from a glycoprotein, apolipoprotein(a) [apo(a), not to be confused with apoA-I, apoA-II, or apoA-IV], covalently coupled with the $apoB_{100}$ of LDL via disulfide linkage. It was shown that humans express a variety of molecular weight isoforms of apo(a) ranging from $M_r$ 300,000 to 700,000 in widely varying concentrations of Lp(a) ranging from undetectable to relatively high levels; isoforms segregate in a codominant Mendelian fashion. Serum levels of Lp(a) have been shown to be genetically determined and to be relatively insensitive to standard dietary and pharmacologic means used to treat hyperlipidemias. Multiple studies since the 1960s have shown that elevated Lp(a) levels constitute a risk factor for atherosclerosis independent of overt hypercholesterolemia and markedly exacerbate the risk of elevated LDL. On the other hand, absence of measurable Lp(a) from plasma is not associated with any known disease.

In 1987, Eaton et al. obtained partial purification of Apo(a) from a single female donor; this revealed a protein of 280,000 $M_r$ (Eaton et al., 1987). Partial amino acid sequence of this protein revealed it to have striking sequence homology with plasminogen, a serine protease which circulates as a zymogen, and which upon activation, catalyzes fibrin clot dissolution. Plasminogen has five domains referred to as "Kringle" domains I–V and a protease domain (the term Kringle derives from the resemblance of the tertiary structure of the Kringle domains with the Scandinavian pastry of the same name). Sequence analysis revealed that apo(a) had several fragments with a high degree of homology with plasminogen Kringle IV and fragments with homology with Kringle V and with the protease domain. It was also shown that apo(a) had no intrinsic protease activity and could not be proteolytically activated to a functional protease, as occurs with plasminogen in response to tissue plasminogen activator and urokinase. The failure to be a substrate for tPA occurs because of a single amino acid substitution in the sequence recognized by these plasminogen-activating proteases. Based on the high degree of homology with Kringle IV of plasminogen, McLean et al. used a probe derived from the known sequence of plasminogen Kringle IV to screen a human liver cDNA library; this identified a cDNA sequence of ~14,000 bp including 5′ and 3′ untranslated regions and encoding a mature protein of 4529 amino acids and a 19-amino acid signal sequence (McLean et al., 1987). Surprisingly, the apo(a) cDNA sequence revealed 37 contiguous repeats of the Kringle IV homology region with extremely high intramolecular sequence fidelity constituting most of the molecule from the 5′ (N-terminal) direction followed by a single copy of the Kringle V domain and lastly the protease domain. The fact that the entire cDNA sequence revealed homology with three domains of plasminogen led these researchers to propose that apo(a) arose by duplication of the plasminogen gene. Subsequent studies have demonstrated that the apo(a) gene and the plasminogen gene comap to a gene cluster on the long arm of human chromosome 6. The molecular size variation between alleles of apo(a) has been shown to most likely result from variation in the interallelic number of Kringle IV repeats.

While it is clear that our understanding of Lp(a) has been dramatically advanced by the contribution from molecular biology, the biological role of Lp(a)/apo(a) and the mechanism relating elevations in Lp(a) to atherogenesis remain unknown. In view of the fact that absence of measurable Lp(a) in plasma is correlated with no known adverse outcome, it may be that there is no essential role for Lp(a)/apo(a). Such conclusions must be drawn with caution, however, and this issue must be considered unresolved at this time. With regard to atherogenesis, the homology between apo(a) and plasminogen is very intriguing and provocative. This has led to studies which have demonstrated that Lp(a) can bind to fibrin and fibrinogen, can compete with plasminogen for tissue plasminogen activator [while, as noted above, tPA cannot activate apo(a)], and can compete with plasminogen for binding to endothelium and platelets; based on these data, it has been proposed that Lp(a) may impede the fibrinolytic function of plasmin, perhaps increasing the risk of thrombosis with coronary artery occlusion. Regarding a direct proatherogenic effect, uncertainty remains regarding the in vivo metabolism of Lp(a). In vitro, while conflicting data exist, Lp(a) appears to be ligand for the LDL receptor, although isoforms from various patients appear to compete with LDL for receptor binding with differing affinities (Hofmann et al., 1988). Therefore, in vivo metabolism of Lp(a) via LDL receptors is tenable (see below). Alternative pathways which could promote enhanced macrophage uptake, including the possibility that Lp(a) is taken up via the scavenger receptor, have also begun to be investigated. Native Lp(a) is not avidly taken up by the macrophage scavenger receptor; however, in vitro modifications including malondialdehyde conjugation (Scanu et al., 1991;

Haberland *et al.*, 1987) and oxidative modification (Naruszewicz *et al.*, 1992) greatly enhance uptake via the scavenger receptor in a manner similar to that seen with LDL modified by like means.

Transgenic animal experiments have recently begun to aid in clarification of fundamental questions regarding Lp(a). Expression of apo(a) appears to be relatively restricted to several primate species; good animal models have been lacking. Transgenic mice have been produced which express a human apo(a) isoform of $M_r$ 550,000 and achieve plasma levels of apo(a) comparable to the median values found in Caucasians (Lawn *et al.*, 1992). Caveats encountered in these transgenic animals were that most of the plasma apo(a) was not associated with lipoproteins, and that apo(a) apparently does not associate with mouse LDL but does with human LDL. When these mice were fed an atherogenic diet, they were found to have significantly more atherosclerotic lesions covering a greater area and of greater severity than seen in control mice fed the atherogenic diet or in transgenic animals fed the normal chow diet. Immunostaining the aorta with antibody to human apo(a) stained lesioned areas but not uninvolved areas; staining appeared to be associated with macrophage foam cells. This study raises questions regarding the dose–response relationship between apo(a) expression and atherogenesis, and also regarding the potential roles of free apo(a) versus lipoprotein-bound apo(a) [i.e., Lp(a)] in atherogenesis which will require further research to resolve. It does show, however, that expression of the human transgene is associated with development of early atherosclerotic lesions and provides the basis for development of animal models to study this system. *In vivo* metabolism of Lp(a) has been examined in transgenic mice which overexpress the human LDL receptor (Hofmann *et al.*, 1988). It was shown that control mice cleared both LDL and Lp(a) slowly; in the transgenic animals, the clearance of both ligands was markedly accelerated and to a similar degree. It was concluded that the LDL receptor may play a significant role in the *in vivo* metabolism of Lp(a). Recently, a transgenic mouse expressing human apoB has been developed and crossed with the transgenic mouse expressing apo(a). Unlike mice expressing apo(a) alone, the double transgenic mice contain floating Lp(a) particles analogous to those found in humans. These mice should provide a particularly useful model in studying Lp(a) metabolism and its role in atherosclerosis.

## E. Lipoprotein modifications affecting atherogenesis

A significant body of evidence suggests that oxidation of LDL is a central event in the initiation and progression of the atherogenic process (reviewed in Steinberg *et al.*, 1989; Witztum *et al.*, 1991). *In vitro* studies have demonstrated that oxidation of LDL converts it to a form that is taken up by scavenger receptors that are preferentially or exclusively expressed on macrophages (Parthasarathy *et al.*, 1986). Unlike LDL receptors, these so-called "scavenger receptors" are not downregulated by high levels of intracellular cholesterol and can therefore mediate massive levels of cholesterol accumulation (Goldstein *et al.*, 1979). Furthermore, oxidized LDL appears to be capable of modulating the expression of genes involved in the recruitment of circulating monocytes to sites of lesion formation, such as MCP-1 (Cushing *et al.*, 1990). Immunologic and biochemical studies have demonstrated the presence of oxidized LDL within human, rabbit, and murine lesions (Yla-Herttuala *et al.*, 1989; Palinski *et al.*, 1994). Furthermore, treatment of hypercholesterolemic rabbits with the antioxidant probucol has been demonstrated to reduce the rate of atherosclerotic lesion development (Carew *et al.*, 1987; Kita *et al.*, 1987). These observations imply that an understanding of the pathogenesis of atherosclerosis will require the identification and characterization of the genes involved in promoting or preventing the oxidation and uptake of LDL.

The oxidation hypothesis suggests that oxidation of LDL occurs in the extracellular space of the artery wall, presumably in microdomains adjacent to cells where the LDL particle is relatively inaccessible to water-soluble antioxidants. In cell culture, endothelial cells, smooth muscle cells, and macrophages have all been demonstrated to be capable of mediating cell-dependent oxidation, with macrophages being particularly active. Undoubtedly, different mechanisms are involved with each cell type, and it is likely that even a single cell type, such as the monocyte/macrophage, has more than one pathway capable of initiating oxidation. Oxidants and oxidant-generating systems that have been proposed to contribute to LDL modification include superoxide anion (Heinecke *et al.*, 1984), the combination of hydrogen peroxide plus a peroxidase activity (Wieland *et al.*, 1993), and 15 lipoxygenase (Parthasarathy *et al.*, 1989). It is important to note, however, that there is little evidence to firmly support any of these mechanisms *in vivo*. Molecular approaches are just now being applied to the problem of determining what roles these specific enzyme systems may play in promoting modifications of LDL in the artery wall. By analogy to the uses of transgenic and knockout technologies in the study of the function of the LDL receptor and apoE, efforts to overexpress or delete specific oxidant-generating systems are likely to provide new insights into how these systems contribute to the atherogenic process. Because many oxidant generating systems may contribute to LDL modifications, it may be necessary to study combinations of gene knockouts or transgenes, similar to the approaches required to determine the mechanisms involved in the clearance of triglyceride-rich lipoproteins.

## F. The macrophage scavenger receptors

As noted in the introduction, in 1979 Goldstein and Brown and their colleagues demonstrated that chemically modified LDL was taken up by macrophages much more rapidly than native LDL, and that this uptake could lead to cholesterol accumulation and the development of macrophage foam cells. They termed the macrophage receptor which mediated this uptake the "scavenger receptor," based on its ability to take up a wide variety of ligands whose only common feature appeared to be a polyanionic character, and they hypothesized its relationship to atherogenesis [see Brown *et al.* (1983) for review]. Work throughout the 1980s has led to the description of a natural ligand for this receptor— oxidized LDL—and has presented a wealth of data both *in vitro* and *in vivo* consistent with the importance of oxidized LDL to the process of atherogenesis [see Steinberg *et al.* (1989) and Witztum *et al.* (1991) for review]. In 1988, the scavenger receptor was purified to homogeneity from bovine liver membranes (from the Kupfer cell component) and bovine lung (from the alveolar macrophage component) using ligand affinity chromatography followed by monoclonal antibody immunoaffinity purification [see Krieger (1992) for review]. This yielded a 220-kDa homotrimeric integral membrane glycoprotein which yielded three 77-kDa monomers upon disulfide reduction.

Study of the scavenger receptor pathway was ushered into the molecular biology era with the publication in 1990 of the cloned sequences of two related cDNAs encoding variant forms of the bovine scavenger receptor (dubbed type I and type II scavenger receptors) [Kodama *et al.* 1990; Rohrer *et al.* 1990; see Krieger (1992) for review]. Based on partial amino acid sequence of the purified receptor, oligonucleotide probes were synthesized and used to screen a bovine lung cDNA library. Two cDNAs were isolated, encoding proteins of 453 amino acids (type I receptor) and 349 amino acids (type II receptor) corresponding to the receptor monomers. Sequence analysis revealed that these proteins differed only in their carboxy termini (Figure 5.6). These proteins are integral membrane proteins with a single transmembrane domain and with the N-terminus oriented cytoplasmically and the C-terminus oriented extracellularly. Both receptor isoforms are broken down into structural/functional domains. They share identical domains I–V (numbered from the cytoplasmic N-termini) and differ only in their C-termini. Domain I is the N-terminal domain and consists of the cytoplasmic N-terminal 50 amino acids. Domain II is the membrane-spanning domain and consists of 26 amino acids. Proceeding in the C-terminal direction, domain III is the first extracellular domain and consists of 32 amino acids with two asparagine residues constituting sites of N-linked glycosylation. Domain IV consists of 163 amino acids with 23 seven amino acid repeats which in the native (trimeric) receptors form an amphipathic α-helical-coiled coil; there are five sites of N-linked glycosylation in domain IV. The next and last domain common to both receptor isoforms is domain V. This domain has 72 amino acids arranged as sequential trimers characteristic of collagen and, therefore, in the native (trimeric) receptor is predicted to assemble into a collagen-like triple helix. Domain VI in the type I scavenger receptor consists of the C-terminal 110

Type I                                    Type II

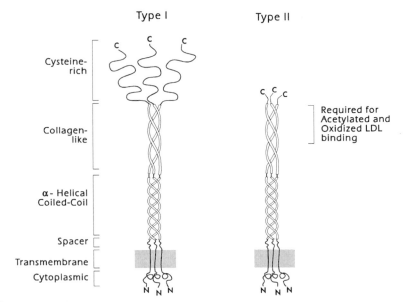

**Figure 5.6.** Predicted structure of the type I and type II macrophage scavenger receptors. Both receptors are derived from a common primary transcript that is alternatively spliced to yield two distinct mRNAs. These mRNAs encode proteins that are distinguished by the presence or absence of a C-terminal cysteine-rich domain of unknown function. The N-terminal regions of the type I and II scavenger receptors, which encode the cytoplasmic, transmembrane spaces, α-helical-coiled coil, and collagen-like domains, are identical. Oxidized and acetylated LDL interact with both forms of the receptor via amino acid residues present in the collagen-like domains.

amino acids and contains six cysteine residues; domain VI in the type II receptor consists of the C-terminal 6 amino acids and contains no cysteines. Subsequently, highly conserved cDNAs encoding type I and type II scavenger receptors have been cloned from mouse, human, and rabbit. In 1990, Freeman *et al.* (1990) used PCR to isolate type I and type II scavenger receptor cDNAs from the mouse macrophage cell line, P388D$_1$. They showed that the two mRNAs are simultaneously expressed in these cells. Using mice with abundant genetic polymorphisms generated by a cross between two different strains of mice, they mapped the type I and type II loci using RFLP analysis and probes specific for the two receptor types which could distinguish between the two parent strains of mice on the basis of different RFLPs. With breeding, the strain-specific scavenge receptor loci of both receptor types always cosegregated. It thus appears that there is one scavenger receptor locus (which they termed the *Scvr* locus), and that the two receptor types arise via alternate RNA splicing to give rise to the two different mRNAs. This was confirmed 3 years later when Emi *et al.* reported the cloning of the human scavenger receptor gene from a human genomic DNA library (Emi *et al.*, 1993). The human gene is about 80 kb in length and consists of 11 exons interspersed by 10 introns. Functional domains of the type I and type II scavenger receptors were found to correlate with exon organization with exons 1 through 8 being common to both receptor isoforms. It was confirmed that the mRNAs encoding the two receptor isoforms were indeed generated by alternative splicing of the primary transcript, with the 3' end of the mRNA being coded for by exons 10 and 11 in the type I receptor and by exon 9 in the type II receptor. They also reported mapping the gene to the short arm of human chromosome 8.

The high-affinity binding of a wide array of ligands is shared by both the type I and the type II receptors. From this observation, it has been concluded that domain VI (the C-terminal-most domain) does not mediate or determine specificity of ligand binding and that, rather, binding is mediated by domain V, the collagen-like triple helix. While the molecular basis for binding of the wide array of ligands which bind to the scavenger receptor with high affinity is not completely understood, it has been proposed that the scavenger receptor functions as a sort of "molecular flypaper" with neutral and cationic amino acids in domain V (the collagen-like domain) forming the basis for high-affinity binding to a diverse set of polyanionic ligands (Krieger, 1992).

The type I and II scavenger receptors are exclusively expressed in cells of the monocyte macrophage lineage (Wu *et al.*, 1992), and therefore provide a model for investigating the molecular mechanisms responsible for macrophage-specific gene expression. Promoter and enhancer elements that direct the expression of the scavenger receptor have recently been cloned and characterized (Wu *et al.*, 1994; Moulton *et al.*, 1994). Positive regulation is dependent on the interactions of several ets-domain transcription factors and members of the AP-1 gene family. These transcription factors are likely to regulate a diverse set of genes required for the development and functional competence of the macrophage. Because the scavenger receptor is selectively expressed in macrophage-like cells, the scavenger-receptor promoter and enhancer elements may be useful in targeting the expression of other genes to macrophages in atherosclerotic lesions of transgenic mice to assess their activities in promoting or retarding lesion development. On the basis of *in vitro* ligand crosscompetition experiments, it has been proposed that macrophages express other receptors which can mediate binding and uptake of oxidized LDL in addition to the type I and type II scavenger receptors. In 1989, Sparrow *et al.* reported that macrophages express a receptor which recognizes oxidized LDL but not acetyl LDL based on the ability of oxidized LDL to essentially quantitatively compete for acetyl LDL binding and uptake while there was a component of oxidized LDL binding and uptake which could not be competed for by acetyl LDL. These results imply the existence of additional receptors that are capable of binding and internalizing oxidized LDL, but not acetylated LDL. Consistent with this hypothesis, ligand-blotting experiments have led to the identification of a macrophage-specific membrane protein of 94-kDa $M_r$ that binds oxidized LDL, but not acetylated LDL (D. Steinberg, personal communication). A similar protein has also been identified recently in Kupffer cells (de Rijke et al., 1994). In addition, expression-cloning approaches resulted in the isolation of a cDNA clone from a murine macrophage library that encoded a protein capable of binding oxidized LDL (Endemann et al., 1993). Sequence analysis of this clone revealed that it corresponded to CD36, an 88-kDa myeloid-specific membrane protein previously demonstrated to interact with thrombospondin. Due to the relatively similar molecular weights, it is possible that CD36 corresponds to the 94-kDa protein identified by ligand-blotting experiments from mouse peritoneal macrophages. Oxidized LDL has also been demonstrated to be highly immunogenic. Autoantibodies capable of forming immune complexes with oxidized LDL have been demonstrated in atherosclerotic lesions of humans, rabbits, and mice (Yla-Herttuala et al., 1994a,b; Palinski et al., 1994). These observations suggest that oxidized LDL could be taken up by macrophages as part of immune complexes, e.g., via Fc receptors. It is clear that a combination of transgenic and gene knockout experiments will be required to determine the relative importance of these various receptors in the development of foam cells in vivo.

## IV. SUMMARY AND FUTURE DIRECTIONS

Molecular biology has revolutionized our understanding of lipid metabolism and atherosclerosis. The cloning of cDNAs encoding the structural, enzymatic, and regulatory proteins that mediate the synthesis, transport, and catabolism of lipids has permitted a detailed analysis of the molecular basis for many of the steps involved in these processes. The isolation of genomic clones encoding these proteins has led to the definition of the promoter sequences and transcription factors that determine their patterns and levels of expression. The combination of molecular cloning techniques with mouse embryology and genetics has further permitted the development of transgenic and gene knockout approaches to the study of gene function in vivo. These approaches have already had a dramatic impact on the evaluation of specific gene products and the development of mouse models of atherosclerosis.

Although significant inroads have been made in the treatment of hypercholesterolemia by the development of drugs that inhibit enzymes involved in cholesterol synthesis or block the resorbtion of bile acids, the application of molecular approaches to the study of lipid metabolism has suggested a number of new approaches to pharmacologic intervention. In particular, the study of the mechanisms that control the transcription rates of genes involved in cholesterol metabolism may permit the development of drugs that influence their expression. For example, drugs that block the effect of cholesterol on the proteolysis of SREBP would appear to be of potential benefit. Finally, molecular biology represents the cornerstone of the nascent science of gene therapy, an approach that is currently envisioned primarily for the treatment of otherwise untreatable conditions, such as homozygous FH, but which may take on broader applications as technology is improved. For example, based on the observation that overexpression of apoA-I in mice inhibits the development of atherosclerosis, gene therapy approaches that increased the expression of apoA-I in humans with polygenic causes for premature atherosclerosis might be of benefit.

Many genes involved in the atherogenic process remain to be clearly identified. For example, the specific enzyme systems that generate prooxidant (or antioxidant) molecules involved in the modification of LDL *in vivo* have not been established. Similarly, the relative roles of the macrophage scavenger receptor and other potential receptors for modified LDL in foam cell formation remain to be clearly defined. Although not the topic of this review, a great deal of effort in the atherosclerosis field is now focused on the biology of the artery wall and how this biology is perturbed by the presence of foam cells and T cells that elaborate a diverse array of cytokines, growth factors, and other hormone-like molecules. Many of these proteins also represent potential targets for therapeutic intervention.

The complications of atherosclerosis continue to be the leading causes of death in Western societies. The advances in our understanding of the mechanisms that underlie the development of this disease give reason for optimism that new and increasingly effective therapeutic strategies can be developed to reduce this toll. Although achieving these goals will require a substantial investment in fundamental and applied research, the recent history of medical progress leaves no doubt that this investment will pay lasting dividends.

## Acknowledgments

We are grateful to Dr. Joseph Witztum for helpful suggestions and to Margarita Reyes for assistance with the preparation of the manuscript. T.P.K. is supported by an American Heart Association Bugher Foundation Fellowship.

## References

Ashcom, J. D., Tiller, S. E., Dickerson, K., Cravens, J. L., Argraves, W. S., and Strickland, D. K. (1990). The human alpha2-macroglobulin receptor: Identification of a 420-kd cell surface glycoprotein specific for the activated conformation of alpha2-macroglobulin. *J. Cell Biol.* 110:1041–1048.

Assmann, G., von Eckardstein, A., and Funke, H. (1993). High density lipoproteins, reverse transport of cholesterol, and coronary artery disease. *Circulation* 87 (**Suppl. III**):III-28–III-34.

Beisiegel, U., Weber, W., Ihrke, G., Herz, J., and Stanley, K. K. (1989). The LDL-receptor-related protein, LRP, is an apolipoprotein E-binding protein. *Nature* **341**:162–164.

Beisiegel, U., Weber, W., and Bengtsson-Olivecrona, G. (1991). Lipoprotein lipase enhances the binding of chylomicrons to low density lipoprotein receptor-related protein. *Proc. Natl. Acad. Sci. USA* **88**:8342–8346.

Breslow, J. L. (1985). Human apolipoprotein molecular biology and genetic variation. *Annu. Rev. Biochem.* **54**:699–727.

Breslow, J. L. (1989). Familial disorders of high density lipoprotein metabolism. In "Metabolic Basis of Inherited Disease" (C. R. Scriver, A. L. Beaudet, W. S. Sly, and D. Valle, Eds.) pp. 1251–1266. McGraw–Hill, New York.

Breslow, J. L. (1993). Transgenic mouse models of lipoprotein metabolism and atherosclerosis. *Proc. Natl. Acad. Sci. USA* **90**:8314–8318.

Briggs, M. R., Yokoyama, C., Wang, X., Brown, M. S., and Goldstein, J. L. (1993). Nuclear protein that binds sterol regulatory element of low density lipoprotein receptor promoter. I. Identification of the protein and delineation of its target nucleotide sequence. *J. Biol. Chem.* **268**:14490–14496.

Brown, M. S., and Goldstein, J. L. (1983). Lipoprotein metabolism in the macrophage: Implications for cholesterol deposition in atherosclerosis. *Annu. Rev. Biochem.* **52**:223–261.

Brown, M. S., and Goldstein, J. L. (1986). A receptor-mediated pathway for cholesterol homeostasis. *Science* **232**:34–47.

Capecchi, M. R. (1989). Altering the genome by homologous recombination. *Science* **244**:1288–1292.

Carew, T. E., Schwenke, D. C., and Steinberg, D. (1987). Antiatherogenic effect of probucol unrelated to its hypocholesterolemic effect: Evidence that antioxidants *in vivo* can selectively inhibit low density lipoprotein degradation in macrophage-rich fatty streaks and slow the progression of atherosclerosis in the Wantanabe heritable hyperlipidemic rabbit. *Proc. Natl. Acad. Sci. USA* **84**:7725–7729.

Chappell, D. A., Fry, G. L., Waknitz, M. A., Iverius, P., Williams, S. E., and Strickland, D. K. (1992). The low density lipoprotein receptor-related protein/alpha2-macroglobulin receptor binds and mediates catabolism of bovine milk lipoprotein lipase *J. Biol. Chem.* **267**:25764–25767.

Chappell, D. A., Fry, G. L., Waknitz, M. A., Muhonen, L. E., Pladet, M. W., Iverius, P. H., and Strickland, D. K. (1993). Lipoprotein lipase induces catabolism of normal triglyceride-rich lipoproteins via the low density lipoprotein receptor-related protein/alpha2-macroglobulin receptor in vitro. *J. Biol. Chem.* **268**:14168–14175.

Choi, S., and Cooper, A. D. (1993). A comparison of the roles of low density lipoprotein (LDL) receptor and the LDL receptor-related protein/alpha2-macroglobulin receptor in chylomicron remnant removal in the mouse in vitro. *J. Biol. Chem.* **268**:15804–15811.

Chowdhury, J. R., Grossman, M., Gupta, S., Chowdhury, N. R., Baker, J. R., and Wilson, J. M. (1991). Long-term improvement of hypercholesterolemia after ex vivo gene therapy in LDLR-deficient rabbits. *Science* **254**:1802–1805.

Cooper, A. D. (1992). Hepatic clearance of plasma chylomicrons remnants. *Sem. Liv. Dis.* **12**:386–396.

Cushing, S. D., Berliner, J. A., Valente, A. J., Territo, M. C., Navab, M., Parhami, F., Gerrity, R., Schwartz, C. J., and Fogelman, A. M. (1990). Minimally modified low density lipoprotein induces monocyte chemotactic protein 1 in human endothelial cells and smooth muscle cells. *Proc. Natl. Acad. Sci. USA* **87**:5134.

Cybulsky, M. I., and Gimbrone, M. A., Jr. (1991). Endothelial expression of a mononuclear leukocyte adhesion molecule during atherogenesis. *Science* **251**:788–791.

Davidson, N. O. (1993). Apolipoprotein B mRNA editing: A key controlling element targeting fats to proper tissue. *Ann. Med.* **25**:539–543.

de Rijke, Y. B., and Van Berkel, T. J. C. (1994). Rat liver Kupffer and endothelial cells express different binding proteins for modified low density lipoproteins. *J. Biol. Chem.* **269**:824–827.

Eaton, D. L., Fless, G. M., Kohr, W. J., McLean, J. W., Xu, Q. T., Miller, C. G., Lawn, R. M., and Scanu, A. M. (1987). Partial amino acid sequence of apolipoprotein(a) shows that it is homologous to plasminogen. *Proc. Natl. Acad. Sci. USA* **84**:3224–3228.

Emi, M., Asaoka, H., Matsumoto, A., Itakura, H., Kurihara, Y., Wada, Y., Kanamori, H., Yazaki, Y., Takahashi, E., Lepert, M., Lalouel, J., Kodama, T., and Mukai, T. (1993). Structure, organization, and chromosomal mapping of the human macrophage scavenger receptor gene. *J. Biol. Chem.* **268**:2120–2125.

Endemann, G., Stanton, L. W., Madden, K. S., Bryant, C. M., White, R. T., and Protter, A. A. (1993). CD36 is a receptor for oxidized low density lipoprotein. *J. Biol. Chem.* **268**:11811–11816.

Fazio, S., Lee, Y., Ji, Z., and Rall, S. C. (1993). Type III hyperlipoproteinemic phenotype in transgenic mice expressing dysfunctional apolipoprotein E. *J. Clin. Invest.* **92**:1497–1503.

Freeman, M., Ashkenas, J., Rees, D. J., Kingsley, D. M., Copeland, N. G., Jenkins, N. A., and Krieger, M. (1990). An ancient, highly conserved family of cysteine-rich protein domains revealed by cloning type I and type II murine macrophage scavenger receptors. *Proc. Natl. Acad. Sci. USA* **87**:8810–8814.

Friedmann, T. (1992). A brief history of gene therapy. *Nature Genet.* **2**:93–98.

Gilbert, W. (1978). Why genes in pieces? *Nature* **271**:501.

Glomset, J. A. (1968). The plasma lecithin: Cholesterol acyltransferase reation. *J. Lipid Res.* **9**:155–167.

Goldstein, J. L., and Brown, M. S. (1977). The low-density lipoprotein pathway and its relation to atherosclerosis. *Annu. Rev. Biochem.* **46**:897–930.

Goldstein, J. L., Ho, Y. K., Basu, S. K., and Brown, M. S. (1979). Binding site on macrophages that mediates uptake and degradation of acetylated low density lipoprotein, producing massive cholesterol deposition. *Proc. Natl. Acad. Sci. USA* **76**:333–337.

Goldstein, J. L., and Brown, M. S. (1989). Familial hypercholesterolemia. *In* "Metabolic Basis of Inherited Disease" (C. R. Scriver, A. L. Beaudet, W. S. Sly, and D. Valle, Eds.), pp. 1215–1250. McGraw–Hill, New York.

Goldstein, J. L., and Brown, M. S. (1990). Regulation of the mevalonate pathway. *Nature* **343**:425–430.

Grossman, M., Raper, S. E., Kozarsky, K., Stein, E. A., Engelhard, J. F., Muller, D., Lupien, P. J., and Wilson, J. M. (1994). Successful ex vivo therapy directed to liver in a patient with familial hypercholesterolemia. *Nature Genet.* **6**:335–341.

Haberland, M. E., and Fogelman, A. M. (1987). The role of altered lipoproteins in the pathogenesis of atherosclerosis. *Am. Heart J.* **113**:573–577.

Hayek, T., Azrolan, N., Verdery, R. B., Walsh, A., Chajek-Shaul, T., Agellon, L. B., Tall, A. R., and Breslow, J. L. (1993). Hypertriglyceridemia and cholesterol ester transfer protein interact to dramatically alter high density lipoprotein levels, particle sizes, and metabolism. *J. Clin. Invest.* **92**:1143–1152.

Hedrick, C. C., Castellani, L. W., Warden, C. H., Puppione, D. L., and Lusis, A. J. (1993). Influence of mouse apolipoprotein A-II on plasma lipoproteins in transgenic mice. *J. Biol. Chem.* **25**:20676–20682.

Heinecke, J. W., Rosen, H., and Chait, A. (1984). Iron and copper promote modification of low density lipoprotein by human arterial smooth muscle cells in culture. *J. Clin. Invest.* **74**:1890–1894.

Herz, J., Hamann, U., Rogne, S., Myklebost, O., Gausepohl, H., and Stanley, K. K. (1988). Surface location and high affinity for calcium of a 500-kd liver membrane protein closely related to the LDL-receptor suggest a physiological role as lipoprotein receptor. *EMBO J.* **7:**4119–4127.

Herz, J., Goldstein, J. L., Strickland, D. K., Ho, Y. K., and Brown, M. S. (1991). 39-kDa protein modulates binding of ligands to low density lipoprotein receptor-related protein/alpha2-macroglobulin receptor. *J. Biol. Chem.* **266:**21232–21238.

Herz, J., Clouthier, D. E., and Hammer, R. E. (1992). LDL receptor-related protein internalizes and degrades uPA-PAI-1 complexes and is essential for embryo implantation. *Cell* **71:**411–421.

Hobbs, H. H., Russell, D. W., Brown, M. S., and Goldstein, J. L. (1990). The LDL receptor locus in familiar hypercholesterolemia: Mutational analysis of a membrane protein. *Annu. Rev. Genet.* **24:**133–170.

Hodges, P., and Scott, J. (1992). Apolipoprotein B mRNA editing: A new tier for the control of gene expression. *Trends Biochem. Sci.* **17:**77–81.

Hofmann, S. L., Russell, D. W., Brown, M. S., Goldstein, J. L., and Hammer, R. E. (1988). Overexpression of low density lipoprotein (LDL) receptor eliminates LDL from plasma in transgenic mice. *Science* **239:**1277–1281.

Hua, X., Yokoyama, C., Wu, J., Briggs, M. R., Brown, M. S., Goldstein, J. L., and Wang, X. (1993). SREBP-2, a second basic-helix-loop-helix-leucine zipper protein that stimulates transcription by binding to a sterol regulatory element. *Proc. Natl. Acad. Sci. USA* **90:**11603–11607.

Hussain, M. M., Maxfield, F. R., Mas-Olivas, J., Tabas, I., Ji, Z., Innerarity, T. L., and Mahley, R. W. (1991). Clearance of chylomicron remnants by the low density lipoprotein receptor-related protein/alpha2-macroglobulin receptor. *J. Biol. Chem.* **266:**13936–13940.

Inazu, A., Brown, M. L., Hesler, C. B., Agellon, L. B., Kiozumi, J., Takata, K., Maruhama, Y., Mabuchi, H., and Tall, A. R. (1990). Increased high-density lipoprotein levels caused by a common cholesterol-ester transfer protein gene mutation. *N. Engl. J. Med.* **323:**1234–1238.

Ishibashi, S., Brown, M. S. Goldstein, J. L., Gerard, R. D., Hammer, R. E., and Herz, J. (1993). Hypercholesterolemia in low density lipoprotein receptor knockout mice and its reversal by adenovirus-mediated gene delivery. *J. Clin. Invest.* **92:**883–893.

Ishibashi, S., Goldstein, J. L., Brown, M. S., Herz, J., and Burns, D. K. (1994a). Massive xanthomatosis and atherosclerosis in cholesterol-fed low density lipoprotein receptor-negative mice. *J. Clin. Invest.* **93:**1885–1893.

Ishibashi, S., Herz, J., Maeda, N., Goldstein, J. L., and Brown, M. S. (1994b). The two-receptor model of lipoprotein clearance: tests of the hypothesis in "knockout" mice lacking the low density lipoprotein receptor, apolipoprotein E, or both proteins. *Proc. Natl. Acad. Sci. USA,* **91:**4431–4435.

Kahn, J. A., and Glueck, C. J. (1978). Familial hypobetalipoproteinemia. *JAMA* **240:**47–48.

Kerjaschki, D., and Farquhar, M. (1982). The pathogenic antigen of Heymann nephritis is a membrane glycoprotein of the renal proximal tubule brush border. *Proc. Natl. Acad. Sci. USA* **79:**5557–5561.

Khoo, J. C., Mahoney, E. M., and Witztum, J. L. (1981). Secretion of lipoprotein lipase by macrophages in culture. *J. Biol. Chem.* **256:**7105–7108.

Kita, T., Nagano, Y., Yodoke, M., Ishi, K., Kume, N., OOshima, A., Yoshida, H., and Kawai, C. (1987). Probucol prevents the progression of atherosclerosis in Watanabe heritable hyperlipidemic rabbit, an animal model for familial hypercholesterolemia. *Proc. Natl. Acad. USA* **84:**5928–5931.

Kodama, T., Freeman, M., Rohrer, L., Zabrecky, J., Matsudaira, P., and Krieger, M. (1990). Type I macrophage scavenger receptor contains alpha-helical and X collagen-like coiled coils. *Nature* **343:**531–535.

Kounnas, M. Z., Argraves, W. S., and Strickland, D. K. (1992). The 39-kDa receptor-associated protein interacts with two members of the low density lipoprotein receptor family, alpha2-macroglobulin receptor and glycoprotein 330. *J. Biol. Chem.* **267**:21162–21166.

Kowal, R. C., Herz, J., Goldstein, J. L., Esser, V., and Brown, M. S. (1989). Low density lipoprotein receptor-related protein mediates uptake of cholesteryl esters derived from apolipoprotein E-enriched lipoproteins. *Proc. Natl. Acad. Sci. USA* **86**:5810–5814.

Krieger, M. (1992). Molecular flypaper and atherosclerosis: Structure of the macrophage scavenger receptor. *Trends Biochem. Sci.* **17**:141–146.

Kuhn, T. S. (1962). "The Structure of Scientific Revolutions." Univ. of Chicago Press, Chicago.

Lawn, R. M., Wade, D. P., Hammer, R. E., Chiesa, G., Verstuyft, J. G., and Rubin, E. M. (1992). Atherogenesis in transgenic mice expressing human apolipoprotein(a). *Nature* **360**:670–672.

Leiper, J. M., Bayliss, J. D., Pease, R. J., Brett, D. J., Scott, J., and Shoulders, C. C. (1994). Microsomal triglyceride transfer protein, the abetalipoproteinemia gene product, mediates the secretion of apolipoprotein B-containing lipoproteins from heterologous cells. *J. Biol. Chem.* **269**:21951–21954.

Li, H., Reddick, R. L., and Maeda, N. (1993). Lack of apo A-I is not associated with increased susceptibility to atherosclerosis in mice. *Arterio. Thrombo.* **13**:1814–1821.

Mahley, R. W., and Rall, S. C. (1989). Type III hyperlipoproteinemia (Dysbetalipoproteinemia): The role of apolipoprotein E in normal and abnormal lipoprotein metabolism. *In* "Metabolic Basis of Inherited Disease." (C. R. Scriver, J. B. Stanbury, J. B. Wyngaarden, and D. S. Frederickson, Eds.), pp. 1195–1213. McGraw–Hill, New York.

Matsunaga, T., Hiasa, Y., Yanagi, H., Maeda, T., Hattori, N., Yamakawa, K., Yamanouchi, Y., Tanaka, I., Obara, T., and Hamaguchi, H. (1991). Apolipoprotein A-I deficiency due to a codon 84 nonsense mutation of the apolipoprotein A-I gene. *Proc. Natl. Acad. Sci. USA* **88**:2793–2797.

McLean, J. W., Tomlinson, J. E., Kuang, W. J., Eaton, D. L., Chen, E. Y., Fless, G. M., Scanu, A. M., and Lawn, R. M. (1987). cDNA sequence of human apolipoprotein(a) is homologous to plasminogen. *Nature* **330**:132–137.

Moulton, K. S., Semple, K., Wu, H., and Glass, C. K. (1994). Cell-specific expression of the macrophage scavenger receptor gene is dependent on PU.1 and a composite AP-1/ets motif. *Mol. Cell. Biol.* **14**:4408–4418.

Nakashima, Y., Plump, A. S., Raines, E. W., Breslow, J. L., and Ross, R. (1994). Apo E-deficient mice develop lesions of all phases of atherosclerosis throughout the arterial tree. *Arterio. Thrombo.* **14**:133–140.

Naruszewicz, M., Selinger, E., and Davignon, J. (1992). Oxidative modification of lipoprotein(a) and the effect of beta-carotene. *Metabolism* **41**:1215–1224.

Osborne, T. (1991). Single nucleotide resolution of the sterol regulatory region in the promoter for 3-hydroxy-3-methylglutaryl coenzyme A reductase. *J. Biol. Chem.* **266**:13947–13951.

Palinski, W., Ord, V. A., Plump, A. S., Breslow, J. L., Steinberg, D., and Witztum, J. L. (1994). Apoprotein E-deficient mice are a model of lipoprotein oxidation in atherogenesis: Demonstration of oxidation-specific epitopes in lesions and high titers of autoantibodies to malondialdehyde-lysine in serum. *Arterio. Thrombo.* **14**:605–616.

Parthasarathy, S., Printz, D. J., Boyd, D., Joy, L., and Steinberg, D. (1986). Macrophage oxidation of low density lipoprotein generates a modified form recognized by the scavenger receptor. *Arteriosclerosis* **6**:505–510.

Parthasarathy, S., Wieland, E., and Steinberg, D. (1989). A role for endothelial cell lipoxygenase in the oxidative modification of low density lipoprotein. *Proc. Natl. Acad. Sci. USA* **86**:1046–1050.

Piedrahita, J. A., Zhang, S. H., Hagaman, J. R., Oliver, P. M., and Maeda, N. (1992). Generation of mice carrying a mutant apolipoprotein E gene inactivated by gene targeting in embryonic stem cells. *Proc. Natl. Acad. Sci. USA* **89**:4471–4475.

Plump, A. S., Smith, J. D., and Hayek, T. (1992). Severe hypercholesterolemia and atherosclerosis in apolipoprotein E-deficient mice created by homologous recombination in ES cells. *Cell* **71**:343–353.

Rauh, G., Keller, C., Kormann, B., Spengel, F., Schuster, H., Wolfram, G., and Zollner, N. (1992). Familial defective apolipoprotein B100 clinical characteristics of 54 cases. *Atherosclerosis* **92**:233–241.

Raychowdhury, R., Niles, J. L., McCluskey, R. T., and Smith, J. A. (1989). Autoimmune target in Heymann Nephritis is a glycoprotein with homology to the LDL receptor. *Science* **244**:1163–1165.

Reddick, R. L., Zhang, S. H., and Maeda, N. (1994). Atherosclerosis in mice lacking apo E; Evaluation of lesional development and progression. *Arterio. Thrombo.* **14**:141–147.

Rohrer, L., Freeman, M., Kodama, T., Penman, M., and Krieger, M. (1990). Coiled-coil fibrous domains mediate ligand binding by macrophage scavenger receptor type II. *Nature* **343**:570–572.

Ross, R. (1993). The pathogenesis of atherosclerosis: a perspective for the 1990s. *Nature* **362**:801–809.

Russell, D. W., Yamamoto, T., Schneider, W. J., Slaughter, C. J., Brown, M. S., and Goldstein, J. L. (1983). cDNA cloning of the bovine low density lipoprotein receptor: Feedback regulation of a receptor mRNA. *Proc. Natl. Acad. Sci. USA* **80**:7501–7505.

Scanu, A. M., Lawn, R. M., and Berg, K. (1991). Lipoprotein(a) and atherosclerosis. *Ann. Int. Med.* **115**:209–218.

Schmitz, G., Bruning, T., Williamson, E., and Nowicka, G. (1990). The role of HDL in reverse cholesterol transport and its disturbances in Tangier disease and HDL deficiency with xanthomas. *Eur. Heart J.* **11**:197–211.

Schultz, J. R., Verstuyft, J. G., Gong, E. L., Nichols, A. V., and Rubin, E. M. (1992). ApoAI and ApoAII transgenic mice: A comparison of atherosclerotic susceptibility. *Circulation* **86(Suppl. I)**:I-472.

Sharkey, M. F., Miyanohara, A., Elam, R. L., Friedmann T., and Witztum, J. L. (1990). Post transcriptional regulation of retroviral vector-transduced low density lipoprotein receptor activity. *J. Lipid Res.* **31**:2167–2178.

Sharp, D., Blinderman, L., Combs, K. A., Kienzle, B., Ricci, B., Wagner-Smith, K., Gil, C. M., Turck, C. W., Bouma, M.-E., Rader, D. J., Aggerbeck, L. P., Gregg, R. E., Gordon, D. A., and Wetterau, J. R. (1993). Cloning and gene defects in microsomal triglyceride transfer protein associated with abetalipoproteinemia. *Nature* **365**:65–69.

Shimano, H., Yamada, N., Katsuki, M., Shimada, M., Gotoda, T., Harada, K., Murase, T., Fukazawa, C., Takaku, F., and Yazaki, Y. (1992a). Overexpression of apolipoprotein E in transgenic mice: Marked reduction in plasma lipoproteins except high density lipoprotein and resistance against diet-induced hypercholesterolemia. *Proc. Natl. Acad. Sci. USA* **89**:1750–1754.

Shimano, H., Yamada, N., Katsuki, M., Yamamoto, K., Gotoda, T., Harada, K., Shimada, M., and Yazaki, Y. (1992b). Plasma lipoprotein metabolism in transgenic mice overexpressing apolipoprotein E. *J. Clin. Invest.* **90**:2084–2091.

Shoulders, C. C., Brett, D., Bayliss, J. D., Narcisi, T. M. E., Jarmuz, A., Grantham, T. T., Leoni, P. R. D., Bhattacharya, S., Pease, R. J., Cullen, P. M., Levi, S., Byfield, P. G. H., Purkiss, P., and Scott, J. (1993). Abetalipoproteinemia is caused by defects of the gene encoding the 97 kDa subunit of a microsomal triglyceride transfer protein. *Hum. Mol. Genet.* **2**:2109–2116.

Steinberg, D., Parthasarathy, S., Carew, T. E., Khoo, J. C., and Witztum, J. L. (1989). Beyond cholesterol. Modifications of low-density lipoprotein that increase its atherogenicity [see comments]. *N. Engl. J. Med.* **320**:915–924.

Strickland, D. K., Ashcom, J. D., Williams, S., Burgess, W. H., Migliorin, M., and Argraves, W. S. (1990). Sequence identity between the alpha2-macroglobulin receptor and low density lipoprotein receptor-related protein suggests that this molecule is a multifunctional receptor. *J. Biol. Chem.* **265**:17401–17404.

Strickland, D. K., Ashcom, J. D., Williams, S., Battey, F., Behre, E., McTigue, K., Battey, J. F., and Argraves, W. S. (1994). Primary structure of apha2-macroglobulin receptor-associated protein. *J. Biol. Chem.* **266**:13364–13369.

Sudhof, T. C., Goldstein, J. L., Brown, M. S., and Russell, D. W. (1985). The LDL receptor gene: A mosaic of exons shared with different proteins. *Science* **228**:815–822.

Szanto, A., Balasubramaniam, S., Roach, P. D., and Nestel, P. J. (1992). Modulation of the low-density-lipoprotein-receptor-related protein and its relevance to chylomicron-remnant metabolism. *Biochem.* **288**:791–794.

Teng, B., Burant, C. F., and Davidson, N. O. (1993). Molecular cloning of apolipoprotein B messenger RNA editing protein. *Science* **260**:1816–1819.

Utermann, G. (1989). The mysteries of lipoprotein(a). *Science* **246**:904–910.

van den Maagdenberg, A. M. J. M., Hofker, M. H., Krimpenfort, P. J. A., de Bruijn, I., van Vlijmen, B., van der Boom, H., Havekes, L. M., and Frants, R. R. (1993). Transgenic mice carrying the apolipoprotein E3-Lieden gene exhibit hyperlipoproteinemia. *J. Biol. Chem.* **268**:10540–10545.

Virchow, R. (1940). "Cellular Pathology." Edwards Brothers, Ann Arbor.

Wang, X., Briggs, M. R., Hua, X., Yokoyama, C., Goldstein, J. L., and Brown, M. S. (1993). Nuclear protein that binds sterol regulatory element of low density lipoprotein receptor promoter. II. Purification and characterization. *J. Biol. Chem.* **268**:14497–14504.

Wang, X., Sato, R., Brown, M. S., Hua, X., and Goldstein, J. L. (1994). SREBP-1, a membrane-bound transcription factor released by sterol-regulated proteolysis. *Cell* **77**:53–62.

Warden, C. H., Hedrick, C. C., Qiao, J., Castellani, L. W., and Lusis, A. J. (1993). Atherosclerosis in transgenic mice overexpressing apolipoprotein A-II. *Science* **261**:469–472.

Wieland, E., Parthasarathy, S., and Steinberg, D. (1993). Peroxidase-dependent metal-independent oxidation of low density lipoprotein in vitro: a model for in vivo oxidation? *Proc. Natl. Acad. Sci. USA* **90**:5929–5933.

Willnow, T. E., Sheng, Z., Ishibashi, S., and Herz, J. (1994). Inhibition of hepatic chylomicron remnant uptake by gene transfer of a receptor antagonist. *Science* **264**:1471–1474.

Witztum, J. S., and Steinberg, D. (1991). Role of oxidized low density lipoprotein in atherogenesis. *J. Clin. Invest.* **88**:1785–1792.

Wu, H., Moulton, K., and Glass, C. K. (1992). Macrophage scavenger receptors and atherosclerosis. *T.C.M.* **2**:220–225.

Wu, H., Moulton, K., Horvai, A., Parik, S., and Glass, C. K. (1994). Combinatorial interactions between AP-1 and ets-domain proteins contribute to the developmental regulation of the macrophage scavenger gene. *Mol. Cell Biol.* **14**:2129–2139.

Yamamoto, T., Davis, C. G., Brown, M. S., Schneider, W. J., Casey, M. L., Goldstein, J. L., and Russell, D. W. (1984). The human LDL receptor: A cysteine-rich protein with multiple Alu sequences in its mRNA. *Cell* **39**:27–38.

Yla-Herttuala, S., Palinski, W., Rosenfeld, M. E., Parthasarathy, S., Carew, T. E., Butler, S., Witztum, J. L., and Steinberg, D. (1989). Evidence for the presence of oxidatively modified low density lipoprotein in atherosclerotic lesions of rabbit and man. *J. Clin. Invest.* **84**:1086–1095.

Yla-Herttuala, S., Palinski, W., Butler, S. W., Picard, S., Steinberg, D., and Witztum, J. L. (1994a). Rabbit and human atherosclerotic lesions contain IgG that recognizes epitopes of oxidized LDL. *Arterio. Thrombo.* **14**:32–40.

Yokode, M., Hammer, R. E., Ishibashi, S., Brown, M. S., and Goldstein, J. L. (1990). Diet-induced hypercholesterolemia in mice: Prevention by overexpression of LDL receptors. *Science* **1273**:1275.

Yokoyama, C., Wang, X., Briggs, M. R., Admon, A., Wu, J., Hua, X., Goldstein, J. L., and Brown, M. S. (1993). SREBP-1, a basic-helix–loop–helix-leucine zipper protein that controls transcription of the low density lipoprotein receptor gene. *Cell* **75**:187–197.

Young, S. G. (1990). Recent progress in understanding apolipoprotein B. *Circulation* **82**:1574–1594.

Zhang, S. H., Reddick, R. L., Piedrahita, J. A., and Meada, N. (1992). Spontaneous hypercholesterolemia and arterial lesions in mice lacking apolipoprotein E. *Science* **258**:468–471.

# 6 Molecular Genetics of Phenylketonuria: From Molecular Anthropology to Gene Therapy

**Randy C. Eisensmith\* and Savio L. C. Woo\*,†**
\*Department of Cell Biology
†Howard Hughes Medical Institute
Baylor College of Medicine
Houston, Texas 77030

## I. GENERAL BACKGROUND

### A. Discovery of phenylketonuria

Classical phenylketonuria (PKU) was first recognized as a distinct metabolic disorder causing severe, irreversible mental retardation by the Norwegian biochemist and physician Asbjörn Fölling in 1934. Asked to investigate the cause of a distinctive "mousy" odor in two mentally retarded siblings, Fölling observed increased levels of phenylpyruvate in their urine. An examination of an additional 430 mentally retarded children revealed 8 other children with similar symptoms, including 2 other sib pairs. Because of the structural similarity between phenylpyruvate and the essential amino acid phenylalanine, Fölling hypothesized that phenylalanine was the most likely source of the urinary phenylpyruvate. This hypothesis was based in part on the earlier observations of Kotake et al. (1922), who had demonstrated that rabbits fed excessive amounts of phenylalanine excreted phenylpyruvic acid. Fölling pursued this hypothesis by examining the effects of both oral phenylalanine and dietary protein loads on urinary phenylpyruvate levels in one affected individual. Both of these procedures increased the excretion of phenylpyruvate in the urine. Based on these observations, Fölling published two articles in which he proposed the existence of an autosomally transmitted recessive disorder of phenylalanine metabolism, which he called "imbecillitas phenylpyrouvica" (Fölling, 1934a,b). A year after Fölling's initial observations, Penrose confirmed that this metabolic disorder was indeed transmitted as an autosomal recessive trait (Penrose, 1935). The term phenylketonuria was subsequently introduced to describe this metabolic disorder by Penrose and Quastel (1937). Through his identification of PKU as an inherited metabolic disorder, Fölling had supplied a prime example of an "inborn error of metabolism" originally proposed by Sir Archibald Garrod (1908).

## B. The biochemical defect in phenylketonuria

Fölling's observations clearly implicated some defect of phenylalanine metabolism in phenylketonuric individuals. However, the precise biochemical deficiency responsible for classical PKU was not identified until George Jervis (1947) showed that the administration of phenylalanine caused a rapid elevation in serum tyrosine levels in normal individuals, but not in individuals with PKU. Udenfriend and Cooper (1952) subsequently demonstrated an enzymatic system in the soluble fraction of rat liver that was capable of converting phenylalanine to tyrosine. Jervis (1953) then showed that postmortem liver tissue obtained from normal individuals could sustain this enzymatic reaction, while tissue obtained from PKU patients could not. These observations provided the first direct evidence that PKU was caused by a profound deficiency in a hepatic enzyme system that converted phenylalanine to tyrosine.

## C. Phenylalanine hydroxylation in humans

Although the reports by Udenfriend and Cooper (1952) and Jervis (1953) defined the role of hepatic phenylalanine hydroxylation in PKU, Mitoma (1956) observed that two different protein products were necessary for phenylalanine hydroxylation to occur in the liver. Shortly thereafter, Kaufman (1958a) demonstrated that a reduced pteridine was also required for this reaction to proceed. Kaufman (1959, 1963) quickly identified the two protein components of this reaction as a phenylalanine hydroxylase and a pteridine reductase.

Subsequent studies have revealed that phenylalanine hydroxylation in humans is a complex reaction involving at least six different enzymes and their requisite cofactors (Figure 6.1). The central reaction, the oxidation of the L-phenylalanine to L-tyrosine, is catalyzed by phenylalanine hydroxylase (phenylalanine 4-monooxygenase; PAH; EC 1.14.16.1). One atom of molecular oxygen is incorporated into tyrosine while the second atom is reduced to $H_2O$. Thus, PAH is a mixed-function oxygenase. Although early studies suggested that PAH may be present in nonhepatic tissues in humans (Ayling et al., 1974) and more recent studies have demonstrated the presence of small amounts of PAH mRNA in leukocytes (Sarkar and Sommer, 1989), authentic expression of PAH is most likely limited to the liver in humans (Crawfurd et al., 1981).

In addition to L-phenylalanine, molecular oxygen, and PAH, the co-

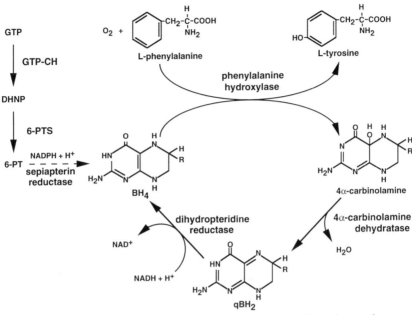

**Figure 6.1.** The phenylalanine hydroxylation system in man, including the synthetic and regenerative pathways for the pteridine cofactor. GTP, guanosine triphosphate; GTP-CH, GTP-cyclohydrolase I; DHNP, D-*erythro*-dihydroneopterin triphosphate; 6-PTS, 6-pyruvoyl tetrahydrobiopterin synthase; 6-PT, 6-pyruvoyl tetrahydrobiopterin; qBH$_2$, quinonoid dihydrobiopterin; BH$_4$, L-*erythro*-tetrahydrobiopterin.

factor L-*erythro*-tetrahydrobiopterin (BH$_4$) is also required for the phenylalanine hydroxylation reaction to proceed. The pteridine cofactor is first oxidized to 4$\alpha$-carbinolamine during the phenylalanine hydroxylation reaction. In a secondary reaction, the 4$\alpha$-carbinolamine is converted to quinonoid dihydrobiopterin (qBH$_2$) and H$_2$O by 4$\alpha$-carbinolamine dehydratase (Lazarus *et al.*, 1983), originally identified as phenylalanine hydroxylase stimulating protein (Kaufman, 1970). For the phenylalanine hydroxylation reaction to proceed catalytically, sufficient levels of BH$_4$ must be maintained *in vivo*. Immediate requirements are met through the regeneration of BH$_4$ from qBH$_2$. This reaction is catalyzed by the enzyme quinonoid dihydropteridine reductase, more commonly referred to as dihydropteridine reductase (DHPR; EC 1.6.99.7). This enzyme requires a reduced pyridine nucleotide cofactor as an electron donor.

Long-term requirements for BH$_4$ are met through its biosynthesis from guanosine triphosphate (GTP) via a complex and as yet incompletely characterized multistep pathway (Figure 6.1). The first step in this pathway, the conversion of GTP to D-*erythro*-dihydroneopterin triphosphate, is catalyzed by the enzyme GTP-cyclohydrolase I (GTP-CH; EC 3.5.4.16). The dihydroneopterin triphosphate is then converted to 6-pyruvoyl tetrahydrobiopterin (6-PT) by the enzyme 6-pyruvoyl tetrahydrobiopterin synthase (6-PTS). In the presence of NADPH and a third enzyme, sepiapterin reductase, 6-pyruvoyl tetrahydrobiopterin is ultimately converted into $BH_4$, although the number of steps and the identities of the intermediates remain unclear.

## D. Dietary therapy for phenylketonuria

As information accumulated regarding the clinical and genetic characteristics (Jervis, 1939; Jervis et al., 1940), the biochemical basis (Jervis, 1947, 1953), and the frequency of PKU (Jervis, 1954), Woolf proposed the use of a low-phenylalanine diet as a possible therapy for PKU (Woolf and Vulliamy, 1951). Several landmark studies were soon published describing significant reductions in serum phenylalanine and urinary phenylpyruvate levels in young PKU patients following dietary restriction (Bickel et al., 1954; Armstrong and Tyler, 1955; Woolf et al., 1955). These same reports also described some apparent improvements in the mental developmental and behavioral performance of treated PKU patients.

Nearly 40 years of experience has confirmed the dramatic benefits of dietary therapy in the treatment of PKU. Individuals carefully administered diets consisting of modified protein hydrolysates or free amino acids supplemented with additional vitamins and nutrients can generally thrive both physically (McBurnie et al., 1991) and intellectually (Koch et al., 1984). However, despite these successes, dietary therapy is not without certain limitations. Some selective cognitive impairments are apparent even in well-treated patients (Koch et al., 1984), as are behavioral disturbances (Smith et al., 1988). Furthermore, termination of treatment, even in adolescents or adults, is accompanied by measurable declines in neuropsychological function (Smith et al., 1978; Woolf, 1979). Finally, noncompliance during pregnancy can cause microcephaly, mental retardation, impaired somatic growth, and congenital malformations in the offspring of women with PKU, a syndrome referred to as "maternal PKU" (Lenke and Levy, 1980).

One clear finding of many of these studies is the inverse relationship between age of onset of therapy and the ultimate intellectual outcome. The most profound consequences of PKU can be prevented or greatly ameliorated by dietary treatment, but only if affected individuals are treated early in life. These observations, apparent from even the earliest trials (Bickel et al., 1954; Armstrong and Tyler, 1955; Woolf et al., 1955), generated considerable interest in the development and implementation of neonatal screening programs for PKU and related hyperphenylalaninemias.

## E. Newborn screening for phenylketonuria

The earliest screening programs, initiated in California (Centerwall, 1957) and slightly later in Great Britain (Gibbs and Woolf, 1959; Boyd, 1961), were based on a ferric chloride test that detects urinary phenylpyruvate. Unfortunately, the harmful elevations of phenylalanine in serum precede by several weeks the increased urinary excretion of phenylpyruvate (Armstrong and Binkley, 1956; Armstrong and Low, 1957). Thus, the development of a semiquantitative test for the measurement of serum phenylalanine levels in neonates was a critical advance in the treatment of PKU (Guthrie, 1961; Guthrie and Susi, 1963). This "Guthrie test" permitted relatively inexpensive yet comprehensive neonatal screening programs to be instituted in many Western countries.

## F. Clinical aspects of phenylketonuria

Because of the complex nature of the phenylalanine hydroxylation system in humans, defects in several different protein components could conceivably lead to a loss of phenylalanine hydroxylation and consequently to elevated levels of serum phenylalanine or hyperphenylalaninemia (HPA). Early experiments detected both DHPR (Mitoma et al., 1957) and $BH_4$ (Kaufman, 1958b) in postmortem liver tissue from classical PKU patients, suggesting that most cases of this disorder were caused by a reduction or a complete absence of PAH activity. However, variant forms of HPA can be caused by deficits in DHPR (Kang et al., 1970; Kaufman et al., 1975a, 1978), GTP-CH (Niederweiser et al., 1984; Dhondt et al., 1985), and 6-PTS (Niederweiser et al., 1985, 1986), while some instances of transient HPA have been attributed to a deficiency of $4\alpha$-carbinolamine dehydratase (Dhondt et al., 1988; Blaskovics and Guidici, 1988). Collectively, these $BH_4$ deficiencies comprise only 1 or 2% of all cases of HPA, and detailed descriptions of the clinical aspects and genetic bases of the $BH_4$-deficient hyperphenylalaninemias are well beyond the scope of this chapter. However, their presence can sometimes confound the proper diagnosis of HPA subtypes in newborns. Differential diagnosis requires a careful evaluation of biopterin and its precursors, intermediates, and metabolites in blood and urine after the initial determination of serum phenylalanine levels. The remainder of this chapter will focus on hyperphenylalaninemias caused by PAH deficiency.

The PAH-deficient hyperphenylalaninemias (McKusick's MIM Cat. 261600) encompass a broad range of biochemical and clinical phenotypes. Serum phenylalanine levels in untreated children with classical PKU can rise from a normal value of $62 \pm 15$ μM (mean $\pm$ SD; Gregory et al., 1986) to greater than 1200 μM. These dramatic increases are almost always accompanied by profound and irreversible mental retardation. Recent studies have employed magnetic resonance imaging to demonstrate reversible changes in brain myelination (Thompson *et al.*, 1991; Bick *et al.*, 1991) that are related to phenylalanine levels in treated PKU patients. Changes in neurotransmitter receptor density have also been observed in animal models for PKU (Matsuo and Hommes, 1988; Hommes, 1993). It is not yet clear if these effects are wholly responsible for the mental impairment. Additional symptoms frequently reported among untreated individuals with classical PKU include the distinctive mousy odor that stimulated Fölling's initial investigations, hypopigmentation and other dermatological conditions, behavioral disturbances, and convulsive seizures [see Scriver *et al.*, (1989) for a comprehensive review of clinical manifestations in classical PKU]. In contrast to individuals with classical PKU, some individuals only mildly deficient in hepatic PAH activity display only slight increases in serum phenylalanine and can achieve near-normal or normal IQ levels even in the absence of treatment.

This heterogeneity of biochemical and clinical phenotypes in the PAH-deficient hyperphenylalaninemias has been reflected in an often bewildering array of clinical classification systems. For the purposes of the present discussion, the classification scheme proposed by Güttler (1980) will be employed. "Classical PKU" will be used to refer to the most severely affected class of hyperphenylalaninemic patients, those with serum phenylalanine levels greater than 1200 $\mu$M and who are profoundly retarded in the absence of treatment. "Mild PKU" will refer to those patients with serum phenylalanine levels ranging from 800 to 1200 $\mu$M. These patients still require dietary therapy to avoid various degrees of mental impairment, but can tolerate more dietary phenylalanine than the classical group while maintaining similarly low levels of serum phenylalanine. "HPA" will be used to describe those individuals with serum phenylalanine levels ranging from 250 to 800 $\mu$M and who are clinically normal even without dietary therapy.

## II. CHARACTERIZATION OF THE HUMAN PHENYLALANINE HYDROXYLASE GENE

### A. The cloning of human PAH cDNAs

The cloning of the human PAH cDNA began with the purification of PAH protein from rat liver and the production of a monospecific antibody. The anti-rat PAH antibody was used to isolate PAH mRNA from rat liver by polysome immunoprecipitation, and the rat PAH mRNA was reverse transcribed to produce a rat PAH cDNA (Robson *et al.*, 1982). The identity of this cDNA clone was verified through the use of hybrid-selected translation to produce PAH protein, and further confirmed by comparing the amino acid sequence predicted

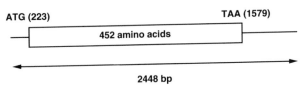

ATG (223)    TAA (1579)

452 amino acids

2448 bp

**Figure 6.2.** Schematic diagram of the human PAH cDNA clone hPAH247. The 2448-bp cDNA contains a 1356-bp open reading frame which encodes a protein containing 452 amino acid residues.

from the cDNA sequence with a partial amino acid sequence obtained from the purified rat enzyme (Robson *et al.*, 1982, 1984). This rat PAH cDNA clone then served as a specific hybridization probe to identify and isolate several human PAH cDNA clones from a human liver cDNA library. The longest of these clones, designated hPAH247, was 2448-bp long and contained an open reading frame beginning at position 223 and ending at position 1579 (Figure 6.2). This open reading frame encoded a protein of 452 amino acids, with a predicted molecular weight of 51,862 (Kwok *et al.*, 1985). The amino acid composition of this protein, as deduced from the nucleotide sequence of hPAH247, was nearly identical to that reported for the human protein (Friedman and Kaufman, 1973; Shiman and Gray, 1980). A second human PAH cDNA clone, hPH7, has also been isolated using a similar approach (Speer *et al.*, 1986).

## B. Human PAH is encoded by a single gene

Before the application of molecular genetic methodologies, the number of genes encoding PAH in humans was not known. In rat, electrophoretic analysis detected two discrete species of PAH subunit in some strains (Mercer *et al.*, 1984; Dahl and Mercer, 1986). Similar analyses suggested that the human PAH protein was also multimeric, with two subunit species very close in size to those of rat (Abita *et al.*, 1983). In both species it was unclear whether the two electrophoretic variants were created by post-translational modification of a single protein produced at one genetic locus, by the expression of two different alleles at a single genetic locus, or by two separate genetic loci.

This question was quickly and easily answered for the human PAH protein by expressing the full-length human PAH cDNA clone hPAH247 (Ledley *et al.*, 1985b). An expression vector was constructed containing hPAH247 under the control of the driven by the human metallothionine gene promoter. When this vector was transfected into cultured mammalian cells normally lacking PAH activity, properly sized PAH mRNA, immunoreactive PAH protein, and pterin-dependent enzymatic activity similar to authentic human PAH could be detected in cell extracts (Ledley *et al.*, 1985b). These findings conclusively demonstrated that the assembled, functional human PAH protein is the product of a single gene, suggesting that the two electrophoretic variants sometimes observed result from post-translational modification of a single gene product.

Studies using highly specific monoclonal antibodies have conclusively shown that the two distinct species of human PAH are phosphorylated and dephosphorylated forms of the same basic subunit (Smith *et al.*, 1984). Phosphorylation of rat PAH by cAMP-dependent protein kinase increases its activity several-fold (Abita *et al.*, 1976), and this regulatory mechanism is clearly of physiological significance in the rat (Donlon and Kaufman, 1978). However, the importance of phosphorylation-induced activation in the regulation of human PAH is still unresolved.

## C. Chromosomal localization of the human PAH gene

Initial attempts to localize the human PAH gene relied on classical linkage studies between PKU and certain polymorphic protein markers. In these studies, moderate linkage between PKU and the phosphoglucomutase locus PGM-1 and the amylase loci AMY-1 and AMY-2 suggested that the PAH gene was located on chromosome 1 (Berk and Saugstad, 1974; Kamaryt *et al.*, 1978). However, subsequent investigations using improved methods for identifying heterozygotes in PKU families failed to confirm this assignment (Paul *et al.*, 1979a).

Since human PAH is encoded by a single gene, the location of the PAH gene could be determined simply by performing Southern analyses on genomic DNA isolated from human/rodent cell hybrids containing different combinations of human chromosomes, using the human PAH cDNA as a specific hybridization probe. Such studies demonstrated that the human PAH gene was on chromosome 12 (Lidsky *et al.*, 1984). Through deletion chromosome mapping and *in situ* hybridization of metaphase chromosome preparations from human lymphoblastoid cell lines of normal karyotype, the PAH gene was further localized to band region 12q22-q24.1 (Lidsky *et al.*, 1985b).

## D. Structure of the human PAH gene

Southern analysis also proved useful in determining the structural organization of the human PAH gene. Preliminary observations suggested that the chromosomal PAH gene is over 65 kilobases (kb) in length and contains multiple intervening sequences (Lidsky *et al.*, 1985c). The large size of this gene necessitated the use of cosmid vectors to further examine its molecular structure. A human genomic DNA library was therefore constructed using cosmid vectors and probed with the human PAH cDNA. Four cosmid clones containing overlapping PAH genomic sequences were isolated. Further mapping of these clones

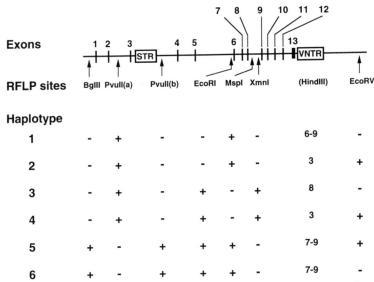

**Figure 6.3.** Structure of the human PAH gene. The positions of eight RFLP sites, one polymorphic STR, and one polymorphic VNTR are shown relative to the 13 exons contained within the PAH gene. Each specific combination of RFLP, STR, and VNTR sites constitutes a unique haplotype. The six most common RFLP haplotypes reported among European populations are shown at the bottom of the figure.

showed that the human PAH gene spans about 90 kb and contains 13 exons (DiLella *et al.*, 1986a). Exons 6–13 are compactly clustered within a 20-kb fragment of the gene, while exons 1–5 are generally separated by large introns which range in size from 3 to 23 kb (Figure 6.3). The mature messenger RNA is about 2.4 kb (DiLella *et al.*, 1986a).

## E. Homologies among the pterin-dependent hydroxylases

PAH is but one member of a class of closely related pterin-dependent enzymes, the aromatic L-amino acid hydroxylases. However, PAH shares more with tryptophan hydroxylase (TPH) and tyrosine hydroxylase (TYH) than an absolute requirement for the reduced pterine cofactor $BH_4$. The nucleotide sequences are highly conserved between the three cDNAs (Grennet *et al.*, 1987) and, consequently, there are strong homologies in the amino acid sequences of these three proteins, both within and between different species (Ledley *et al.*, 1985a; Grennet *et al.*, 1987; Morales *et al.*, 1990; Nagatsu and Ichinose, 1991; Onishi *et al.*, 1991). Several antibodies have been produced that recognize all three proteins (Grennet *et al.*, 1987; Haan *et al.*, 1987; Jennings and Cotton, 1990; Jennings *et al.*, 1991).

The high degree of cDNA sequence homology gave rise to the hypothesis that the genes for PAH, TYH, and TPH arose from a common ancestor. According to this hypothesis, two hydroxylases with different substrate specificities were generated through duplication of a single protohydroxylase gene followed by sequence divergence. This process was then repeated on one of these two genes to yield the third hydroxylase gene. Since the cDNA sequences of TPH and PAH are more closely related to each other than either is to TYH, these two genes are proposed to have diverged most recently (Grenett et al., 1987). The observation that a single gene encodes both TPH and PAH in Drosophila (Morales et al., 1990) provides some support for this hypothesis, as does the finding that the intron/exon boundaries are highly conserved in all three genes (Stoll and Goldman, 1991).

The putative domain structure is also similar in these three enzymes (Dahl and Mercer, 1986; Grennet et al., 1987). The carboxy-terminal domain containing the basic determinants for hydroxylation activity (Iwaki et al., 1986), including the $BH_4$-binding site (Jennings and Cotton, 1990; Jennings et al., 1991), is more highly conserved than the amino-terminal domain, which contains the determinants for phosphorylation-mediated activation (Campbell et al., 1986; Pigeon et al., 1987) and substrate specificity (Iwaki et al., 1986). Truncated versions of rat PAH and mouse TPH lacking significant portions of the N-terminal region can still perform the basic hydroxylation reaction, but with an apparent loss of substrate specificity (Iwaki et al., 1986; Stoll and Goldman, 1991). In addition, the PAH found in Chromobacterium violaceum can function well despite lacking the region homologous to the N-terminal portion of the proteins from Drosophila or mammalian species (Onishi et al., 1991).

## III. MOLECULAR GENETICS OF PAH-DEFICIENT PKU AND RELATED HPAS

### A. Polymorphic haplotypes at the human PAH locus

### 1. RFLP-based haplotypes at the human PAH locus

Southern analyses of the PAH locus demonstrated the presence of several restriction fragment length polymorphisms (RFLPs) on both normal and mutant chromosomes. Restriction enzymes used to detect RFLPs include BglII, PvuII (two sites), EcoRI, MspI, XmnI, HindIII, and EcoRV (Woo et al., 1983; Lidsky et al., 1985a). The relative locations of these eight RFLP sites within the human PAH gene are shown at the top of Figure 6.3. Each specific combination of RFLP sites constitutes a unique haplotype. By Southern analysis, seven of these RFLPs are dimorphic, and one (HindIII) is trimorphic. These eight RFLPs can therefore define 384 unique haplotypes, about 70 of which have actually been observed

(Eisensmith and Woo, 1992b). The 6 most common RFLP haplotypes reported among European populations are shown at the bottom of Figure 6.3. The finding of so few haplotypes on a majority of all normal and mutant PAH chromosomes implies a high degree of linkage disequilibrium between certain polymorphic sites (Chakraborty *et al.*, 1987).

## 2. Extended haplotypes at the human PAH locus

Polymerase chain reaction (PCR)-based methods have recently been applied to the detection of many of the polymorphic sites used for determination of RFLP haplotypes at the PAH locus (Dworniczak *et al.*, 1991d,e; Wedemeyer *et al.*, 1991; Goltsov *et al.*, 1992a,b). The determination of flanking sequence information for the *Bgl*II (Dworniczak *et al.*, 1991e), *Pvu*IIa (Dworniczak *et al.*, 1991b,d), *Pvu*IIb (R. Eisensmith, unpublished observation), *Msp*I (Wedemeyer *et al.*, 1991), *Xmn*I (Goltsov *et al.*, 1992b), and *Hind*III (Goltsov *et al.*, 1992a) sites has allowed the specific amplification of fragments containing these polymorphisms. During the course of these studies, detailed examination of the amplification fragment containing the polymorphism detected by *Hind*III digestion demonstrated that this RFLP is caused by the presence of a variable number of tandem repeats (VNTR) situated between two constant *Hind*III sites. This VNTR system, located 3kb downstream of the final exon of the PAH gene, contains various multiples of a 30-bp repeat (Figure 6.3). Comparison of this system in Caucasian PKU families demonstrates the presence of at least 10 alleles that differ in the number of repeated units (Byck *et al.*, 1992; Goltsov *et al.*, 1992a; R. Eisensmith, unpublished observation). The incorporation of this VNTR into the existing RFLP haplotype scheme increases the potential number of haplotypes at the PAH locus to 1280.

In addition to the polymorphisms detectable by restriction endonuclease digestion, experiments using oligonucleotide probes for di-, tri-, and tetranucleotide repeats suggest that there are a number of short tandem repeats (STRs) within the PAH locus, some of which are also polymorphic. The most well characterized of these STRs is a TCTA repeat located in intron 3 (Figure 6.3). An examination of this STR system in about 100 Caucasian and Chinese PKU families identified at least nine alleles, each differing from the next by a single 4-bp repeat (Goltsov *et al.*, 1993). The inclusion of this STR polymorphism into the PAH haplotype system dramatically increases the possible number of haplotypes to 11,520. This number will undoubtedly continue to increase as the more than 10 silent polymorphisms now known are included and as other polymorphic systems are characterized. However, it is also likely that only a small number of these potential haplotypes will actually be observed.

Although the development of PCR-based methods for haplotyping of the PAH locus led to a dramatic increase in the number of possible haplotypes, and hence the informativity of haplotype analysis (see Section III.A.3), complete RFLP haplotyping of all eight of the sites shown in Figure 6.3 by PCR is not yet possible. The polymorphic EcoRV site is located about 20 kb from regions of known sequence in the PAH gene and the determination of sequence information flanking this site has not yet been reported. More troublesome, however, is the polymorphic EcoRI site in the PAH gene. This site is located more than 1 kb from each end of a repetitive LINE-1 element, and specific amplification of this region has not yet been achieved. Despite these technical problems, PCR-based methods are proving to be valuable aids in clinical applications of haplotype analysis.

## 3. Clinical applications of haplotype analysis

Comparisons between standardized linkage disequilibria and physical map distances for all of the different combinations of the RFLP, VNTR, and STR sites used to define the extended haplotypes at the PAH locus failed to distinguish any obvious hotspots for recombination (Daiger et al., 1986; Eisensmith et al., 1992; Goltsov et al., 1993). These RFLPs are therefore tightly linked to specific PAH genes and can be used to identify and follow the transmission of normal or mutant chromosomes in PKU families. Initially, the most important consequence of this tight linkage or association was that haplotype analysis could be used to perform prenatal diagnosis or carrier screening in PKU families (Woo et al., 1983; Lidsky et al., 1985a,c; Speer et al., 1986). The frequencies of RFLP haplotypes have now been established on normal and mutant chromosomes in many populations (see Table 6.1 and references therein). Because of the high degree of heterozygosity at many of the RFLP sites, RFLP haplotype analysis is very informative for the performance of prenatal diagnostic or carrier screening tests in most populations. For example, in the general European population, RFLP haplotype analysis is informative in about 86% of all PKU families (Lidsky et al., 1985a). In contrast, among Asians, RFLP haplotype analysis is informative in only about 32% of all PKU families (Daiger et al., 1989b; Chen et al., 1989). Consequently, RFLP haplotype analysis is much less useful for carrier screening or prenatal diagnosis in Asian populations.

The low degree of informativity of RFLP haplotype analysis among Asian populations can be largely remedied through the use of additional polymorphic systems in the PAH locus. For example, based on the allele frequencies derived from a study of Chinese PKU families, the STR system alone is more than twice as informative as the original RFLP haplotype system (75 vs 32%; Goltsov et al., 1993), and the informativity of haplotype analysis based upon a combination of RFLP, VNTR, and STR polymorphisms approaches 80% among Asian PKU families. A similar situation exists in Caucasian populations, where analysis of extended haplotypes based on combinations of RFLP, VNTR, and

STR polymorphisms is informative in about 95% of all PKU families (Eisensmith *et al.*, 1994). Further increases in the level of accuracy of diagnostic tests based on haplotype analysis are still possible, provided that other similarly polymorphic systems are present in or near the PAH locus.

Although haplotype analysis is now extremely informative, this technique is not without limitations. Most significantly, prenatal diagnostic and carrier screening tests based upon haplotype analysis are only informative in families in which PKU has already occurred. Since the overwhelming majority of new cases of PKU (over 95%) are the offspring of previously unrecognized carriers, they are undetectable by this method. Any diagnostic procedures that will be informative in individuals without a prior family history of PKU must rely on the detection of the many molecular lesions of the PAH gene that are responsible for PKU and related HPAs.

## 4. RFLP haplotypes among normal and mutant chromosomes

Following the initial discovery of RFLP haplotypes and their use in prenatal diagnosis, many investigators examined the associations between PAH haplotypes and normal or mutant chromosomes in various ethnic groups. The first of these studies clearly demonstrated no significant associations between any single RFLP site or RFLP haplotype and PKU (Chakraborty *et al.*, 1987). This result implied that chromosomes bearing PKU-causing mutations are of several different haplotypes. The distribution of PAH haplotypes and their associations with normal and mutant chromosomes in European and Asian populations are summarized in Table 6.1.

Of the 71 RFLP haplotypes recorded at the PAH locus (Eisensmith and Woo, 1992b), only 7 (haplotypes 1–7) account for over 80% of all mutant PAH chromosomes in most European populations (Eisensmith *et al.*, 1992). Haplotypes 1, 4, and 7 are relatively common among normal chromosomes in Europeans, Asians, and Polynesians, suggesting that the formation of these three haplotypes preceded racial divergence (Hertzberg *et al.*, 1989). Haplotypes 2 and 3 are common among mutant chromosomes in northern (Chakraborty *et al.*, 1987; Apold *et al.*, 1990; Svensson *et al.*, 1991) and central European (Aulehla-Scholz *et al.*, 1988; Herrmann *et al.*, 1988; Lichter-Konecki *et al.*, 1988b; Reiss *et al.*, 1988) populations, but are relatively rare among normal chromosomes. In eastern European countries, the predominant mutant chromosome is haplotype 2 (Daiger *et al.*, 1989a; Zygulska *et al.*, 1991), while in southern European populations, the major mutant chromosome is haplotype 6 (Kalaydjieva *et al.*, 1990; Dianzani *et al.*, 1990; Lichter-Konecki *et al.*, 1989; Stuhrmann *et al.*, 1989). Haplotypes 5 and 7 are more often present on normal chromosomes in several European populations, especially in Germany, but haplotype 7 was strongly associated with mutant chromosomes in Norway. Haplotypes 1 and 4

**Table 6.1.** Relative Frequencies and Distributions of PAH Haplotypes in European and Asian Populations

| Region/country | Haplotype | | | | | | | | | | | | | | | | No. of alleles | |
|---|---|---|---|---|---|---|---|---|---|---|---|---|---|---|---|---|---|---|
| | 1 | | 2 | | 3 | | 4 | | 5 | | 6 | | 7 | | 1–7 | | | |
| | N | M | N | M | N | M | N | M | N | M | N | M | N | M | N | M | N | M |
| Northern Europe | | | | | | | | | | | | | | | | | | |
| Denmark[a] | 35* | 18 | 4 | 20* | 3 | 38**** | 32* | 13 | 10* | 0 | 0 | 3 | 10* | 2 | 94 | 94 | 66 | 66 |
| Norway[b] | 28 | 21 | 7 | 11 | 4 | 17*** | 27*** | 9 | 14 | 6 | 2 | 0 | 8 | 20* | 92 | 84 | 85 | 86 |
| Sweden[c] | 31*** | 15 | 8 | 23**** | 2 | 14**** | 19 | 27 | 8 | 6 | 12**** | 1 | 7 | 4 | 76 | 90*** | 132 | 136 |
| Western Europe | | | | | | | | | | | | | | | | | | |
| France[d] | 26 | 31 | 6 | 18† | 3 | 9 | 15 | 8 | 8* | 0 | 2 | 3 | 12 | 4 | 72 | 73 | 66 | 74 |
| Germany[e,f,g,h] | 24 | 22 | 8 | 32**** | 3 | 13**** | 20 | 19 | 12**** | 2 | 2 | 2 | 14**** | 2 | 83 | 92**** | 246 | 246 |
| Scotland[i] | 32 | 30 | 3 | 9 | 6 | 18 | 13 | 6 | 10 | 3 | 3 | 0 | 10 | 0 | 77 | 66 | 31 | 33 |
| Switzerland[j] | 42 | 50 | 6 | 11 | 0 | 5 | 25 | 18 | 11 | 0 | 0 | 5 | 11 | 0 | 95 | 89 | 36 | 38 |
| Eastern Europe | | | | | | | | | | | | | | | | | | |
| Bulgaria[j] | 24 | 16 | 2 | 38**** | 2 | 0 | 15* | 10 | 2 | 2 | 2 | 22*** | 10* | 0 | 57 | 88*** | 42 | 50 |
| Czechoslovakia[k] | 14 | 0 | 33 | 66* | 5 | 0 | 14 | 23 | 5 | 0 | 0 | 0 | 0 | 0 | 71 | 91 | 21 | 22 |
| Hungary[k] | 32 | 13 | 16 | 55**** | 8 | 5 | 16 | 10 | 0 | 2 | 0 | 0 | 5 | 0 | 77 | 85 | 38 | 40 |
| Poland[l] | 27 | 9 | 9 | 57**** | 0 | 2 | 11 | 11 | 9 | 2 | 0 | 5 | 7 | 5 | 63 | 91**** | 44 | 44 |
| Southern Europe | | | | | | | | | | | | | | | | | | |
| Greece[m] | 30 | 47 | 10 | 18 | 0 | 0 | 20 | 29 | NA | NA | NA | NA | NA | NA | ≥60 | ≥94* | 10 | 17 |
| Italy[n] | 29 | 40 | 0 | 6 | 3 | 3 | 24* | 9 | 9 | 4 | 2 | 18*** | 5 | 0 | 72 | 80 | 63 | 68 |
| Turkey[o,p] | 14 | 25 | 7 | 1 | 2 | 1 | 31* | 17 | 10 | 4 | 2 | 36**** | 11* | 2 | 78 | 86 | 84 | 91 |
| Asia | | | | | | | | | | | | | | | | | | |
| China[q] | 3 | 0 | 0 | 3 | 3 | 3 | 77 | 83 | 0 | 0 | 0 | 0 | 0 | 6 | 86 | 95 | 35 | 44 |
| Japan[q] | 0 | 9 | 0 | 0 | 0 | 0 | 78 | 73 | 0 | 0 | 11 | 0 | 11 | 18 | 100 | 100 | 9 | 11 |

| S. E. Asians[r] | 8 | ND | 0 | ND | 1 | ND | 72 | ND | 0 | ND | 0 | ND | 11 | ND | 92 | ND | 74 | ND |
| Polynesians[r] | 21 | ND | 0 | ND | 1 | ND | 58 | ND | 0 | ND | 1 | ND | 16 | ND | 97 | ND | 630 | ND |

*Note.* NA, not available; these data were combined into a single category representing haplotypes 5–12 rather than reported separately. ND, not determined; PKU was not present among the S. E. Aisans or Polynesians examined in these studies.

[a]Chakraborty et al., 1987.
[b]Apold et al., 1990.
[c]Svensson et al., 1991.
[d]Rey et al., 1988.
[e]Aulehla-Scholz et al., 1988.
[f]Hermann et al., 1988.
[g]Lichter-Konecki et al., 1988b.
[h]Riess et al., 1988.
[i]Sullivan et al., 1989.
[j]Kalaydjieva et al., 1990.
[k]Daiger et al., 1989a.
[l]Zygulska et al., 1991.
[m]Hofman et al., 1989.
[n]Dianzani et al., 1990.
[o]Lichter-Konecki et al., 1989.
[p]Stuhrmann et al., 1989.
[q]Daiger et al., 1989b.
[r]Hertzberg et al., 1989.

* Significant disequilibrium between normal or mutant chromosome and haplotype ($p < 0.05$ by Fisher's exact test).
** Significant disequilibrium between normal or mutant chromosome and haplotype ($p < 0.01$ by Fisher's exact test).
*** Significant disequilibrium between normal or mutant chromosome and haplotype ($p < 0.005$ by Fisher's exact test).
**** Significant disequilibrium between normal or mutant chromosome and haplotype ($p < 0.001$ by Fisher's exact test).

are common among both mutant and normal chromosomes throughout most European populations (Eisensmith et al., 1992). In contrast to the heterogeneity observed between different European populations, nearly all normal (Daiger et al., 1989b; Hertzberg et al., 1989) and mutant (Daiger et al., 1989b) chromosomes in Asian populations are haplotype 4.

## 5. Haplotype/phenotype correlations in PKU and related HPAs

Despite these differences in the distribution of RFLP haplotypes among normal and mutant chromosomes in different populations, there are significant correlations between some RFLP haplotypes and PKU or HPA phenotypes within specific populations (Güttler et al., 1987; Rey et al., 1988; Verelst et al., 1988; Lichter-Konecki et al., 1988b; John et al., 1988; Trefz et al., 1988; Svensson et al., 1991). For example, nearly all individuals in northern European populations who carry any combination of mutant PAH haplotypes 2 or 3 alleles exhibit a severe or classical form of PKU, while patients who carry mutant PAH alleles of haplotypes 1 or 4 exhibit a wide range of phenotypes. Studies of families in which PKU and HPA occur in different siblings indicate that classical PKU and milder forms of HPA are phenotypic manifestations due to the presence of various combinations of different PAH alleles (Ledley et al., 1986b; Avigad et al., 1991; Okano et al., 1991b; Güttler et al., 1993b; Svensson et al., 1993a). The reasons why the segregation of different PAH alleles can produce significantly different phenotypes will be discussed under section III.D.

## B. Mutations in the human PAH gene

## 1. PKU is not caused by deletion of the entire PAH gene

Even before the complete organization of the human PAH gene was known, Southern hybridization was employed to examine the role of the PAH gene in PKU. Genomic DNAs from cell lines derived from two classical PKU patients (GM934 and GM2406) and two normal individuals were digested with various restriction enzymes and probed using the hPAH247 cDNA. Identical hybridization patterns were obtained from all of these samples, confirming that the overall organization of the PAH genes in these patients was similar to that of the normal gene (Woo et al., 1983). Densitometric analyses further demonstrated that the hybridization signals generated by the PKU DNA samples were not due to separate, nonoverlapping deletions in the two PAH genes. Thus, deletion of the entire PAH gene was not the cause of PKU in these individuals.

## 2. Heterogeneity of PAH genotypes

Prior to mutational analysis of the PAH gene, evidence from three sources suggested that PAH genotypes are heterogeneous. First, both CRM-positive and

-negative phenotypes could be detected in the livers of PKU patients (Choo *et al.*, 1979, 1980; Bartholomé and Dresel, 1982; Yamashita *et al.*, 1985). Second, adequate amounts of PAH mRNA could be detected in the livers of some patients, but not in others (DiLella *et al.*, 1985). Third, there was an extremely broad spectrum of clinical phenotypes observed in PAH-deficient patients, which will be further discussed later in this chapter. Based on these observations, several authors proposed that the phenotypic variation present in PAH deficiencies reflected an underlying heterogeneity at the molecular level (Scriver *et al.*, 1988). Subsequent analyses of mutant PAH genes have detected nearly 190 different single-base substitutions and deletions in the PAH gene, confirming this hypothesis.

## 3. Molecular lesions in the human PAH gene

Initial mutation detection studies focused on chromosomes of mutant haplotypes 2 and 3. The relatively high frequency of these haplotypes among mutant chromosomes in northern European populations (Table 6.1) suggested close associations with single mutations, as had previously been observed at the β-globin locus (for review, see Orkin and Kazazian, 1984). This hypothesis was attractive since it could also explain the strong correlations observed between these two RFLP haplotypes and severe PKU phenotypes (Güttler *et al.*, 1987). This prediction was confirmed through direct molecular analysis of mutant haplotype 2 and 3 chromosomes.

The PAH genes from northern Europeans bearing mutant haplotype 2 or 3 chromosomes were isolated by molecular cloning and sequence analysis was performed. These studies revealed the presence of a C-to-T transition in exon 12 of the PAH gene from an individual homozygous for mutant haplotype 2 (DiLella *et al.*, 1987) and a G-to-A transition at the consensus splice-donor site at the exon 12/intron 12 boundary region of the PAH gene from an individual homozygous for mutant haplotype 3 (DiLella *et al.*, 1986b). In the mutation terminology employed by the PAH Gene Mutation Analysis Consortium, the C-to-T missense mutation associated with haplotype 2 is referred to as R408W, where R is the single-letter amino acid code for the original amino acid, 408 is the codon containing the mutation, and W is the single-letter amino acid code for the mutant amino acid. The mutation that occurs in the intron 12 splicing signal is referred to as IVS12nt1, indicating that it is a single-base substitution occurring at the first nucleotide of the twelfth intervening or intronic sequence in the PAH gene.

Shortly after the discovery of these two PAH mutations, population screening studies were performed to determine the relative frequencies of these alleles in different populations. Such studies demonstrated that the R408W mutation was never found on normal chromosomes and was rarely found on chromosomes of haplotypes other than haplotype 2 in northern and eastern

European populations, suggesting that this mutation is both responsible for PKU and is distinct from other mutations that are present on chromosomes bearing other haplotypes. Similar results have been observed for the IVS12nt1 mutation and mutant haplotype 3 chromosomes in northern Europeans. The absence of these mutations from most other mutant chromosomes implies that these other chromosomes must bear novel PKU mutations, confirming the heterogeneous nature of this disorder at the molecular level. Mutation/haplotype associations will be discussed in greater detail under section IV.

With the introduction of PAH mutation detection methods based on the polymerase chain reaction (DiLella et al., 1988) and related technologies, such as chemical cleavage (Dianzani et al., 1991), single-strand conformational polymorphism (SSCP) analysis (Labrune et al., 1991), or denaturing gradient gel electrophoresis (DGGE) analysis (Guldberg et al., 1993a), dramatic progress has been made in the detection of PAH mutations within the past few years. At present, nearly 190 different single-base substitutions and deletions have been reported in the PAH gene (PAH Gene Mutation Analysis Consortium, April 1994). The identity and position of known PAH mutations are listed in Table 6.2. Also included in this table are any novel mutation/haplotype associations and the population in which they were initially observed.

Most of the disease-causing changes in the PAH gene are single-base substitutions, including missense (120/189; 63.5%), splicing (19/189; 10.1%), and nonsense (15/189; 7.9%) mutations and silent polymorphisms (11/189; 5.8%). At least 20 of the missense mutations occur at highly mutagenic CpG dinucleotides. Deletions or insertions of various types account for the remaining mutations. Deletions of single exons are most common when the exon-skipping effects of splicing (Marvit et al., 1987) and other mutations (Okano et al., 1994) near intron/exon boundaries are included (23/189, 12.2%). Other large deletions include the loss of approximately 7 kb containing exon 3 and flanking intronic regions from chromosomes of Yemenite Jews (Avigad et al., 1990) and the loss of a region apparently containing exons 1 and 2 from chromosomes of Scottish PKU patients (Sullivan et al., 1985). The presence of a third large deletion, purportedly involving the 3' end of the gene, has been inferred from Southern analyses of several Japanese PKU patients (Trefz et al., 1990), although more recent evidence suggests that this deletion may actually involve exons 5 and 6 (Okano et al., 1994). Smaller deletions include a 22-bp deletion in exon 6 (Kleiman et al., 1992), a 15-bp in-frame deletion in exon 11 (Jaruzelska et al., 1993b), an 11-bp deletion in exon 7 causing a frameshift that produces a premature stop at codon 278 (Dworniczak et al., 1992), 3 single-codon deletions (Svensson et al., 1990; Caillaud et al., 1991; Guldberg et al., 1993a), 4 two-base deletions (Guldberg et al., 1993a,b; Bénit et al., 1994; P. Guldberg to the PAH Gene Mutation Analysis Consortium, April 1994), and 10 single-base deletions (Eigel et al., 1991; Kalaydjieva et al., 1992; Guldberg et al., 1993a; P. Guldberg to the PAH Gene Mutation Analysis Consortium, March 1993; Ramus and Cotton, 1993; Bénit et al., 1994; P. Guldberg and L. Tyfield to the PAH Gene Mutation Analysis Consortium, April 1994). Only three insertions have been observed; two are caused by the formation of novel cryptic splice sites induced by missense mutations within introns (Dworniczak et al., 1991a; Takahashi et al., 1994), while the third is caused by a 4-bp insertion within exon 10 (H. K. Ploos van Amstel to the PAH Gene Mutation Analysis Consortium, June 1994). No disease-causing mutations have yet been detected in the promoter region (Svensson et al., 1993b) or polyadenylation site of the gene, although four polymorphic changes in the promoter region have been described (Tyfield et al., 1993; Svensson et al., 1993b).

## 4. Distribution of mutations with the human PAH gene

Comparisons of the primary structures of PAH proteins from different species (Morales et al., 1990) or of PAH and related hydroxylases in mammalian species (Grenett et al., 1987) have demonstrated a high degree of homology in the regions of the protein encoded by exons 4 through 11 of the human PAH protein, with the highest degree of homology in the regions encoded by exons 7 to 9. This central region has been hypothesized to be important for the common functions of these proteins, i.e., the binding of oxygen and cofactor, and the catalytic conversion of substrate, while those features unique to individual proteins, such as substrate specificity, may lie outside this region. The importance of this central region is reflected not only by the high degree of homology, but also by the large number of PKU-causing mutations found in this region. Approximately 20% of all known mutations affect exon 7 and more than 80% affect the region encoded by exons 5 through 12 (see Table 6.2). Despite the preponderance of mutations in exon 7 of the PAH gene, calculations of mutation frequency corrected for the relative size of the exon and its sequence composition do not suggest that this exon is a hotspot for mutation within the PAH gene (Dworniczak et al., 1992); this clustering instead reflects the critical nature of the region of the protein encoded by this exon for proper enzyme function. This analysis provides further support for the studies of Jennings and co-workers who used monoclonal anti-idiotype antibodies to demonstrate that the cofactor binding domain lies within the region of the protein encoded by exon 7 (Jennings et al., 1991).

One obvious consequence of this mutation clustering in regions of high homology between PAH and related proteins is that a majority of the PAH mutations observed occur in amino acid residues that are highly conserved between different proteins and species. For example, while only 38% (172/452) of the amino acid residues in human PAH are conserved in TYH and rabbit TPH (Grenett et al., 1987), 58% (69/120) of the missense mutations occur at these

**Table 6.2.** Mutations and Polymorphisms in the PAH Gene of PKU or HPA Patients

| Exon/ Intron | Mutation Designation | Nucleotide Change | Codon Change | Haplotype | Population | Comments | References |
|---|---|---|---|---|---|---|---|
| PRO | -348 t→c | -348 t→c | - | 1 | European | g | Svensson et al., 1993b |
| PRO 5'UTR | -224 g→a -146 c→t | -224 g→a -146 c→t | - | 9 | European | g, h | Svensson et al., 1993b |
| 5'UTR | -71 a→c | -71 a→c | - | 34 | English | g | Tyfield et al., 1993 |
| E1 | M1I | ATG→ATA | Met→Ile | 7 | Norwegian | a, b | Eiken et al., 1992a |
| E1 | M1V | ATG→GTG | Met→Val | 2 | French-Canadian | a | John et al., 1989 |
| | | | | 2 | French | | Lyonnet et al., 1992 |
| E1 | ΔctS16fs | TCT→TGA | Ser→Ter | ND | German | c | Guldberg et al., to Consortium, 1994 |
| E1 | Q20X | CAG→TAG | Gln→Ter | ND | German | | Guldberg et al., to Consortium, 1994 |
| E1/E2 | ΔE1/E2 | - | - | ND | Scottish | d | Sullivan et al., 1985 |
| I1 | IVS1nt5 | +5 g→t | - | 4 | Danish | e | Guldberg et al., 1993a |
| E2 | F39L | TTC→TTG | Phe→Leu | 1 | Anglo-Australian | | Forrest et al., 1991 |
| E2 | ΔF39 | ΔTTC | - | 21 | Danish | | Guldberg et al., 1993a |
| E2 | S40L | TCA→TTA | Ser→Leu | ND | German | | Guldberg et al., to Consortium, 1994 |
| E2 | L41F | CTC→TTC | Leu→Phe | ND | Sicilian | | Guldberg et al., 1993c |
| E2 | G46S | GGT→AGT | Gly→Ser | 5 | Danish | | Guldberg et al., 1993a |
| E2 | A47V | GCA→GTA | Ala→Val | ND | Danish | f | Guldberg et al., to Consortium, 1994 |
| E2 | L48S | TTG→TCG | Leu→Ser | 3, 4 | German Turks | | Konecki et al., 1991 |
| E2 | ΔtF55fs | TTT→TTG | Phe→Leu | 1 | German | c | Eigel et al., 1991 |
| E2 | E56D | GAG→GAT | Glu→Asp | 10 | Chinese | | Li et al., 1994 |
| E2/I2 | E56E/IVS2nt1 | Gg→Aa | Glu→Glu | 7, 42 | Arab | e | Kleiman et al., 1991 |
| I2 | IVS2nt5 | +5 g→t | - | 28 | Danish | e | Guldberg et al., 1993a |
| I2 | IVS2nt6 | +6 t→g | - | ND | German | e | Guldberg et al. to Consortium, 1994 |
| I2 | IVS2nt19 | +19 t→c | - | 5, 6, 11, ND | European | g | Lichter-Konecki et al., to Consortium, 1994 |
| E3 | ΔE3 | - | - | ND | Yemenite Jewish | | Avigad et al., 1990 |
| E3 | T63P/ H64N | ACC→CCC/ CAC→AAC | Thr→Pro/ His→Asn | 1 | Danish | h | Guldberg et al., 1993a |
| E3 | I65T | ATT→ACT | Ile→Thr | 9 | French-Canadian | f | John et al., 1992 |
| | | | | 5, B | English | | Tyfield et al., 1993 |
| | | | | 1 | Spanish | | Desviat et al., 1993 |
| | | | | 5.9, 9.8, 21.12 | Australian | | Ramus et al. to Consortium, 1994 |
| E3 | S67P | TCT→CCT | Ser→Pro | 4 | Welsh | | Tyfield et al. to Consortium, 1993 |
| E3 | R68S | AGA→AGT | Arg→Ser | 1 | European | | Horst et al., 1991 |
| E3 | D84Y | GAT→TAT | Asp→Tyr | 4 | E. German | | Kunert et al. to Consortium, 1992 |
| E3 | S87R | AGC→AGA | Ser→Arg | ND | Danish | f | Guldberg et al. to Consortium, 1994 |

Notes. PRO, promoter region; 5'UTR-5' untranslated region; ND, not determined; a, translation initiation codon; b, de novo mutation; c, frameshift mutation generating premature termination codon; d, unspecified deletion; e, mutation affecting proper splicing; f, mutation associated with HPA; g, polymorphism; h, linked single base substitutions; i, mutant protein has increase in apparent $K_{m(Pha)}$; j, mutation involving CpG dinucleotide; k, possible example of recurrent mutation; l, first reported as Ser→Arg (Forrest et al., 1991); m, mutation creates cryptic splice, resulting in inframe insertion; n, first reported as L430L (Forrest et al., 1991).

Table 6.2. (*Continued*)

| Exon/ Intron | Mutation Designation | Nucleotide Change | Codon Change | Haplotype | Population | Comments | References |
|---|---|---|---|---|---|---|---|
| E3 | T92I | ACA→ATA | Thr→Ile | ND | Sicilian | | Guldberg et al., 1993c |
| E3 | ΔI94 | ΔATC | ΔIle | 2 | Portugese | f, i | Caillaud et al., 1991 |
| E3 | L98S | TTG→TCG | Leu→Ser | ND | Pakistani | f | Guldberg et al., 1993c |
| E3 | A104D | GCC→GAC | Ala→Asp | 1 | European | f | Horst et al., 1991 |
| E3 | R111X | CGA→TGA | Arg→Ter | 4 | Chinese | j | Wang et al., 1989 |
| | | | | 4 | European | k | Guldberg et al., 1993a, b |
| E4 | W120X | TGG→TAG | Trp→Ter | ND | German | | Guldberg et al. to Consortium, 1994 |
| E4 | T124I | ACC→ATC | Thr→Ile | 28 | Danish | | Guldberg et al., 1993a |
| E4 | D143G | GAT→GGT | Asp→Gly | 11 | Norwegian | | Apold et al. to Consortium, 1992 |
| I4 | IVS4nt-5 | -5 c→g | - | ND | Sicilian | e | Guldberg et al., 1993c |
| I4 | IVS4nt-1 | -1 g→a | - | 4 | Chinese | e | Wang et al., 1991c |
| E5 | G148S | GGT→AGT | Gly→Ser | 1 | E. German | | Kunert et al. to Consortium, 1992 |
| E5 | Y154N | TAC→AAC | Tyr→Asn | ND | American | | Guldberg et al. to Consortium, 1994 |
| E5 | R158Q | CGG→CAG | Arg→Gln | 4 | European | j | Dworniczak et al., 1989 |
| | | | | 16, 28 | French | | Abadie et al., 1993 |
| E5 | R158W | CGG→TGG | Arg→Trp | 4 | Taiwanese | j | Takarada et al., 1993b |
| E5 | F161S | TTT→TCT | Phe→Ser | 4 | Chinese | | Li et al., 1992 |
| E5 | I164T | ATT→ACT | Ile→Thr | ND | French | | Rey et al. to Consortium, 1993 |
| E5 | H170R | CAT→CGT | His→Arg | ND | Norwegian | | Apold et al. to Consortium, 1993 |
| I5 | ΔgIVS5nt1 | +1 g→t | - | 1 | Bulgarian | e | Kalaydjieva et al., 1992 |
| I5 | IVS5nt1g→a | +1 g→a | - | ND | German | e | Guldberg et al. to Consortium, 1994 |
| E6 | G171A | GGG→GCG | Gly→Ala | ND | French | | Bénit et al. to Consortium, 1993 |
| E6 | Q172X | CAG→TAG | Gln→Ter | ND | German | | Guldberg et al. to Consortium, 1994 |
| E6 | P173T | CCC→ACC | Pro→Thr | 4 | Japanese | | Shirahase & Shimada, 1992 |
| E6 | I174T | ATC→ACC | Ile→Thr | 1 | Danish | | Guldberg et al., 1993a |
| E6 | R176L | CGA→CTA | Arg→Leu | ND | Danish | f | Guldberg et al. to Consortium, 1994 |
| E6 | R176X | CGA→TGA | Arg→Ter | ND | Sicilian | j | Guldberg et al., 1993c |
| E6 | E178G | GAA→GGA | Glu→Gly | ND | Danish | f | Guldberg et al. to Consortium, 1994 |
| E6 | W187X | TGG→TGA | Trp→Ter | ND | Sicilian | | Guldberg et al., 1993c |
| E6 | V190A | GTG→GCG | Val→Ala | 3 | E. German | | Kunert et al. to Consortium, 1992 |
| E6 | L194P | CTG→CCG | Leu→Pro | ND | N. Irish | | Zschocke et al. to Consortium, 1994 |
| E6 | Δ22bpL197fs | TTG→TTG | Leu→Leu | ND | Arab | c | Kleiman et al., 1992 |
| E6 | Y204C | TAT→TGT | Tyr→Cys | 4 | Chinese | | Wang et al., 1991b |
| | | | | 3 | Chinese | | R. Eisensmith, unpublished observation |
| E6 | Y204X | TAT→TAG | Tyr→Ter | ND | Sicilian | | Guldberg et al., 1993c |
| E6 | P211T | CCA→ACA | Pro→Thr | ND | Sicilian | | Guldberg et al., 1993c |

(*continued*)

residues. The disproportionate frequency of mutation at these highly conserved residues reinforces the idea that they are extremely important for proper enzyme function.

Despite the methodological advances that have made direct sequencing of the entire coding region and the immediately adjacent intronic regions of the PAH gene relatively easy, there are still some PKU patients with no detected alterations in these portions of the gene (Guldberg et al., 1993a,b; R. Eisensmith,

**Table 6.2.** (*Continued*)

| Exon/ Intron | Mutation Designation | Nucleotide Change | Codon Change | Haplotype | Population | Comments | References |
|---|---|---|---|---|---|---|---|
| E6 | ΔcP211fs | CCA→CAC | Pro→His | ND | American | c | Guldberg *et al.* to Consortium, 1993 |
| E6 | L213P | CTT→CCT | Leu→Pro | ND | American | | Guldberg *et al.* to Consortium, 1994 |
| E6 | C217G | TGT→GGT | Cys→Gly | ND | Dutch | | Ploos van Amstel *et al.* to Consortium, 1994 |
| E6 | G218V | GGC→GTC | Gly→Val | 1 | Danish | | Guldberg *et al.*, 1993a |
| E6 | E221G | GAA→GGA | Glu→Gly | 4 | Turkish | | Konecki *et al.*, 1991 |
| E6 | ΔagE221/D222fs | GAA→GAA | Glu→Glu | 4 | Danish | c | Guldberg *et al.*, 1993a |
| E6 | I224M | ATT→ATG | Ile→Met | 4 | Japanese | | Shirahase & Shimada, 1992 |
| E6 | P225T | CCC→ACC | Pro→Thr | ND | Romanian | | Guldberg *et al.* to Consortium, 1994 |
| E6 | V230I | GTT→ATT | Val→Ile | ND | Danish | f, j | Guldberg *et al.* to Consortium, 1994 |
| E6 | S231P | TCT→CCT | Ser→Pro | ND | Italian | | Dianzani *et al.*, 1992 |
| E6 | Q232Q | CAA→CAG | Gln→Gln | ND | Anglo-Australian | j | Forrest *et al.*, 1991 |
| | | | | 3, 4, 7 | Caucasian | | Kalaydjieva *et al.*, 1991b |
| I6 | IVS6nt-2 | -2 a→g | - | 1 | Bulgarian | e | Kalaydjieva *et al.*, 1992 |
| E7 | T238P | ACT→CCT | Thr→Pro | 4 | European | | Dworniczak *et al.*, 1992 |
| E7 | G239S | GGT→AGT | Gly→Ser | ND | French | | Rey *et al.* to Consortium, 1993 |
| E7 | R241H | CGC→TGC | Arg→His | 5 | Polish | j | Jaruzelska *et al.* to Consortium, 1992 |
| E7 | R241C | CGC→TGC | Arg→Cys | 4 | Taiwanese | j | Takarada *et al.*, 1993c |
| E7 | L242F | CTC→TTC | Leu→Phe | ND | European | | Dworniczak *et al.*, 1992 |
| E7 | R243Q | CGA→CAA | Arg→Gln | 4 | Chinese | j | Wang *et al.*, 1991b |
| | | | | 9 | Chinese | | R. Eisensmith, unpublished observation |
| E7 | R243X | CGA→TGA | Arg→Ter | 4 | European | j | Wang *et al.*, 1990 |
| E7 | P244L | CCT→CTT | Pro→Leu | 12 | Spanish | | Desviat *et al.*, 1992 |
| E7 | V245A | GTG→GCG | Val→Ala | 3 | Danish | f | Guldberg *et al.* to Consortium, 1994 |
| E7 | V245E | GTG→GAG | Val→Glu | 3.8 | N. Irish | | Guldberg *et al.*, 1993a |
| E7 | V245V | GTG→GTA | Val→Val | 3, 4, 16, 17, 28 | European | g | Dworniczak *et al.*, 1990 |
| | | | | 6 | European | | Lichter-Konecki *et al.* to Consortium, 1993 |
| | | | | 4 | Chinese | | R. Eisensmith, unpublished observation |
| E7 | G247V | GGC→GTC | Gly→Val | 4 | Chinese | | Li *et al.*, 1992 |
| E7 | L248P | CTG→CGG | Leu→Pro | ND | Dutch | | Ploos van Amstel *et al.* to Consortium, 1994 |
| E7 | L249F | CTT→TTT | Leu→Phe | 1 | Portugese | | Caillaud *et al.*, 1992 |
| E7 | R252G | CGG→GGG | Arg→Gly | 7 | European | | Dworniczak *et al.*, 1992 |
| E7 | R252Q | CGG→TGG | Arg→Gln | ND | French | j | Rey *et al.* to Consortium, 1993 |
| E7 | R252W | CGG→TGG | Arg→Trp | ND | French | j | Abadie *et al.*, 1989 |
| | | | | 1, 7 | French | | Berthelon *et al.*, 1991 |
| | | | | 2, 6, 11 | Swedish | | Svensson *et al.*, 1993a |
| E7 | L255S | TTG→TCG | Leu→Ser | ND | African-American | | Hofman *et al.*, 1991 |

unpublished observation; B. Dworniczak, personal communication). Apart from methodological limitations or sequencing errors, these patients may harbor as yet unidentified mutations in the noncoding regions of the gene, including promoter, cap site, or polyadenylation site mutations, or mutations causing cryptic splice sites deep within introns. Since several studies have reported the identification of 99 and 94% of all PAH mutations in PKU and HPA patients from Denmark (Guldberg *et al.*, 1993a) and Sicily (Guldberg *et al.*, 1993c), these mutations are presumably rare in these populations.

**Table 6.2.** (*Continued*)

| Exon/ Intron | Mutation Designation | Nucleotide Change | Codon Change | Haplotype | Population | Comments | References |
|---|---|---|---|---|---|---|---|
| E7 | L255V | TTG→GTG | Leu→Val | 18, 21 | Chinese | | Li et al., 1992 |
| E7 | G257C | GGC→TGC | Gly→Cys | ND | German | | Guldberg et al. to Consortium, 1994 |
| E7 | A259T | GCC→ACC | Ala→Thr | 3 | Norwegian | | Apold et al. to Consortium, 1992 |
| E7 | A259V | GCC→GTC | Ala→Val | 42 | French | | Labrune et al., 1991 |
| E7 | R261P | CGA→CCA | Arg→Pro | ND | Belgian | | Guldberg et al. to Consortium, 1994 |
| E7 | R261Q | CGA→CAA | Arg→Gln | 1 | French | j | Abadie et al., 1989 |
| | | | | 1, 2 | European, N. African | | Berthelon et al., 1991 |
| | | | | 28 | Danes | | Tyfield et al., 1991 |
| E7 | R261X | CGA→TGA | Arg→Ter | ND | Italian | j, k | Bosco et al. 1991 |
| | | | | 3 | German, Turkish | | Dworniczak et al., 1991c |
| | | | | 2 | Japanese | | Shirahase et al., 1991 |
| E7 | F263L | TTC→TTG | Phe→Leu | ND | Caucasian | | Takarada et al., 1993a |
| E7 | H264L | CAC→CTC | His→Leu | ND | European | | Cardoso et al., to Consortium, 1993 |
| E7 | Δtl269fs | ATC→ACA | Ile→Thr | ND | English | c | Tyfield et al. to Consortium, 1994 |
| E7 | R270K | AGA→AAA | Arg→Lys | ND | American | | Guldberg et al. to Consortium, 1993 |
| E7 | R270S | AGA→AGT | Arg→Ser | 1 | Arabs | | Kleiman et al., 1993 |
| E7 | G272X | GGA→TGA | Gly→Ter | 7 | Norwegian | | Apold et al., 1990 |
| E7 | S273P | TCC→TTC | Ser→Phe | 7 | Belgian, French | | Melle et al., 1991 |
| E7 | Δ11bpL274fs | AAG→AAC | Leu→Asn | 9 | German | c | Dworniczak et al., 1992 |
| E7 | M276I | ATG→ATT | Met→Ile | ND | Anglo-Australian | | Forrest et al., 1991 |
| E7 | M276V | ATG→GTG | Met→Val | 4 | Japanese | | Goebel-Schreiner & Schreiner, 1993 |
| E7 | Y277C | TAT→TGT | Tyr→Cys | ND | Portugese | | Leandro et al., 1993 |
| E7 | Y277D | TAT→GAT | Tyr→Asp | ND | French | | Labrune et al., 1991 |
| E7 | T278A | ACC→GCC | Thr→Ala | ND | Portugese | | Leandro et al., 1993 |
| E7 | T278I | ACC→ATC | Thr→Ile | ND | Japanese | | Takahashi et al., 1994 |
| E7 | T278N | ACC→AAC | Thr→Asn | ND | Sicilian | | Guldberg et al., 1993c |
| E7 | E280K | GAA→AAA | Glu→Lys | 4 | French | j, k | Abadie et al., 1989 |
| | | | | 38 | Algerian, French | | Lyonnet et al., 1989 |
| | | | | 1 | Caucasian | | Okano et al., 1990a |
| E7 | P281L | CCg→CTg | Pro→Leu | 1, 4 | Caucasian | j | Okano et al., 1991 |
| I7 | IVS7nt1 | +1 g→a | - | ND | Italian | e | Dianzani et al., 1991 |
| | | | | 4 | Danish | | Guldberg et al., 1993a |
| I7 | IVS7nt2 | +2 t→a | - | 7 | Chinese | e | Wang et al., 1992 |
| I7 | IVS7nt5 | +5 g→a | - | ND | German | e | Guldberg et al. to Consortium, 1994 |
| E8 | D282N | GAC→AAC | Asp→Asn | 1 | Danish | | Guldberg et al., 1993a |
| E8 | R297H | CGC→CAC | Arg→His | ND | American | j | Guldberg et al. to Consortium, 1994 |

(*continued*)

# C. Expression analysis of PAH mutations

The exclusive association of single-base substitutions with mutant PAH chromosomes suggests, but does not prove, that these changes are the cause of the PKU or HPA phenotype. This is especially true for rare or private substitutions identified through examination of only a portion of the gene. Some of these may be benign polymorphisms linked to other expressed mutations. Ideally, supporting evidence should come from more comprehensive genetic and biochemical stud-

**Table 6.2.** (Continued)

| Exon/ Intron | Mutation Designation | Nucleotide Change | Codon Change | Haplotype | Population | Comments | References |
|---|---|---|---|---|---|---|---|
| E8 | F299C | TTT→TGT | Phe→Cys | ND | Caucasian | | Okano et al., 1989 |
| | | | | 8 | Norwegians | | Eiken et al, 1992b |
| E8 | A300S | GCC→TCC | Ala→Ser | 1 | Bulgarian | | Kalaydjieva et al., 1992 |
| E8 | Q304Q | CAG→CAA | Gln→Gln | ND | American | e?, g | Guldberg et al. to Consortium, 1993 |
| I8 | IVS8nt1 | +1 g→a | - | ND | American | e | Guldberg et al. to Consortium, 1993 |
| E9 | I306V | ATT→GTT | Ile→Val | 4 | Danish | f | Economou-Petersen et al., 1992 |
| E9 | A309D | GCC→GAC | Ala→Asp | 7 | French-Canadian | | Rozen et al. to Consortium, 1994 |
| E9 | A309V | GCC→GTC | Ala→Val | ND | Sicilian | | Guldberg et al., 1993c |
| E9 | L311P | CTG→CCG | Leu→Pro | 1 | German | | Lichter-Konecki et al., 1988a |
| | | | | 10 | German | | Riess et al., 1988 |
| | | | | 7 | Greek | | Hofman et al., 1989 |
| E9 | P314H | CCT→CAT | Pro→His | ND | Belgian | | Guldberg et al. to Consortium, 1994 |
| E9 | A322G | GCC→GGC | Ala→Gly | 12 | Swedish | f | Svensson et al., 1992 |
| E9 | A322T | GCC→ACC | Ala→Thr | 1 | Bulgarian | j | Kalaydjieva et al., 1991c |
| I9 | IVS9nt-6 | -6 g→t | - | ND | Japanese | e, m | Takahashi et al., 1994 |
| E10 | W326X | TGG→TAG | Trp→Ter | 4 | Chinese | | Wang et al., 1992 |
| E10 | F331L | TTT→CTT | Phe→Leu | 1 | French | | Bénit et al. to Consortium, 1993 |
| E10 | L333F | CTC→TTC | Leu→Phe | ND | Berber | | Abadie et al., 1993a |
| E10 | Q336X | CAA→TAA | Gln→Ter | 11 | Danish | | Guldberg et al., 1993a |
| E10 | D338Y | GAC→TAC | Asp→Tyr | 4 | French-Canadian | | Rozen et al. to Consortium, 1994 |
| E10 | K341R | AAG→AGG | Lys→Arg | ND | English | | Cadiou et al. to Consortium, 1994 |
| E10 | A342T | GCA→ACA | Ala→Thr | 5 | Danish | | Guldberg et al., 1993a |
| E10 | Y343C | TAT→TGT | Tyr→Cys | ND | Sicilian | | Guldberg et al., 1993c |
| E10 | A345T | GCT→ACT | Ala→Thr | 7 | Chinese | | Li et al., 1994 |
| E10 | ΔgG346fs | GGG→GGC | Gly→Gly | 16 | Danish | c | Guldberg et al., 1993a |
| E10 | L347F | CTC→TTC | Leu→Phe | ND | English | | Cadiou et al. to Consortium, 1994 |
| E10 | L348V | CTG→GTG | Leu→Val | 9 | Caucasian | | Eisensmith et al., 1991 |
| E10 | S349P | TCA→CCA | Ser→Pro | ND | Anglo-Australian | l | Forrest et al., 1991 |
| | | | | 1 | French-Canadian | | John, 1991 |
| | | | | 4 | Moroccan Jewish | | Weinstein et al., 1992 |
| E10 | insgtcaS349fs | TCA→TCG | Ser→Ser | ND | Dutch | c | Ploos van Amstel et al. to Consortium, 1994 |
| E10 | ΔgG352fs | GGT→GGG | Gly→Gly | ND | N. African | c | Bénit et al., 1994 |
| | | | | 2 | Italian-Canadian | | Rozen et al. to Consortium, 1993 |
| I10 | IVS10nt1 | +1 g→a | - | ND | Belgian | e | Guldberg et al. to Consortium, 1994 |
| I10 | IVS10nt3 | +3 a→g | - | ND | Danish | e?, f | Guldberg et al. to Consortium, 1994 |

ies. Genetic studies should demonstrate segregation between the substitution and the disease phenotype in PKU or HPA kindreds. In addition, the substitution should never be observed on a large number of normal alleles. Furthermore, if gene scanning techniques, such as SSCP or DGGE, were used to detect the putative mutation, it should be the only substitution apart from obvious polymorphisms observed in the entire gene. Such genetic evidence is usually adequate for frequently occurring mutations. However, in all cases, and especially for the steadily increasing number of substitutions observed on only a few chro-

Table 6.2. (*Continued*)

| Exon/ Intron | Mutation Designation | Nucleotide Change | Codon Change | Haplotype | Population | Comments | References |
|---|---|---|---|---|---|---|---|
| I10 | IVS10nt546 | +546 g→a | - | 6 | Irish, S. European | e | Dasovich et al., 1991 |
| | | | | 6, 10, 36 | S. European | e, m | Dworniczak et al., 1991a |
| | | | | 6, 10, 36 | S. European | | Kalaydjieva et al., 1991a |
| | | | | 6, 34 | English | | Tyfield et al., 1993 |
| I10 | IVS10nt554 | +554 c→t | - | ND | European | e | Jaruzelska et al., 1992 |
| E11 | Y356X(C→A) | TAC→TAA | Tyr→Ter | 3, 4, 7, 9 | Chinese | e | Wang et al., 1992 |
| E11 | Y356X(C→G) | TAC→TAG | Tyr→Ter | 12 | Danish | e? | Guldberg et al., 1993a |
| E11 | S359X | TCA→TGA | Ser→Ter | ND | Italian | | Dianzani et al., 1992 |
| E11 | ΔgK363fs | AAG→AAC | Lys→Asn | ND | American | c | Guldberg et al. to Consortium, 1993 |
| E11 | ΔL364 | ΔCTT | ΔLeu | 5 | Swedish | | Svensson et al., 1990 |
| E11 | Δ15bpL364 | ΔCTT...GAG | ΔLeu...Glu | 4 | Polish | | Jaruzelska et al., 1993b |
| E11 | P366H | CCC→CAC | Pro→His | ND | Sicilian | | Guldberg et al., 1993c |
| E11 | T372S | ACA→TCA | Thre→Ser | ND | Dutch | | Ploos van Amstel et al. to Consortium, 1994 |
| E11 | ΔgcA373fs | GCC→CAT | Ala→His | 1, 12 | French | c | Bénit et al., 1994 |
| E11 | ΔtY377fs | TAC→ACA | Tyr→Thr | ND | Iceland | c | Guldberg et al. to Consortium, 1994 |
| E11 | T380M | ACG→ATG | Thr→Met | ND | Sicilian | f | Guldberg et al., 1993c |
| | | | | 4.3 | N. Irish | | Zschocke et al. to Consortium, 1994 |
| E11 | L385L | CTG→CTC | Leu→Leu | ND | Anglo-Australian | g, n | Forrest et al., 1991 |
| | | | | 3, 7 | Caucasian | | Kalaydjieva et al., 1991b |
| E11 | Y386C | TAT→TGT | Tyr→Cys | ND | American | | Guldberg et al. to Consortium, 1994 |
| E11 | V388L | GTG→CTG | Val→Leu | ND | American | | Guldberg et al. to Consortium, 1994 |
| E11 | V388M | GTG→CTG | Val→Met | ND | Japanese | | Takahashi et al., 1992 |
| | | | | ND | Portugese | | Leandro et al., 1993 |
| | | | | 1 | Danish | | Guldberg et al., 1993a |
| E11 | ΔagA389fs | GCA→GAC | Ala→Ala | ND | Sicilian | c | Guldberg et al., 1993c |
| E11 | E390G | GAG→GGG | Glu→Gly | ND | Berber | | Abadie et al., 1993 |
| E11 | D394A | GAT→GCT | Asp→Ala | ND | Sicilian | | Guldberg et al., 1993c |
| E11 | D394H | GAT→CAT | Asp→His | ND | German | | Guldberg et al. to Consortium, 1994 |
| E11 | A395G | GCC→GGC | Ala→Gly | ND | American | | Guldberg et al. to Consortium, 1994 |
| E11 | A395P | GCC→CCC | Ala→Pro | ND | Danish | | Guldberg et al., 1993a |
| E11 | V399V | GTA→GTT | Val→Val | 4 | Chinese | g | Huang et al., 1990 |
| E11 | ΔaR400fs | AGG→GGA | Arg→Gly | 1.8 | Australian | e | Ramus & Cotton, 1993 |
| I11 | IVS11nt1 | +1 g→a | - | ND | American | e | Guldberg et al. to Consortium, 1994 |
| E12 | A403V | GCT→GTT | Ala→Val | ND | Sicilian | | Guldberg et al., 1993c |

(*continued*)

mosomes, examination of the effect of the substitution on protein function is desirable. Unfortunately, since expression of human PAH is limited to the liver (Murthy and Berry, 1975), accurate biochemical characterization of mutant PAH proteins would require liver tissue samples from individuals homozygous for a given mutant allele. However, such samples are difficult to obtain not only for ethical reasons, but also for genetic ones; most individuals with PKU or HPA are compound heterozygous for two different mutant alleles and may produce two or more species of PAH proteins, depending on the stability of the mutant subunit

Table 6.2. (*Continued*)

| Exon/ Intron | Mutation Designation | Nucleotide Change | Codon Change | Haplotype | Population | Comments | References |
|---|---|---|---|---|---|---|---|
| E12 | ΔcP407fs | CCT→CCC | Pro→Pro | ND | French | c | Bénit et al., 1994 |
| E12 | R408Q | CGG→CAG | Arg→Gln | 12 | Scandinavian | f, j | Svensson et al., 1992 |
| E12 | R408W | CGG→TGG | Arg→Trp | 2 | Danish | j | DiLella et al., 1987 |
|  |  |  |  | 1 | French-Canadian | k | John et al., 1990 |
|  |  |  |  | 44 | Taiwanese |  | Tsai et al., 1990 |
|  |  |  |  | 5 | Polish |  | Zygulska et al., 1991 |
|  |  |  |  | 41 | Taiwanese |  | Lin et al., 1992 |
|  |  |  |  | 34 | Portugese |  | Caillaud et al., 1992 |
| E12 | R413P | CGC→CCC | Arg→Pro | 4 | Oriental |  | Wang et al., 1991a |
| E12 | Y414C | TAC→TGC | Tyr→Cys | 4 | Caucasian | f | Okano et al., 1991a |
| E12 | Y414Y | TAC→TAT | Tyr→Tyr | ND | Danish | g | Guldberg & Güttler, 1993 |
| E12 | D415N | GAC→AAC | Asp→Asn | 1 | Danish | f | Economou-Petersen et al., 1992 |
| E12 | T418P | ACC→CCC | Thr→Pro | 4 | Chinese |  | Li et al., 1994 |
| I12 | IVS12nt1 | +1 g→a | - | 3 | Danish | e | DiLella et al., 1986 |
| E13 | A447D | GCC→GAC | Ala→Asp | ND | American |  | Guldberg et al. to Consortium, 1993 |
| ? | - | - | - | ND | Japanese |  | Trefz et al., 1990 |

*in vivo*. To overcome these limitations, *in vitro* expression analyses are often performed to demonstrate that the observed genetic alteration does in fact alter protein function.

The first example of expression analysis of a PAH mutation was performed on the R408W mutation (DiLella *et al.*, 1987). In this study, an expression vector containing the mutant PAH cDNA was created by site-directed mutagenesis. Plasmids containing the normal and mutant cDNA sequences were then introduced into cultured mammalian cells and the levels of PAH mRNA, immunoreactive PAH protein, and PAH enzyme activity were determined. In these experiments, both the normal and the mutant constructs produced similar levels of PAH mRNA, but the mutant construct failed to produce immunoreactive PAH protein or PAH enzyme activity. These results demonstrate that the effects of the R408W mutation are compatible with the severe PKU phenotype found in patients homozygous for this mutation (DiLella *et al.*, 1987).

In vitro expression analysis has now been performed on about 30 missense (DiLella *et al.*, 1987; Lichter-Konecki *et al.*, 1988a; Okano *et al.*, 1989, 1990b, 1991a,c; Dworniczak *et al.*, 1991b; Eisensmith *et al.*, 1991; Wang *et al.*, 1991a,b; John *et al.*, 1992; Li *et al.*, 1992, 1994; Svensson *et al.*, 1992; Kleiman *et al.*, 1993; Weinstein *et al.*, 1993), splicing (Marvit *et al.*, 1987; Dworniczak *et al.*, 1991a), or nonsense mutations (Svensson *et al.*, 1993a) and two single-codon deletions (Caillaud *et al.*, 1991; Svensson *et al.*, 1993a). Fifteen of these mutations, including the Δ194 single-codon deletion (Caillaud *et al.*, 1991), result in the expression of mutant PAH proteins with significant amounts of residual enzyme activity. The remaining characterized mutations result in the expression of mutant PAH proteins with little or no residual activity. Of the uncharacterized mutations, many are nonsense or splicing mutations that presumably yield inactive proteins, although this assumption remains to be verified.

Many mutant PAH proteins are unstable when expressed *in vitro*. In most cases, the degree of residual activity of these mutant PAH proteins is directly proportional to the amount of immunoreactive protein produced. Thus, these mutant proteins appear to have specific activities similar to that of the wild-type protein, but less inherent stability or less resistance to the effects of proteases. Whether this situation is true *in vivo et situ* remains to be determined. Expression of several of these mutant PAH proteins in primary hepatocytes obtained from the PAH-deficient Pah[enu1] mouse (McDonald *et al.*, 1990) yields results nearly identical to those from COS or other cultured mammalian cell lines (R. Eisensmith, unpublished observation), suggesting that the instability observed *in vitro* is not caused by the nonhepatic nature of the cell types used in these studies.

## D. Genotype/phenotype correlations in PKU and related HPAs

The demonstration of altered enzyme activity of mutant proteins in expression analysis is only a further indication of the segregation between mutations and disease phenotype. It does not prove conclusively that a mutation causes disease. This is especially true when the levels of residual enzyme activity obtained in expression studies are significantly higher than those obtained from biopsied liver tissue. Despite this limitation, significant correlations have been demonstrated between the results of *in vitro* expression studies and biochemical indices used for the diagnosis of hyperphenylalaninemia phenotypes (Okano *et al.*, 1991b; Svensson *et al.*, 1993a; Okano *et al.*, 1994). Significant correlations have also been shown between the results of *in vitro* expression studies and IQ at age 9 in early-treated patients (Trefz *et al.*, 1993). In these studies, a rudimentary formula consisting of a simple average of the *in vitro* levels of PAH enzyme activity associated with the proteins produced by each mutant allele was used to estimate the *in vivo* PAH activity in patients of different genotypes. Okano and colleagues (1991b) demonstrated significant correlations between this "predicted PAH activity" and pretreatment serum phenylalanine levels, phenylalanine tolerance, or serum phenylalanine levels following the administration of an oral protein load, while Svensson and co-workers (1993a) showed a strong correlation between this measure and phenotypic classifications based on dietary phenylalanine tolerance. The data from these two studies are summarized in Table 6.3. "Severe" mutations were those resulting in the expression of mutant PAH proteins having little or no enzymatic activity when measured *in vitro*. These findings were in agreement with the results of previous *in vivo* and *in*

**Table 6.3.** Genotype/Phenotype Correlation in Danish, German, and Swedish PKU and HPA Patients

| Mutant allele | R243X, R252W, G272X, E280K, P281L, ΔL364, R408W, or IVS12nt1 | R158Q | R261Q | Y414C | R408Q or A322G |
|---|---|---|---|---|---|
| R243X, R252W, G272X, E280K, P281L, ΔL364, R408W, or IVS12nt1 | C (74) | C (6) | C (5) M (4) | M (36) H (2) | M (1) H (3) |
| R158Q | | C (5) | | H (2) | |
| R261Q | | | M (4) | M (2) | |
| Y414C | | | | H (5) | H (1) |
| R408Q or A322Q | | | | | H (1) |

*Note.* C, classical PKU (see text for definition of phenotypes); M, mild PKU; H, hyperphenylalaninemia. Numbers in parentheses denote the number of patients in the respective categories.

*vitro* studies that demonstrated that classical PKU patients have little or no detectable PAH activity (Kang *et al.*, 1970; Bartholomé *et al.*, 1975; Kaufman *et al.*, 1975b; Trefz *et al.*, 1979, 1981). Mutations of this type were associated with a classical PKU phenotype when present in the homozygous state or in compound heterozygosity with other severe mutations. In contrast, "mild" mutations give rise to mutant proteins that have sizable amounts of residual enzyme activity when expressed *in vitro*. Individuals bearing these milder mutations in compound heterozygosity with severe mutations exhibit a milder form of PKU, while individuals bearing milder mutations in the homozygous state or in compound heterozygosity with other mild mutations can exhibit mild PKU or HPA, depending on the levels of residual activity associated with each mutant protein (Okano *et al.*, 1991b; Svensson *et al.*, 1993a; Avigad *et al.*, 1991). Thus, through comparisons of data from *in vitro* expression studies with data on the biochemical and clinical phenotypes, it is clear that the heterogeneity of PKU and HPA phenotypes can be largely explained by variability within the PAH locus.

The fact that a simple average of the levels of PAH activity derived from expression studies of mutant proteins is strongly correlated with the biochemical and clinical phenotypes of patients bearing these mutations suggests that the effects of deviant gene dosage or negative allelic complementation were not major influences on the manifestation of genotype in these patients. If negative allelic complementation was common in PKU, then some particular genotypic combinations should exist whose phenotypes are poorly predicted from the simple relationship with predicted PAH activity, which assumes equal contribution of both alleles. However, after examining the relationship between genotype and phenotype in over 200 PKU and HPA patients, no evidence of such genotypic combinations was found (Okano *et al.*, 1991b; Svensson *et al.*, 1993a). Similarly, no direct evidence for deviant gene dosages among affected individuals was found in these studies (Okano *et al.*, 1991b; Svensson *et al.*, 1993a).

It is not altogether surprising that no direct evidence has yet been found in support of negative allelic complementation. Negative interallelic complementation assumes the production and interaction of two stable mutant proteins, yet most mutant PAH proteins appear to be unstable. Based on the low relative frequencies of mutant proteins with significant levels of PAH immunoreactivity, patients bearing mutations that produce stable mutant proteins will be extremely rare in most populations, excluding true heterozygotes. Thus, detailed examination of negative allelic complementation awaits the development of more sophisticated expression systems capable of producing equivalent amounts of different mutant peptides. In addition to these two hypotheses, the relationship between PAH genotype and HPA phenotype could be altered by the effects of other genes that are also important determinants of phenylalanine homeostasis (Langenbeck *et al.*, 1988). Although such modifier genes have not yet been identified, the effects of such genes could certainly contribute to the scattering of biochemical values in patients with little or no PAH activity.

Although useful in establishing genotype/phenotype correlations, expression analysis is not entirely without limitations. The *in vitro* activities of several mutant PAH proteins are almost certainly higher than their corresponding activities *in vivo*. Thus, without well-designed and carefully controlled comparative studies between *in vivo* and *in vitro* systems, the results of expression studies cannot be used to assess the absolute hepatic PAH activities in patients carrying mutant PAH alleles. In addition, while the phenotypes of large numbers of patients are well predicted from *in vitro* studies of mutant PAH proteins, there are certainly examples in which PAH activities predicted from *in vitro* studies do not correlate, or correlate only weakly, with the biochemical or clinical phenotypes of treated or untreated patients (Güttler *et al.*, 1993a; Ramus *et al.*, 1993; Verelst *et al.*, 1993). Genotype/phenotype correlations tend to be strongest when sufficiently large numbers of patients of widely differing phenotypes are examined.

Deviations of phenotype from those predicted by genotype are both problematic and interesting. The existence of some individuals whose phenotypes deviate significantly from those predicted by their genotypes effectively precludes genotype determination as the primary diagnostic test for hyperphenylalaninemia. This, however, is not the intended application of these types of study. Their true importance is as a means of identifying patients of different genotypic classes so that they can be compared both genetically and phenotypically. This rational beginning should permit many interesting and important clinical issues to be examined in greater detail. Such studies may expand our understanding of the role of environmental conditions, such as dietary compliance, early discontinuation of treatment, or maternal serum phenylalanine levels, or genetic factors, such as gene dosage, allelic complementation, or multi-loci effects, on the ultimate manifestation of disease phenotype. Patients whose phenotypes are poorly predicted from their PAH genotype can be identified and studied to determine the possible roles of these or other factors. This is especially true in families containing individuals with apparently identical PAH genotypes but significantly different clinical phenotypes (Ledley et al., 1986b; DiSilvestre et al., 1991). In the meantime, the excellent predictive abilities of genotype determination in most of the patients examined thus far suggest that genotype analysis can be a valuable adjunct to phenotype determination in newborns or to haplotype analysis in families with an affected proband. Such testing should permit further refinement of initial diagnosis, optimization of therapy, and determination of long-term prognosis.

## IV. POPULATION GENETICS OF PHENYLKETONURIA

### A. Frequency of PKU and related HPAs in human populations

Despite the apparent disadaptive effects of homozygosity for mutant PAH alleles, PKU and related HPAs exist in many different human populations. The incidence of PKU ranges from a high of approximately 1 in 2600 in Turkey (Özalp et al., 1986) to a low of approximately 1 in 120,000 in Japan (Aoki and Wada, 1988). The average incidence in Caucasians is approximately 1 in 10,000 (Bickel et al., 1981), while in Chinese it is approximately 1 in 16,500 (Liu and Zuo, 1986). Several mechanisms have been advanced to account for the relatively high incidence of PKU in humans. These include founder effect/genetic drift, heterozygote selection, reproductive compensation, elevated mutation rate, and the involvement of multiple loci that confer similar disease phenotypes. The involvement of multiple loci in PKU can be rejected, since more than 95% of all cases of HPA and all cases of classic PKU are the result of PAH deficiencies caused by mutation only at the PAH locus. Despite evidence that some PKU genotypes are the result of recurrent mutation (John et al., 1990; Okano et al., 1990a; Tsai et al., 1990; Dworniczak et al., 1991c; Ramus et al., 1992; Wang et al., 1992), elevated mutation rate at the PAH locus can also be dismissed as the primary mechanism responsible for the high frequency of PKU since many PKU and HPA patients within given populations bear a limited number of prevalent mutant alleles. The evidence for and against founder effect/genetic drift, hetero-zygote selection, and reproductive compensation in PKU will be examined below.

The frequency and distribution of many PAH mutations have been determined in various European, North and South American, Australian, and East Asian populations. Any meaningful summary of mutation frequencies in the general Caucasian population is effectively precluded by the significant variations in both the distribution of mutant haplotypes (see Table 6.1) and in the mutation/haplotype associations (see Table 6.2) in different ethnic sub-groups. However, a range of frequencies found in Caucasian populations can be estimated for a given mutation, both on a specific haplotype background and on the total number of mutant alleles; these data are presented in Table 6.4. For example, in the French-Canadian population, all mutant haplotype 2 alleles contain the M1V mutation (John et al., 1989), and this haplotype comprises approximately one-quarter of all mutant haplotypes. Thus, the M1V mutation accounts for about 25% of all mutant alleles in this population. In contrast, in most eastern European populations, all mutant haplotype 2 alleles contain R408W (Eisensmith et al., 1992), and this haplotype can account for two-thirds or more of all mutant haplotypes, yielding a relative frequency for R408W that approaches 70% in some populations.

## B. Evidence for founder effect/genetic drift in phenylketonuria

### 1. Mutation/haplotype associations at the PAH locus

As briefly introduced in previous sections, there are strong associations between specific RFLP (Eisensmith et al., 1992; Scriver et al., 1993), VNTR (Goltsov et al., 1992), and STR haplotypes and PAH mutations in different populations. However, both the distribution of mutant haplotypes (Daiger et al., 1989a; Eisensmith et al., 1992) and the qualities of the mutation/haplotype associations (Eisensmith and Woo, 1991, 1992a; Eisensmith et al., 1992; PAH Gene Muta-tion Analysis Consortium, 1994) can vary widely between different ethnic sub-groups. The distribution of PAH haplotypes and their associations with normal and mutant chromosomes in European and Asian populations have been pre-sented in Table 6.1. Populations in which novel mutation/haplotype associa-tions have been reported are recorded in Table 6.2.

The associations between five major PAH mutations and their respec-tive RFLP haplotypes in several European sample populations are shown in Table 6.5 (Eisensmith et al., 1992). Mutation/haplotype associations can be defined as exclusive, where a specific mutation is present only on a certain haplotype, inclusive, where all chromosomes of a certain haplotype contain a specific mutation, or both. The association between the IVS12nt1 mutation and mutant

**Table 6.4.** Range of Relative Frequencies of Common PAH
Mutations in Caucasians

| Haplotype | Mutation | Percentage of haplotype[a] (range) | Percentage of total alleles[b] (range) |
|---|---|---|---|
| 1 | F39L | 0–15 | 0–5 |
|  | ΔT55 | 0–15 | 0–5 |
|  | R252W | 0–20 | 0–5 |
|  | R261Q | 0–75 | 0–35 |
|  | E280K | 0–10 | 0–5 |
|  | P281L | 0–25 | 0–15 |
|  | A300S | 0–15 | 0–5 |
|  | R408W | 0–50 | 0–40 |
| 2 | M1V | 0–100 | 0–25 |
|  | R261Q | 0–10 | 0–1 |
|  | R408W | 0–100 | 0–70 |
| 3 | L48S | ND[c] | 0–5 |
|  | IVS12nt1 | 30–100 | 0–35 |
| 4 | L48S | 0–40 | 0–15 |
|  | R158Q | 0–80 | 0–10 |
|  | R243X | 0–20 | 0–5 |
|  | E280K | 0–30 | 0–5 |
|  | P281L | 0–5 | 0–5 |
|  | Y414C | 0–85 | 0–20 |
| 6 | IVS10nt546 | 0–100 | 0–40 |
| 7 | G272X | 0–80 | 0–15 |
| 9 | I65T | 0–100 | 0–20 |
|  | L348V | 0–70 | 0–10 |

[a]Range of values reflect the percentage of all mutant alleles
of a given haplotype that bear a specific mutation.

[b]Range of values reflect the percentage of all mutant alleles
that bear a specific mutation.

[c]ND, not determined.

haplotype 3 is both exclusive and inclusive in the Danish population
(Eisensmith *et al.*, 1992), while in other populations it is neither (Eisensmith *et al.*, 1992). The association between the R408W mutation and mutant haplotype 2 is also both exclusive and inclusive in many eastern European populations (Eisensmith *et al.*, 1992), but is neither in many other European, North American, and Australian populations. This mutation has been observed on haplotype 1 chromosomes in Danish (Eisensmith *et al.*, 1992; Guldberg *et al.*, 1993a),

**Table 6.5.** Associations between PAH Mutations and RFLP Haplotypes[a]

| Region/country | R261Q | | R408W | | IVS12nt1 | | R158Q | | IVS10nt546 | |
|---|---|---|---|---|---|---|---|---|---|---|
| | HT 1 | Non-HT 1 | HT 2 | Non-HT 2 | HT 3 | Non-HT 3 | HT 4 | Non-HT 4 | HT 6 | Non-HT 6 |
| **Northern Europe** | | | | | | | | | | |
| Denmark | 3/30[b] | 0/100 | 20/21 | 1/109 | 40/40 | 0/90 | 4/19 | 0/111 | 6/8 | 0/122 |
| Norway | 7/17 | 0/77 | 7/9 | 8/85 | 16/18 | 2/76 | 2/9 | 0/85 | 0/0 | 0/94 |
| Sweden | 2/20 | 0/116 | 30/31 | 2/105 | 18/19 | 0/117 | 2/37 | 0/99 | 0/2 | 0/134 |
| **Western Europe** | | | | | | | | | | |
| France | 10/25 | 0/45 | 8/9 | 1/61 | 9/9 | 0/61 | ND | ND | ND | ND |
| Germany | 12/40 | 0/45 | 52/52 | 0/49 | 32/32 | 0/34 | 12/47 | 0/30 | 5/5 | 2/109 |
| Switzerland | 13/18 | 0/18 | 3/4 | 0/32 | 2/2 | 0/34 | 2/6 | 0/30 | 1/2 | 0/34 |
| **Eastern Europe** | | | | | | | | | | |
| Czechoslovakia | 0/0 | 0/20 | 12/14 | 0/6 | 0/0 | 0/20 | 2/5 | 0/15 | 0/0 | 0/20 |
| Hungary | 1/5 | 0/36 | 22/22 | 0/19 | 3/3 | 0/38 | 4/5 | 0/36 | 0/0 | 0/41 |
| **Southern Europe** | | | | | | | | | | |
| Italy | 2/20 | 0/28 | 1/4 | 0/44 | 1/4 | 0/44 | 0/1 | 0/47 | 2/5 | 0/43 |
| Turkey | 8/22 | 0/12 | 1/1 | 0/4 | 0/1 | 0/3 | 4/19 | 0/5 | 31/31 | 0/28 |

*Note.* HT1, HT2, HT3, HT4, HT6, haplotypes 1, 2, 3, 4, and 6, respectively. ND, not determined.

[a]Only data from countries where haplotyped samples were screened for these mutations are reported in this table.

[b]Fractions indicate the number of positive alleles over the total number of alleles examined.

English (Tyfield et al., 1991), French (Berthelon et al., 1991), Norwegian (Eisensmith et al., 1992; Eiken et al., 1993), and Swedish (Svensson et al., 1993a) populations, and in French-Canadian (John et al., 1990; Treacy et al., 1993) and non-French-Canadian families (Treacy et al., 1993) in the Quebec province of Canada. In addition, this mutation has been reported on haplotype 5 in a Polish population (Zygulska et al., 1991), on haplotype 34 in a Portugese population (Caillaud et al., 1992), and on haplotypes 41 (Lin et al., 1992) and 44 (Tsai et al., 1990) in a Chinese population. There is also a strong association between the IVS10nt546 mutation and mutant haplotype 6 chromosomes in many southern European populations (Dasovich et al., 1991; Dworniczak et al., 1991a; Kalaydjieva et al., 1991a; Eisensmith et al., 1992; Pérez et al., 1992; Desviat et al., 1993). This association is strongly inclusive, since only a few mutant haplotype 6 alleles lack this mutation in southern Europe, but is not completely exclusive, since this mutation has also been observed on several other haplotypes (Dworniczak et al., 1991a; Kalaydjieva et al., 1991a; Eisensmith et al., 1992; Tyfield et al., 1993).

In contrast to mutant haplotypes 2, 3, and 6, mutant haplotypes 1 and 4 each harbor a large number of independent mutations in most populations. This finding likely reflects the fact that these latter two haplotypes are the most common on normal chromosomes in most populations and should therefore have had a higher probability of sustaining multiple mutational events. Despite the molecular heterogeneity of these two haplotypes, there is still one predominant mutation associated with each of these two haplotypes in some European populations (Table 6.5). For example, R261Q is strongly associated with haplotype 1. This association is most inclusive in Switzerland, where 13/18 haplotype 1 chromosomes contain this mutation (Eisensmith et al., 1992). Although this association is also exclusive of other haplotypes in most European populations, this is not true in all populations; associations between R261Q and haplotypes 2 (Berthelon et al., 1991) and 28 (Tyfield et al., 1991) have been observed in a few western European populations. R158Q is the predominant mutation associated with haplotype 4 in many European populations. This association is also both exclusive and nearly inclusive in some eastern European populations (Eisensmith et al., 1992). However, since R158Q is associated with haplotypes 16 and 28 in a few cases (Abadie et al., 1993b), this association is also not absolutely exclusive. As greater numbers of alleles are studied, it is likely that no mutation/haplotype association will remain completely exclusive or inclusive. Mutation, crossover, gene conversion, and other processes will transfer mutations to chromosomes of other haplotypes, while new mutations will occur on haplotype backgrounds that already contain other mutations.

As illustrated above, the same PAH mutation may be present on more than one haplotype background. Multiple haplotype associations have now been observed for at least 20 mutations in the PAH gene (Table 6.2; R. Eisensmith, unpublished observations), and several investigators have invoked recurrent mutation to account for some of these multiple haplotype associations in different populations (John et al., 1990; Okano et al., 1990a; Tsai et al., 1990; Dworniczak et al., 1991c; Ramus et al., 1992; Wang et al., 1992). For example, the presence of the R408W mutation on haplotypes 1 and 2 has been attributed to recurrent mutation (John et al., 1990; R. Eisensmith, unpublished observation), as this mutation involves a highly mutagenic CpG dinucleotide (Cooper and Youssoufian, 1988; Abadie et al., 1989). Similarly, recurrence is the most likely explanation for the presence of the E280K mutation on both haplotypes 4 and 38 (Lyonnet et al., 1989; Okano et al., 1990a); since these two haplotypes differ at all eight RFLP sites, crossover or gene conversion events are unlikely. Although some of the other PAH mutations that are associated with more than one haplotype also occur at CpG dinucleotides (Table 6.2), there are at least a dozen examples of multiple haplotype associations that do not (Dworniczak et al., 1991a; Kalaydjieva et al., 1991a; Konecki et al., 1991; Li et al., 1992; Wang et al., 1992; PAH Gene Mutation Analysis Consortium, April 1994). It is not yet clear whether the associations of these mutations with more than one haplotype are the result of recombination or similar processes that have transferred these mutations to chromosomes of different haplotypes or whether they are the result of mutation, recombination, or related processes that have caused the conversion of the chromosome bearing the mutation from one haplotype to another. The existence of haplotypes that are associated with more than one mutation and mutations that are associated with more than one haplotype make it impossible to infer the presence of a specific mutation based solely on haplotype analysis or vice versa.

## 2. Founder effects in European and Middle Eastern populations

The role of founder effect and genetic drift has been well established for several PAH mutations in specific populations. Perhaps the most well-documented example is the deletion mutation involving exon 3. This mutation was initially detected in an Israeli PKU patient of Yemenite ancestry (Avigad et al., 1990). Further studies identified this allele as the sole cause of PKU in the Yemenite Jewish population in Israel. Genealogical studies traced this mutation back through many generations to individuals living in villages scattered throughout Yemen. A more detailed examination of religious, legal, and other official documents revealed an ancestor common to all carriers of this deletion mutation who lived in San'a, the capital of Yemen, in the mid-18th century. Since PKU has not been observed among Yemenite Muslims, it is likely that the small Yemenite Jewish community represented a founding population for this deletion. The

**Table 6.6.** Relative Frequencies and Distributions of PAH Mutations in Europe

| Region/country | R158Q | | R261Q | | IVS10nt546 | | R408W | | IVS12nt1 | |
|---|---|---|---|---|---|---|---|---|---|---|
| **United Kingdom** | | | | | | | | | | |
| England[a,b,c] | 2/110 | (1.8) | 3/110 | (2.7) | 5/122 | (4.1) | 12/110 | (10.9) | 25/110 | (22.7) |
| Ireland[b] | 0/36 | (0.0) | 0/36 | (0.0) | 1/36 | (2.8) | 13/36 | (36.1) | 1/36 | (2.8) |
| **Scandinavia** | | | | | | | | | | |
| Denmark[d] | 9/308 | (2.9) | 5/308 | (1.6) | 16/308 | (5.2) | 56/308 | (18.2) | 115/308 | (37.3) |
| Norway[b] | 2/94 | (2.1) | 7/94 | (7.4) | 0/94 | (0.0) | 15/94 | (16.0) | 18/94 | (19.0) |
| Sweden[b] | 3/178 | (1.7) | 3/178 | (1.7) | 0/178 | (0.0) | 39/178 | (21.9) | 28/178 | (15.7) |
| **Northwestern Europe** | | | | | | | | | | |
| Belgium[e] | 20/240 | (8.3) | 16/240 | (6.6) | 11/240 | (4.6) | 15/240 | (6.3) | 26/240 | (10.8) |
| France[b,f] | 4/372 | (1.1) | 23/372 | (6.2) | 7/88 | (8.0) | 22/372 | (5.9) | 21/372 | (5.6) |
| Netherlands[g] | 13/98 | (13.3) | 12/98 | (12.2) | 4/98 | (4.1) | 1/98 | (1.0) | 21/98 | (21.4) |
| **Southwestern Europe** | | | | | | | | | | |
| Italy[b] | 0/72 | (0.0) | 2/72 | (2.8) | 9/72 | (12.5) | 1/72 | (1.4) | 2/72 | (2.8) |
| Portugal[h] | ND | | 6/61 | (9.8) | ND | | 1/61 | (1.6) | 0/61 | (1.6) |
| Sicily[i] | 5/106 | (4.7) | 9/106 | (8.5) | 16/106 | (15.1) | 0/106 | (0.0) | 0/106 | (0.0) |
| Spain[j] | ND | | 0/64 | (0.0) | 13/64 | (20.3) | 0/64 | (0.0) | 0/64 | (0.0) |
| **Central Europe** | | | | | | | | | | |
| Czechoslovakia[b] | 2/36 | (5.6) | 1/36 | (2.8) | 0/34 | (0.0) | 22/36 | (61.1) | 0/36 | (0.0) |
| Germany[b,k] | 12/202 | (5.9) | 20/326 | (6.1) | 5/202 | (2.5) | 80/326 | (24.5) | 47/326 | (14.4) |
| Switzerland[b] | 2/50 | (4.0) | 16/50 | (32.0) | 2/50 | (4.0) | 3/50 | (6.0) | 2/50 | (4.0) |

| | | | | | | | | | |
|---|---|---|---|---|---|---|---|---|---|
| Eastern Europe | | | | | | | | | |
| Hungary[b] | 5/70 | (7.1) | 1/70 | (1.4) | 2/70 | (2.9) | 34/70 | (48.6) | 3/70 | (4.3) |
| Poland[b] | ND | | 0/26 | (0.0) | 1/26 | (3.8) | 17/26 | (65.4) | 0/26 | (0.0) |
| Russia[b] | 8/220 | (3.6) | 2/156 | (1.3) | 4/330 | (1.2) | 133/218 | (61.0) | 3/156 | (1.9) |
| Southeastern Europe | | | | | | | | | |
| Bulgaria[l] | 1/34 | (2.9) | 1/34 | (2.9) | 5/34 | (14.7) | 15/34 | (44.1) | 0/34 | (0.0) |
| Croatia[m] | 1/50 | (2.0) | 3/50 | (6.0) | 2/50 | (4.0) | 19/50 | (38.0) | 1/50 | (2.0) |
| Turkey[b] | 4/83 | (4.8) | 8/83 | (9.6) | 31/79 | (39.2) | 1/83 | (1.2) | 0/83 | (0.0) |

*Note.* Fractions indicate the number of positive alleles over the total number of alleles examined. ND, not determined.

[a]Tyfield et al., 1991.
[b]Eisensmith et al., 1992.
[c]Tyfield et al., 1993.
[d]Guldberg et al., 1993a.
[e]Beaudoin et al. to PAH Gene Mutation Analysis Consortium, April 1994 release.
[f]Abadie et al. to PAH Gene Mutation Analysis Consortium, April 1994 release.
[g]Meijer et al. to PAH Gene Mutation Analysis Consortium, April 1994 release.
[h]Callaud et al., 1992.
[i]Guldberg et al., 1993c.
[j]Desviat et al., 1993.
[k]Horst et al., 1993.
[l]Kalaydjieva et al., 1993.
[m]Baric et al., 1992.

strict orthodoxy and closed nature of this community resulted in the expansion of this mutation throughout the Yemenite Jewish population, accounting for its present high frequency (Avigad et al., 1990).

Founder effect/genetic drift also appears to be the most likely mechanism to account for the limited spatial distribution of the G272X mutation within in several European populations (Apold et al., 1993), as well as for the relatively high frequencies of the M1V (John et al., 1989) and R408W (John et al., 1990) alleles in French-Canadians. Detailed genealogical examination of French-Canadian families residing in eastern Quebec revealed 53 potential founders for the M1V mutation, 19 of whom came from the small historical region of Mortagne-Perche in the early 17th century. Thirty-nine of 43 ancestors of these individuals whose origins were known came from Perche or adjacent counties (Lyonnet et al., 1992). Additional genealogical studies of these French-Canadian (John et al., 1990; Treacy et al., 1993) and non-French-Canadian families (Treacy et al., 1993) have indicated that the R408W mutation associated with RFLP haplotype 1 was most likely introduced into these families through Irish and Scottish ancestors (Treacy et al., 1993). This result suggests a founding population in northwest Europe for the R408W mutation on a haplotype 1 background.

Unlike these relatively localized PAH mutations, genealogical records alone cannot establish whether the present distribution of the major mutant PAH alleles in Europe may also be due at least in part to founder effect and genetic drift. Evidence supporting the existence of at least five different founding populations for the major mutant PAH alleles comes primarily from two sources, the strong mutation/haplotype associations observed for these major mutant alleles, presented in Table 6.5, and the strong gradients in their relative frequencies within Europe (Eisensmith et al., 1992), presented in Table 6.6.

An examination of the frequency distribution for the IVS12nt1 mutation (Table 6.6) shows that it is most frequent in Denmark, where it accounts for more than one-third of all mutant alleles. The frequency of this allele drops to between 10 and 20% in neighboring countries of northern and western Europe and is rare or absent in other European populations. These data suggest that this mutation first occurred in a Danish founding population and was subsequently spread into neighboring populations (Figure 6.4). The strong association observed between this mutation and haplotype 3 (Table 6.5) further suggests that this founding event occurred relatively recently on an evolutionary time scale, probably within the past few thousand years. Genealogical studies have shown that many of the Danish patients positive for this allele can trace ancestors back to villages on the west coast of Jutland (F. Güttler, personal communication). These villages served as points of departure for many of the Viking raiding parties that left Denmark, especially those travelling to England and north-

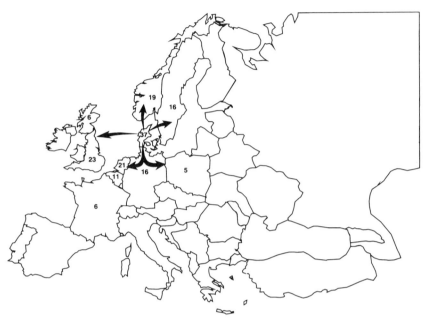

**Figure 6.4.** The frequency distribution of the IVS12nt1 mutation in Europe. Numbers indicate the relative frequencies of the IVS12nt1 mutation in populations where this mutation is prevalent. The arrows indicate the most probable direction of gene flow.

western Europe. Thus, it is not surprising that this mutation is relatively common in England, but less common in other parts of the United Kingdom not colonized by Danes (compare data from England and Ireland in Table 6.6). These observations confirm that this mutation was present in the Danish population more than 1000 years ago and was subsequently spread into neighboring populations by Danish Vikings.

The IVS10nt546 mutation is the most common mutation yet reported in southern European populations (Table 6.6). It accounts for nearly 40% of all mutant alleles in the Turkish population and for between 10 and 20% of all mutant alleles in Spain, Italy, Sicily, and Bulgaria. This allele is rare in western Europe and is mostly absent from Scandinavia and eastern Europe. The frequency distribution of this mutation is certainly suggestive of a Turkish origin with a subsequent spread throughout the Mediterranean basin, as shown in Figure 6.5, although this hypothesis has yet to be supported by additional historical, anthropological, or population genetic studies. Founder effect/genetic drift is also likely to have played a role in the subsequent spread of the IVS10nt546 mutation into some New World populations that accompanied the migration of

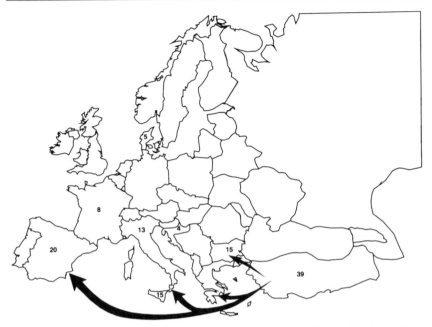

**Figure 6.5.** The frequency distribution of the IVS10nt546 mutation in Europe. Numbers indicate the relative frequencies of the IVS12nt1 mutation in populations where this mutation is prevalent. The arrows indicate the most probable direction of gene flow.

Spanish, Italian, or other settlers from Mediterranean countries. A recent report indicates that the relative frequency of this allele is between 20 and 30% in Chile and Argentina (Pérez et al., 1993).

One obvious feature of the frequency distribution of the R408W mutation is the strong east–west gradient that exists across Europe. This mutation is extremely prevalent in eastern European populations, where it accounts for 50% or more of all mutant alleles (Table 6.6). The high frequency of this mutation in eastern Europe, and its strong association with RFLP haplotype 2 in these populations, led to the suggestion of a Slavic (Eisensmith et al., 1992; Jaruzelska et al., 1993a) or Balto-Slavic (Kalaydjieva et al., 1991c) origin (Figure 6.6). An origin in eastern Europe is also compatible with the high frequency of haplotype 2 among normal chromosomes in these populations. However, this single origin for R408W in eastern Europe is not compatible with the relatively high frequency of this mutation in Ireland (Table 6.6; Eisensmith et al., 1992), nor with its association with haplotype 1 in several northwestern European or North American populations (John et al., 1990; Berthelon et al., 1991; Tyfield et al., 1991; Eisensmith et al., 1992; Eiken et al., 1993; Guldberg et al., 1993a; Svensson et al., 1993a; Treacy et al., 1993). These data suggest a second origin for R408W

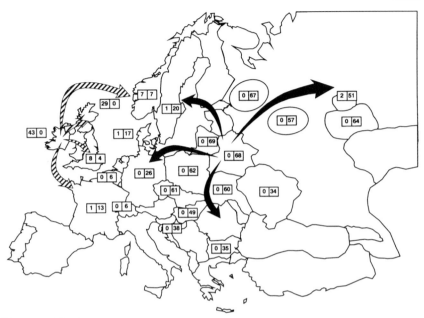

**Figure 6.6.** The frequency distribution of the R408W mutation in Europe. The numbers in the boxes on the left side indicate the relative frequency of the R408W mutation on haplotype 1 chromosomes. The numbers in the boxes on the right indicate the relative frequency of the R408W mutation on haplotype 2 chromosomes. The hatched arrows indicate the most probable direction of gene flow for the R408W mutation on haplotype 1, while the solid arrows indicate the most probable direction of gene flow for the R408W mutation on haplotype 2.

on haplotype 1 within the Irish population (Figure 6.6; Treacy *et al.*, 1993; R. Eisensmith, unpublished observation).

While the association between the R261Q mutation and haplotype 1 is strongly exclusive, it is not inclusive of all mutant haplotype 1 chromosomes. This lack of inclusivity suggests that the event creating the R261Q mutation occurred after the events causing the formation of haplotype 1. Since this mutation was relatively frequent in both Switzerland and Turkey but rare in most other populations (Table 6.6), it could either have occurred in a single common founding population or it could have occurred independently in these two populations. The strong association of this mutation with haplotype 1, especially in Turkey where haplotype 1 is relatively rare among normal alleles, suggests that recurrence is unlikely. This conclusion is further supported by the more recent finding of a strong association between this mutation and a single VNTR allele present on the polymorphic haplotype 1 background in these two populations (Goltsov *et al.*, 1992).

The association of the R158Q mutation with haplotype 4 was very similar to that seen between the R261Q mutation and haplotype 1; it is strongly exclusive but only weakly inclusive. Since this mutation also involves a CpG dinucleotide, it is unlikely that such a strong mutation/haplotype association could be the result of recurrence of this mutation in different European populations, but rather must reflect the occurrence of a single founding event. The relatively high frequency of this mutant allele in the Dutch and Belgian populations (Table 6.6) suggests that this allele may have originated within this region of Europe.

## 3. Population dynamics in Europe

Analysis of multiple polymorphic protein markers has been used by several investigators as a means of tracing human migratory patterns in prehistoric times, and the frequency distributions of the major mutant PAH alleles are largely similar to those observed in studies of other genetic markers in European populations (Menozzi et al., 1978; Sokal et al., 1991). For example, Menozzi and co-workers (1978) identified several axes along which the frequencies of several blood group markers varied. One major axis ran from southeast to northwest, and these authors proposed that this cline reflected the gene flow that accompanied the introduction of early agricultural methods into Europe from the Middle East 5000 to 10,000 years ago. This frequency gradient is roughly similar to that observed for the IVS10nt546 mutant allele, which is relatively common in Mediterranean and Middle Eastern populations, but relatively rare in northern and western European populations. A second cline, running from east to west, was observed by both Menozzi and co-workers (1978) and by Sokal and colleagues (1991) in a study of the ABO protein markers among Europeans. These gradients are very similar to those observed for the R408W mutation on haplotype 2 chromosomes. The underlying causes of this east–west cline may reflect the migrations of early Slavic or Germanic peoples in the middle of the first millennium AD. The frequency distribution of the IVS12nt1 mutation does not conform to any of the patterns previously observed in these two studies, but does bear a strong resemblance to that observed for the ΔF508 CF mutation in studies compiled and reported by DeVoto and co-workers (European Working Group on CF Genetics, 1990), suggesting that human migration rather than selection may have been the most important factor in the spread of these alleles.

## 4. Founder effects in Asian populations

While PKU had been considered a Caucasian disorder for many decades, newborn screening studies for PKU in China have now established an incidence (about 1 in 16,500 births; Liu and Zuo 1986) similar to that in various Caucasian populations. RFLP haplotype analysis (Daiger *et al.*, 1989b) of the PAH locus in Oriental populations has demonstrated significant differences in the genetic basis of PKU in Caucasians and Orientals. This conclusion is further supported by a number of studies of PAH mutations in Asian PKU patients (Tsai *et al.*, 1990; Wang *et al.*, 1989, 1991a,b,c, 1992; Shirahase *et al.*, 1991; Li *et al.*, 1992, 1994; Lin *et al.*, 1992; Okano *et al.*, 1992; Takahashi *et al.*, 1992; Takarada *et al.*, 1993c), which indicate that only about 4 of the 20 or so PAH mutations that account for up to 70% of all mutant alleles in some East Asian populations are present in both races. Of the exceptions, the R261X (Shirahase *et al.*, 1991) and R408W (Tsai *et al.*, 1990) mutations occur on different haplotype backgrounds and are likely to be the result of recurrent mutation. Although the R111X mutation is associated with haplotype 4 in both Europeans and Orientals, recurrence is also likely in this case; the mutation occurs at a CpG dinucleotide and haplotype 4 is the predominant normal chromosome in both populations, making it the most likely target for mutation. The fourth shared mutation, V388M, may be a European allele that was spread to Japan by migration and interbreeding. It can thus be concluded that at least a majority of, if not all, PAH mutations have occurred after the divergence of the Caucasoid and the Mongoloid peoples.

The relative frequencies of these mutations and their distribution in the northern, southern, and total Chinese populations (Wang *et al.*, 1989, 1991a,b,c 1992; Li *et al.*, 1992, 1994), as well as the Japanese (Okano *et al.*, 1992) and Korean (Okano *et al.*, 1992) populations, are shown in Table 6.7. These studies reveal an uneven geographic distribution for certain mutant PAH chromosomes in east Asia. For example, the R413P mutation, which accounts for approximately 9% of all mutant PAH alleles in the total Chinese population, is much more prevalent in northern China (13.8%) than in southern China (2.2%). The frequency of this mutant allele is also very high in Japan (27.3%), despite the much lower incidence of PKU in this population (1:100,000 in Japan vs 1:16,500 in China) (Wang *et al.*, 1991a). In contrast, the IVS4nt1 mutation, which accounts for about 8% of all mutant PAH alleles in the total Chinese population, is much more common among southern Chinese PKU patients (15.2%) than among northern Chinese PKU patients (1.7%) and is absent from the Japanese population (Wang *et al.*, 1991c). These distribution patterns suggest the presence of two or more founding populations for PKU in east Asia.

## 5. Population dynamics in East Asia

Analysis of the distribution of immunoglobulin G heavy-chain allotypes (Gm markers) has shown that the Japanese, Korean, and northern Chinese populations share a Gm pattern typical of a "Northern Mongoloid" group, while a significantly different Gm pattern, typical of a "Southern Mongoloid" group, was

**Table 6.7.** Relative Frequencies of Common PAH Mutations in Orientals

| Mutation | N. Chinese | | S. Chinese | | Total Chinese | | Japanese[a] | | Korean[a] | |
|---|---|---|---|---|---|---|---|---|---|---|
| | N | % | N | % | N | % | N | % | N | % |
| E56D | 0/60 | 0.0 | 1/46 | 2.2 | 1/106 | 0.9 | ND | ND | ND | ND |
| R111X | 5/64 | 7.8 | 8/54 | 14.8 | 12/118 | 10.2 | 6/72 | 8.3 | 0/20 | 0.0 |
| IVS4nt1 | 1/60 | 1.7 | 7/46 | 15.2 | 8/106 | 7.5 | 4/72 | 5.6 | 5/20 | 25.0 |
| F161S | 1/60 | 1.7 | 0/46 | 0.0 | 1/106 | 0.9 | ND | ND | ND | ND |
| Y204C | 9/64 | 14.1 | 5/48 | 10.4 | 14/112 | 12.5 | 6/72 | 8.3 | 4/20 | 20.0 |
| R243Q | 12/64 | 18.8 | 8/46 | 10.4 | 20/110 | 18.2 | 6/72 | 8.3 | 1/20 | 5.0 |
| G247V | 1/60 | 1.7 | 0/46 | 0.0 | 1/106 | 0.9 | ND | ND | ND | ND |
| L255V | 0/60 | 0.0 | 2/46 | 4.3 | 2/106 | 1.9 | ND | ND | ND | ND |
| IVS7nt2 | 1/60 | 1.7 | 0/46 | 0.0 | 1/106 | 0.9 | 1/72 | 1.4 | 0/20 | 0.0 |
| W326X | 0/60 | 0.0 | 2/46 | 4.3 | 2/106 | 1.9 | 0/72 | 0.0 | 0/20 | 0.0 |
| A345T | 1/60 | 1.7 | 0/46 | 0.0 | 1/106 | 0.9 | ND | ND | ND | ND |
| Y356X | 3/60 | 5.0 | 4/46 | 8.7 | 7/106 | 6.6 | 3/72 | 4.2 | 0/20 | 0.0 |
| R413P | 8/60 | 13.3 | 1/46 | 2.2 | 9/106 | 8.5 | 13/72 | 18.1 | 1/20 | 5.0 |
| T418P | 1/60 | 1.7 | 0/46 | 0.0 | 1/106 | 0.9 | ND | ND | ND | ND |
| Total | 43/64 | 67.2 | 38/54 | 70.4 | 80/118 | 67.8 | 39/72 | 54.2 | 11/20 | 55.0 |

[a]from Okano et al., 1992.

observed in southern Chinese populations (Matsumoto, 1988). The Northern Mongoloid pattern was centered around Lake Baikal, while the Southern Mongoloid pattern was centered around the Guangxi area in southwest China (Matsumoto, 1988). A more comprehensive survey of the distribution of these markers within the Chinese population further defined the geographical boundary between these two groups as 30° N latitude, or near the Yangtze river (Zhao and Lee, 1989). Combining the data from the studies of the PAH locus with these other two polymorphic loci, it could be argued that the R413P mutation occurred or was present in the Northern Mongoloid founding population and was then spread to northern China and Japan, where it currently exists at relatively high frequency, while the IVS4nt1 mutation occurred or was present in the Southern Mongoloid population (Figure 6.7). The higher frequency of the R413P mutation in Japan relative to northern China could also be the result of genetic drift in the rather insular Japanese population. The low frequency of the R413P mutation in southern China could reflect the limited intermixing of these two founding populations or the dilution of this mutation by other mutations present in the Southern Mongoloid population. One exception to this hypothesis is the high frequency of the IVS4nt1 mutation in Korea. Based on the Gm haplotype patterns, Koreans are descended from the Northern Mongoloid group, yet they share a major mutation with the southern Chinese population thought to be descended from a Southern Mongoloid group (Okano et al.,

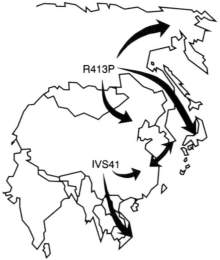

**Figure 6.7.** Proposed origins of the prevalent Asian PAH mutations R413P and IVS4nt1. The high frequency of the R413P mutation in northern China and Japan suggests that it occurred or was already present in a Northern Mongoloid founding population which then spread into northern China, Japan, and northeastern Asia. In contrast, the high frequency of the IVS4nt1 mutation initially observed in southern China suggested that this mutation may have occurred in a Southern Mongoloid founding population which then spread into southern China, Korea, and southeast Asia. The more recent observation that IVS4nt1 is relatively frequent in Koreans, who are genetically more similar to the Northern Mongoloid group than the Southern Mongoloid group, suggests that this mutation may have originated in Koreans after their divergence from other Northern Mongoloid groups, and subsequently introduced into southern China from Korea.

1992). This discrepancy between data from the PAH mutation frequencies and Gm haplotypes may reflect a founding event caused by the movement from southern China to Korea or vice versa that occurred after the divergence of the Northern and Southern Mongoloid peoples.

## C. A role for heterozygote selection in phenylketonuria?

Although there were probably a number of different founding populations in diverse ethnic and racial groups, the presence of so many mutant PAH alleles in different populations implies that founder effect and genetic drift alone cannot be responsible for the high incidence of PKU. Some form of heterozygote selection is still a possibility. One form of selection might come from pressure exerted on a genetic locus closely linked to PAH. The proximity of the γ-interferon locus in region 12q24.1 does provide some speculative basis for this hypothesis, but there is no evidence for this effect. Since selection must act on a phenotype, a second form of selection might come from some compensatory effects brought about by the slight but statistically significant elevations in serum phenylalanine levels present in heterozygotes (Rosenblatt and Scriver, 1968; Gold *et al.*, 1974). A third possible mechanism of selection could involve reproductive compensation, mediated by either a higher rate of reproduction among heterozygotes or a higher survival rate among their offspring. There is some evidence both for (Woolf *et al.*, 1975; Saugstad, 1977) and against (Saugstad, 1973; Paul *et al.*, 1979b) reproductive compensation. One as yet unsubstantiated hypothesis that has been proposed as the means of selection is that the increased serum phenylalanine present in female heterozygotes may protect their offspring against an increased rate of stillbirths caused by ochratoxin A, a mycotoxin found in contaminated grains and lentils (Woolf, 1986).

Unfortunately, it may never be possible to substantiate these selection hypotheses for at least two reasons. First, it may be that some form of selection occurred in the past, but is no longer active. Second, since PKU is less frequent than the other genetic disorders where this mechanism has been proposed, the effects of selection may be too small to detect. In either case, this selective advantage must have existed in regions where there were significantly different climatic, cultural, and dietary conditions, since PKU is present at a relatively high frequency not only in European populations, but in some Asian and Middle Eastern populations as well (Liu and Zuo 1986; Daiger *et al.*, 1989b). Resolving the possible contributions of founder effect, genetic drift, and heterozygote advantage in the origins and subsequent distribution of mutant PAH alleles in human populations will undoubtedly prove to be difficult.

# V. GENE THERAPY FOR PHENYLKETONURIA

## A. Introduction

As discussed in a previous section, the current treatment for PKU is dietary restriction of phenylalanine. Dietary restriction significantly reduces serum phenylalanine levels and can largely reduce or prevent mental impairment if initiated early in the neonatal period. Despite these successes, there are still some limitations or inadequacies associated with dietary therapy for PKU. The simple fact that the treatment restricts normal behavior can reduce compliance, and a reduction in compliance or complete cessation of therapy, even in adolescence or early adulthood, can often be accompanied by a decline in mental or behavioral performance (Smith *et al.*, 1978). Furthermore, dietary noncompliance in pregnant women with PKU can produce developmental abnormalities and mental impairment in their offspring, a syndrome referred to as "maternal PKU" (Lenke and Levy, 1980).

One potential alternative to dietary therapy for PKU is somatic gene therapy, whereby a functional recombinant gene is targeted to the affected organ *in vivo*. Several critical reagents are required for the development of somatic gene therapy for PKU. First, a PAH cDNA clone that produces functional PAH protein must be available. Second, vectors must be developed for the efficient transfer of the PAH cDNA into hepatocytes or other somatic cell targets, where it can be expressed in place of or in addition to the mutant PAH genes. Ideally, the therapeutic gene introduced by these vectors should be maintained in the target cell either through integration into the genome or through extra-chromosomal replication. Either of these mechanisms should permit the trans-mission and expression of this gene in daughter cells. In the absence of stable integration or extrachromosomal replication, the therapeutic gene should be capable of periodic readministration to maintain effective expression levels. Finally, while not essential, an animal model is useful for testing the relative efficiencies of different vectors and strategies for gene transfer.

Early gene transfer experiments described previously demonstrated that PAH mRNA, immunoreactive protein, and pterin-dependent enzyme activity indistinguishable from that found in human liver cells could be produced in cultured mammalian cells through the introduction of an expression vector containing the human PAH cDNA, driven by an appropriate promoter (Ledley *et al.*, 1985b). In this experiment, the PAH cDNA was expressed only tran-siently and at fairly low efficiency. Three novel vector systems have been devel-oped to overcome these limitations, and their potential as transfer agents for the human PAH cDNA has been examined either *in vitro* or *in vivo*. These vectors include recombinant retroviruses (Ledley *et al.*, 1986a; Peng *et al.*, 1988; Liu *et al.*, 1992), DNA/protein complexes (Cristiano *et al.*, 1993b), and recombinant adenoviruses (Fang *et al.*, 1994). Two different strategies have also been explored for the delivery of these vectors to hepatocytes. These include the *ex vivo* and *in vivo* approaches. In the *ex vivo* approach, a single lobe of the liver is resected from the animal and perfused with a collagenase solution. The therapeutic gene is then introduced into isolated liver cells by a recombinant retrovirus or other vector systems followed by autologous reimplantation of the transduced hepato-cytes. In the *in vivo* approach, the recombinant vector is infused directly into the portal or systemic circulation of the animal to be treated. Each of these vectors and approaches have certain advantages and limitations in their application to somatic gene therapy for PKU. Using ethylnitrosourea (ENU) mutagenesis to randomly mutagenize the mouse genome, three different strains of PAH-deficient mice (Pah[enu1], Pah[enu2], and Pah[enu3]) have been created and charac-terized for defects in hepatic phenylalanine hydroxylation (McDonald *et al.*, 1990; Shedlovsky *et al.*, 1993). Two of these strains, Pah[enu1] and Pah[enu2], have been used to assess the efficiency of these vector systems either *in vitro* or *in vivo*.

## B. Recombinant retroviral vectors

Recombinant retroviral vectors are derived from the wild-type Moloney murine leukemia retrovirus. When this retrovirus infects a target cell, an RNA molecule containing the viral genome is released into the cytoplasm along with virally produced reverse transcriptase. The reverse transcriptase transcribes the viral RNA into DNA, which is translocated into the nucleus where it integrates into the target cell genome. This provirus is then transcribed to produce additional RNA copies of the viral genome. The viral genome is translated, the viral proteins are assembled, and the viral genome is packaged into the mature viral particles. These particles are then released from the cell to continue the cycle. Recombinant retroviral vectors are created by replacing all of the retrovirus genes with a therapeutic gene. Transcriptional regulation of the therapeutic gene can be provided either by the retroviral long terminal repeat promoter or by other cellular or viral promoters that can be inserted into the retroviral backbone. The viral gene functions needed to produce recombinant viral particles can be provided in specific viral packaging cells that have been stably transformed with plasmids that express various viral proteins.

Recombinant retroviral vectors were first used to transduce PAH into mouse fibroblast or hepatoma cell lines (Ledley et al., 1986a) or primary hepatocytes from normal mice (Peng et al., 1988). In this first study, the PAH-expressing recombinant retrovirus also contained the neo gene; successful retroviral transduction could therefore be demonstrated through selection with neomycin analogues (Ledley et al., 1986a). In this second study, evidence of successful retroviral transduction came from the detection of PAH mRNA transcripts in extracts of the primary mouse hepatocytes transduced in culture (Peng et al., 1988). More recently, these methods have been used to introduce functional PAH protein into primary hepatocytes isolated from the PAH-deficient mouse strain, Pah^enu1 (McDonald et al., 1990). This strain contains a defect in the PAH gene that leads to a reduction of hepatic PAH activity in homozygotes to approximately 10% of normal levels. Primary hepatocytes from this mutant strain could be isolated and grown in primary culture, where they were transfected by variants of the LNCX retrovirus (Miller and Rosman, 1989) containing the mouse PAH cDNA. As illustrated in Figure 6.8, this procedure introduced significant amounts of PAH enzyme activity into the previously deficient hepatocytes (Liu et al., 1992).

This latter study demonstrated that hepatocytes could be successfully explanted and transduced with the PAH cDNA in culture. The next step in the ex vivo approach requires the reimplantation of the transduced cells. Several studies have examined various methods for reimplanting transduced hepatocytes (Gupta et al., 1991; Ponder et al., 1991). Hepatocytes injected directly into the portal vein or the spleen of mice can migrate into the liver by traversing the

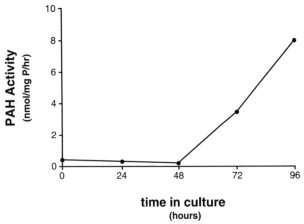

**Figure 6.8.** Recombinant retrovirus-mediated increase in PAH activity in primary hepatocytes isolated from Pah[enu1] mice. A recombinant retrovirus containing the mouse PAH cDNA
under the transcriptional control of the retroviral LTR promoter was used to infect
primary hepatocytes isolated from Pah[enu1] mice after 48 hr in culture. The specific
activity of PAH in these hepatocytes was increased by nearly 20-fold 48 hr after
infection.

portal vasculature and crossing through the fenestrations present in the endothelial cells. Once in the liver, these cells continued to express hepatic markers
and survived for the life of the recipient animals (Ponder et al., 1991). Similar
technologies have also been developed for hepatocyte transplantation in larger
animal models, such as rabbits and dogs (Kay et al., 1992a; Wilson et al., 1992),
and have recently been used in humans (Grossman et al., 1994). Although
transduction of primary hepatocytes by recombinant retroviruses is fairly efficient in vitro, the clinical utility of the ex vivo approach is limited primarily by
the low numbers of cells that can be successfully reimplanted into the liver. An
alternative to this ex vivo approach is the in vivo approach. A number of studies
have now been performed to examine the efficiency of hepatocyte transduction
following the in vivo administration of recombinant retroviral vectors (Ferry et
al., 1991; Kaleko et al., 1991; Kay et al., 1992b, 1993). One such study demonstrated that approximately 1 or 2% of all hepatic parenchymal cells of the mouse
could be successfully transduced after intraportal infusion of a recombinant
retroviral vector into partially hepatectomized animals (Kay et al., 1992b). Despite this low efficiency of transduction in vivo, this approach has been successful
in at least partially correcting the hemophilia B phenotype in factor IX-deficient
dogs (Kay et al., 1993). Whether the efficiency of transduction after in vivo
administration of recombinant retroviral vectors can be increased sufficiently for
future clinical applications in humans remains to be seen. Certainly, recombi-

nant retroviral vectors have many desirable features, including their broad host cell range, their established safety record, and their ability to stably integrate into the host cell genome. Their major limitations remain their inability to integrate into nondividing cell types and their relatively low transduction efficiency, especially when administered in vivo.

## C. Recombinant adenoviral vectors

Recombinant adenoviral vectors are derived from various serotypes of human adenoviruses. These viruses normally enter the target cell through receptor-mediated endocytosis and are sequestered intracellularly in endosomes. Adenoviral proteins then lyse the endosome, delivering the virus into the cytoplasm. The 35-kb linear DNA molecule containing the viral genome is released from the virus and translocated to the nucleus where it replicates extrachromosomally. Some recombinant adenoviral vectors have been created by substituting an expression cassette consisting of a viral or cellular promoter followed by the therapeutic gene and a polyadenylation signal for the adenoviral E1A gene. The resulting E1A-deleted viral vector is replication defective. Other adenoviral vectors may contain additional deletions in regions of the genome encoding other early viral genes.

Although the adenovirus has a natural tropism for the respiratory epithelium, recombinant adenoviral vectors actually have quite a broad host cell range. In addition to lung and other epithelial-derived tissues (Gilardi et al., 1990; Stratford-Perricaudet et al., 1990; Rosenfeld et al., 1991, 1992; Quantin et al., 1992; Bajocchi et al., 1993; Yang et al., 1993), recombinant adenoviral vectors have been successfully used to transduce nonproliferating cell types in vivo including hepatocytes (Levrero et al., 1991; Jaffe et al., 1992; Herz and Gerard, 1993; Ishibashi et al., 1993; Li et al., 1993) and neuronal and glial cells of the central nervous system (Akli et al., 1993; Davidson et al., 1993; Le Gal La Salle et al., 1993). Especially high transduction efficiencies are observed in liver, where 100% of all mouse hepatocytes could be transduced following intraportal infusion of recombinant adenoviral vectors (Li et al., 1993).

In light of these findings, a series of experiments was recently performed to explore the potential of recombinant adenoviruses as vectors for the in vivo delivery of the PAH cDNA to hepatocytes of the PAH-deficient PAH$^{enu2}$ mice. These animals, created by ENU mutagenesis, display many of the same symptoms as PAH-deficient humans, including profound hyperphenylalaninemia, elevation of phenylketones and related compounds in the urine, hypopigmentation, behavioral disturbances, and a maternal PKU phenotype (Shedlovsky et al., 1993). A recombinant adenoviral vector containing the human PAH cDNA under the transcriptional control of the Rous sarcoma virus long terminal repeat was generated and 10$^{10}$ particles of this vector were infused into the portal

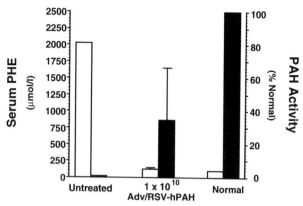

**Figure 6.9** Recombinant adenovirus-mediated correction of PKU in PAH^enu2 mice. The figure shows the relationship between serum PHE levels (□) and liver PAH activity (■) in untreated Pah^enu2 mice, Pah^enu2 mice treated with 10^10 plaque-forming units (PFU) of Adv/RSV-hPAH, and in normal control animals. Each group represents the mean plus or minus the standard deviation from four animals. The infusion of 10^10 PFU of Adv/RSV-hPAH increased hepatic PAH activity from undetectable levels to levels ranging from 10 to 80% of those present in normal controls. Serum phenylalanine levels in these animals were completely normalized by this procedure. The relationship between serum phenylalanine levels and hepatic PAH activity derived from these experiments indicated that only 10–20% of normal PAH activity is sufficient to prevent hyperphenylalaninemia in Pah^enu2 mice.

vasculature of Pah^enu2 mice. This treatment increased hepatic PAH activity from undetectable levels to between 10 and 80% of those present in normal controls and completely normalized serum phenylalanine levels in these animals (Figure 6.9). The relationship between serum phenylalanine levels and hepatic PAH activity derived from these experiments indicated that only 10–20% of normal PAH activity is sufficient to prevent hyperphenylalaninemia in Pah^enu2 mice (Fang et al., 1994).

Unfortunately, the therapeutic effect obtained in these studies did not persist beyond a few weeks, and repeated administration could not duplicate the original effect (Figure 6.10). This result suggested that an immune response against the adenoviral vector was responsible for the failure of repeated treatment, a hypothesis that was supported by the finding of high titers of adenovirus-neutralizing antibodies 4 weeks after infusion of an adenoviral vector (Fang et al., 1994). In addition to this humoral response, the administration of recombinant adenoviral vectors also elicits a T cell-mediated immune response in the host organism that gradually destroys cells transduced by recombinant adenoviruses (Müllbacher et al., 1989; Yang et al., 1994), significantly limiting the long-term expression of genes delivered by existing recombinant adenoviral

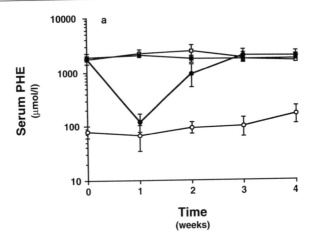

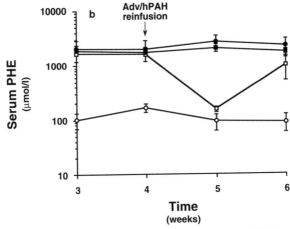

**Figure 6.10** (a) Persistence of the adenovirus-mediated correction of PKU in PAH[enu2] mice. PAH[enu2] mice were infused with either Adv/RSV-hPAH (●; $N = 3$) or the control virus Adv/RSV-hAAT (■; $N = 4$) at $10^{10}$ PFU per mouse or buffer only (□; $N = 5$). Serum PHE levels in normal c57black6 animals (○; $N = 5$) are provided for comparison. Complete normalization of serum PHE was achieved in all animals 1 week after receiving Adv/RSV-hPAH. (b) The effects of reinfusion of Adv/RSV-hPAH. Each treatment group received a second infusion of $10^{10}$ PFU of Adv/RSV-hPAH per mouse at 4 weeks following the initial treatment. This second infusion significantly reduced serum PHE levels at 1 week postinfusion only in those animals previously infused with buffer only. By 2 weeks after the second infusion, the PHE levels in this group of animals also returned to pretreatment levels.

vectors. Modifications of the adenoviral vector that reduce or eliminate expression of the adenoviral genes responsible for evoking this T cell-mediated response may significantly increase the persistence of the vector DNA and, hence, expression of genes delivered by this system; this hypothesis is currently being tested.

Regardless of the degree of persistence of PAH expression in the PAH-deficient mice treated with the recombinant adenoviral vector expressing PAH, these experiments demonstrate the principal advantages of recombinant adenoviral vectors in gene transfer, namely their extremely high efficiency, their broad host cell range, and their ability to transduce nondividing cell types. In addition, the finding that only 10 to 20% of normal hepatic PAH activity is sufficient to restore normal serum phenylalanine levels in these animals establishes a realistically attainable goal for future applications of somatic gene therapy in PKU and in similar metabolic diseases that may be met as more persistent recombinant adenoviral vectors or other gene delivery systems are developed.

## D. DNA/protein complexes

DNA/protein complexes are synthetic delivery systems composed of a negatively charged DNA molecule that has been noncovalently complexed with a positively charged template, such as poly-L-lysine (Wu and Wu, 1987), or to ethidium homodimer (Wagner et al., 1991). The template molecule is then covalently bound to a ligand for a receptor that is present on the surface of the target cell. The electrostatic interactions with the template condense the incorporated DNA into torus- or doughnut-shaped particles that are small enough to enter the cell through clathrin-coated pits containing the high-affinity receptors (Goldstein et al., 1985). A number of receptors expressed on the surface of the hepatocyte may be suitable for receptor-mediated transgene delivery to the liver. DNA/protein complexes have been delivered using the asialoglycoprotein receptor, both in vitro (Markwell et al., 1985; Wu and Wu, 1987; Neda et al., 1991; Cristiano et al., 1993a,b) and in vivo (Wu and Wu, 1988; Wu et al., 1989, 1991; Chowdhury et al., 1993), the transferrin receptor, also both in vitro (Wagner et al., 1990, 1991, 1992; Zenke et al., 1990; Curiel et al., 1991; Cotten et al., 1992; Plank et al., 1992; Michael et al., 1993) and in vivo (Curiel et al., 1992), and more recently, the folate receptor in vitro (Gottschalk et al., 1994; Leamon and Low, 1991, 1993; Turek et al., 1993).

In vitro transduction of cells by DNA/protein complexes is greatly improved by the coadministration of a replication-defective adenovirus (Curiel et al., 1991, 1992; Cotten et al., 1992; Plank et al., 1992; Wagner et al., 1992; Cristiano et al., 1993a; Michael et al., 1993). One effect of the adenovirus is to lyse the endosome before the contents can be routed to the lysosomes or recycled to the cell surface. To reduce cell death, adenovirus has been chemically (Cris-

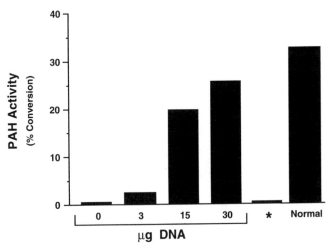

**Figure 6.11.** Transduction of human PAH into primary hepatocytes of Pah$^{enu1}$ mice by DNA/protein complexes. Increasing amounts of PAH activity, expressed as the percentage of phenylalanine in the assay that was converted to tyrosine, could be introduced into primary hepatocytes isolated from PAH-deficent Pah$^{enu1}$ mice by administering DNA/protein complexes containing increasing amounts of a plasmid expressing the human PAH cDNA. The asterisk indicates the absence of significant amounts of PAH activity following the administration of 30 μg of plasmid DNA that had not been incorporated into a DNA/protein complex.

tiano et al., 1993a) or enzymatically (Wagner et al., 1992) coupled to poly-L-lysine. When present either as free virus or as a component of the DNA complex, the transfection rate increased to 100% (Curiel et al., 1991, 1992; Cotten et al., 1992; Plank et al., 1992; Wagner et al., 1992; Cristiano et al., 1993a; Michael et al., 1993). Coadministration of DNA/protein complexes targeting the asialoglycoprotein receptor and a replication-defective adenovirus has been used to introduce a plasmid-expressing human PAH into primary hepatocytes isolated from PAH$^{enu1}$ mice (Figure 6.11). After transduction, near-normal levels of PAH activity were observed (Cristiano et al., 1993b).

    Although this study and others have shown that DNA/protein complexes can efficiently transduce cells in vitro, these vectors have produced only low numbers of transduced hepatocytes or other cell types in vivo (Wu and Wu, 1988; Wu et al., 1989, 1991; Curiel et al., 1992; Chowdhury et al., 1993). The effectiveness of synthetic DNA delivery systems in vivo depends on the route of administration, uptake by a specific cell type, escape from lysosomal degradation, transit through the cytoplasm, transport through the nuclear membrane, recognition and utilization by nuclear enzymes and transcription factors for expression, and persistence of the DNA in an episomal form or by integration into genomic DNA for long-term expression (Hoeben *et al.*, 1992). Improvement in each of these areas is critical to the further development of synthetic DNA vectors, since PKU and other hereditary diseases will probably only be treated by these vectors *in vivo*. Even at their present state of development, synthetic DNA delivery systems have several advantages over other vector systems. First, the absence of viral DNA sequences removes any possible cytopathic effects. Second, the presence of specific receptor ligands provides the possibility to target specific cell types. Third, the absence of a viral backbone removes any restrictions on the size of the DNA fragment that can be incorporated and also precludes any potentially cytopathic effects of viral sequences. Their chief limitations are their low transduction efficiencies *in vivo* and their lack of persistence due to their inability to integrate or replicate in transduced cells.

## VI. CONCLUSIONS

The cloning of the PAH cDNA provided the single most important reagent for understanding the molecular genetics of PKU. The expression of this cDNA conclusively demonstrated that the human PAH protein is the product of a single gene. Since all cases of classic PKU and more than 95% of all cases of HPA are the result of PAH deficiencies, the human PAH cDNA could therefore be used to perform molecular analyses of the PAH gene and to define the molecular basis of PKU and related HPAs. These studies led to the localization of the gene and to the determination of its structure. These studies in turn led to the discovery of RFLP sites in or near the PAH gene that allowed prenatal diagnosis of PKU to be performed for the first time in many PKU families. As mutation detection methods progressed and the mutations responsible for PKU were identified, the PAH genotypes of many affected individuals could be unequivocally determined. These developments not only permitted prenatal diagnosis and carrier screening in all individuals regardless of family history, but, combined with progress in the biochemical characterization of mutant PAH proteins through *in vitro* expression analysis, helped to define the relationship between PAH genotype and disease phenotype. Already such studies have shown that the phenotypic heterogeneity observed in PKU and related HPAs reflects an underlying heterogeneity of PAH genotypes. Genotype determination may also permit many other clinically important issues to be examined. Since the accuracy of these methods is approaching 100%, and since neonatal screening is nearly universal in many human populations, the complete spectrum of mutations responsible for PKU and related HPAs are being defined and absolute mutation frequencies can soon be determined. These developments have facilitated studies of the population genetics of PKU, which have demonstrated the presence of multiple founding populations for PKU in both Europe and East Asia. Additional population studies may lead to some understanding of the role of other mechanisms in maintaining mutant PAH alleles in various human populations. Finally, our knowledge of the molecular basis of PKU and the fact that it is primarily a single gene disorder has made it an attractive model for the development of new treatment modalities based on somatic gene transfer which should also be applicable to a variety of monogenic diseases.

## Acknowledgments

This work was supported in part by NIH Grant HD-17711 to S. L. C. Woo who is also an Investigator with the Howard Hughes Medical Institute.

## References

Abadie, V., Lyonnet S., Maurin, N., Berthelon, M., Caillaud, C., Giraud, F., Mattei, J. F., Rey, J., Rey, F., and Munnich, A. (1989). CpG dinucleotides are mutation hot spots in phenylketonuria. Genomics 5:936–939.

Abadie, V., Jaruzelska, J., Lyonnet, S., Millasseau, P., Berthelon, M., Rey, F., Munnich, A., and Rey, J. (1993a). Illegitimate transcription of the phenylalanine hydroxylase gene in lymphocytes for identification of mutations in phenylketonuria. Hum. Mol. Genet. 2:31–34.

Abadie, V., Lyonnet, S., Melle, D., Berthelon, M., Caillaud, C., Labrune, P., Rey, F., Rey, J., and Munnich, A. (1993b). Molecular basis of phenylketonuria in France. Dev. Brain Dys. 6:120–126.

Abita, J.-P., Blandin-Savoja, F., and Rey, F. (1983). Phenylalanine hydroxylase. Evidence that the enzyme from human liver might not be a phosphoprotein. Biochem. Int. 7:727–737.

Abita, J.-P., Milstien, S., Chang, N., and Kaufman, S. (1976). In vitro activation of rat liver phenylalanine hydroxylase by phosphorylation. J. Biol. Chem. 251:5310–5314.

Akli, S., Caillaud, C., Vigne, E., Stratford-Perricaudet, L. D., Poenaru, L., Perricaudet, M., Kahn, A., and Peschanski, M. R. (1993). Transfer of a foreign gene into the brain using adenovirus vectors. Nature Genet. 3:224–228.

Aoki, K., and Wada, Y. (1988). Outcome of the patients detected by newborn screening in Japan. Acta Paediatr. 30:429–434.

Apold, J., Eiken, H. G., Odland, E., Fredriksen, A., Bakken, A., Lorens, J. G., and Bowman, H. (1990). A termination mutation prevalent in Norwegian haplotype 7 phenylketonuria genes. Am. J. Hum. Genet. 47:1002–1007.

Apold, J., Eiken, H. G., Svensson, E., Kunert, E., Kozak, L., Cechak, P., Güttler, F., Giltay, J., Lichter-Konecki, U., Melle, D., and Jaruzelska, J. M. (1993). The phenylketonuria G272X haplotype 7 mutation in European populations. Hum. Genet. 92:107–109.

Armstrong, M. D., and Brinkley, E. L., Jr. (1956). Studies on phenylketonuria. V. Observations on a newborn infant with phenylketonuria. Proc. Soc. Exp. Biol. Med. 93:418–420.

Armstrong, M. D., and Low, N. L. (1957). Studies on phenylketonuria. VIII. Relation between age, serum phenylalanine level and phenylpyruvic acid excretion. Proc. Soc. Exp. Biol. Med. 94:142–146.

Armstrong, M. D., and Tyler, F. H. (1955). Studies on phenylketonuria. I. Restricted phenylalanine intake in phenylketonuria. J. Clin. Invest. 34:565–580.

Aulehla-Scholz, C., Vorgerd, M., Sautter, E., Leupold, D., Mahlmann, R., Ullrich, K., Olek, K., and Horst, J. (1988). Phenylketonuria: Distribution of DNA diagnostic patterns in German families. Hum. Genet. 78:353–355.

Avigad, S., Cohen, B. E., Bauer, R., Schwartz, G., Frydman, M., Woo, S. L. C., and Shiloh, Y. (1990). A single origin of phenylketonuria in Yemenite Jews. *Nature* **344:**168–170.

Avigad, S., Kleiman, S., Weinstein, M., Cohen, B. E., Schwartz, G., Woo, S. L. C., and Shiloh, Y. (1991). Compound heterozygosity in nonphenylketonuria hyperphenylalaninemia: The contribution of mutations for classical phenylketonuria. *Am. J. Hum. Genet.* **49:**393–399.

Ayling, J. E., Pirson, W. D., At-Janabi, J. M., and Helfand, G. D. (1974). Kidney phenylalanine hydroxylase from man and rat: Comparison with the liver enzyme. *Biochemistry* **13:**78–85.

Bajocchi, G., Feldman, S. H., Crystal, R. G., and Mastrangeli, A. (1993). Direct *in vivo* gene transfer to ependymal cells in the central nervous system using recombinant adenovirus vectors. *Nature Genet.* **3:**229–234.

Barić, I., Mardešič, D., Gjurič, G., Sarnavka, V., Göbel-Schreiner, B., Lichter-Konecki, U., Konecki, D. S., and Trefz, F. (1992). Haplotype distributions and mutations at the PAH locus in Croatia. *Hum. Genet.* **90:**155–157.

Bartholomé, K., and Dresel, A. (1982). Studies on the molecular defect in phenylketonuria and hyperphenylalaninemia using antibodies against phenylalanine hydroxylase. *J. Inherited Metab. Dis.* **5:**7–10.

Bartholomé, K., Lutz, P., and Bickel, H. (1975). Determination of phenylalanine hydroxylase activity in patients with phenylketonuria and hyperphenylalaninemia. *Pediatr. Res.* **9:**899–903.

Bénit, P., Rey, F., Melle, D., Munnich, A., and Rey, J. (1994). Novel frame shift deletions of the phenylalanine hydroxylase gene in phenylketonuria. *Hum. Mol. Genet.* **3:**675–676.

Berk, K., and Saugstad, L. F. (1974). A linkage study of phenylketonuria. *Clin. Genet.* **6:**147–152.

Berthelon, M., Caillaud, C., Rey, F., Labrune, P., Melle, D., Feingold, J., Frézal, J., Briard, M.-L., Farriaux, J.-P., Guibaud, P., Journel, H., Le Marec, B., Maurin, N., Nivelon, J.-L., Plauchu, H., Saudubray, J.-M., Tron, P., Rey, J., Munnich, A., and Lyonnet, S. (1991). Spectrum of phenylketonuria mutations in Western Europe and North Africa, and their relation to polymorphic DNA haplotypes at the phenylalanine hydroxylase locus. *Hum. Genet.* **86:**355–358.

Bick, U., Fahrendorf, G., Ludolph, A. C., Vassallo, P., Weglage, J., and Ullrich, K. (1991). Disturbed myelination in patients with treated hyperphenylalaninemia: Evaluation with magnetic resonance imaging. *Eur. J. Pediatr.* **150:**185–189.

Bickel, H., Gerrard, J., and Hickmans, E. M. (1954). The influence of phenylalanine intake on the chemistry and behavior of a phenylketonuria child. *Acta Paediatr. Scand.* **43:**64–77.

Bickel, H., Bachmann, C., and Beckers, R. (1981). Neonatal mass screening for metabolic disorders. *Eur. J. Pediatr.* **137:**133–139.

Blaskovics, M., and Guidici, T. (1988). A new variant of biopterin deficiency. *N. Engl. J. Med.* **319:**1611–1612.

Bosco, P., Ceratto, N., and Romano, V. (1991). Identification of a new PKU mutation (R261ter) by SSCP analysis. *Clin. Genet.* **40:**392.

Boyd, M. M. (1961). Phenylketonuria; City of Birmingham screening survey. *Br. Med. J.* **1:**771–773.

Byck, S., Morgan, K., John, S. W. M., Blanc, L., Bouchard, G., and Scriver, C. R. (1992). Extended RFLP and HindIII specific haplotypes at the PAH (PKU) locus on chromosomes in France and Québec. *Am. J. Hum. Genet.* **51**[Suppl.]: A571.

Caillaud, C., Lyonnet, S., Rey, F., Melle, D., Frebourg, T., Berthelon, M., Vilarinho, L., Vaz Osorio, R., Rey, J., and Munnich, A. (1991). A 3-base pair in-frame deletion of the phenylalanine hydroxylase gene results in a kinetic variant of phenylketonuria. *J. Biol. Chem.* **266:**9351–9354.

Caillaud, C., Vilarinho, L., Vilarinho, A., Rey, F., Berthelon, M., Santos, R., Lyonnet, S., Briard, M. L., Osorio, R. V., Rey, J., and Munnich, A. (1992). Linkage disequilibrium between phenylketonuria and RFLP haplotype 1 at the phenylalanine hydroxylase locus in Portugal. *Hum. Genet.* **89:**69–72.

Campbell, D. G., Hardie, D. G., and Vulliet, P. R. (1986). Identification of four phosphorylation sites in the N-terminal region of tyrosine hydroxylase. *J. Biol. Chem.* **261**:10489–10492.

Centerwall, W. R. (1957). Phenylketonuria. *J. Am. Med. Assoc.* **165**:392–397.

Chakraborty, R., Lidsky, A. S., Daiger, S. P., Güttler, F., Sullivan, S., DiLella, A. G., and Woo, S. L. C. (1987). Polymorphic DNA haplotypes at the human phenylalanine hydroxylase locus and their relationship with phenylketonuria. *Hum. Genet.* **76**:40–46.

Chen, S.-H., Hsiao, K.-J., Lin, L.-H., Liu, T.-T., Tang, R.-B., and Su, T.-S. (1989). Study of restriction fragment length polymorphisms at the human phenylalanine hydroxylase locus and evaluation of its potential application in prenatal diagnosis of phenylketonuria. *Hum. Genet.* **81**: 226–230.

Choo, K. H., Cotton, R. G. H., Danks, D. M., and Jennings, I. G. (1979). Genetics of the mammalian phenylalanine hydroxylase system. Studies of human liver phenylalanine hydroxylase subunit structure and of mutations in phenylketonuria. *Biochem. J.* **181**:285–294.

Choo, K. H., Cotton, R. G. H., Jennings, I. G., and Danks, D. M. (1980). Observations indicating the nature of the mutation in phenylketonuria. *J. Inherited Metab. Dis.* **2**:79–84.

Chowdhury, N. R., Wu, C. H., Wu, G. Y., Yerneni, P. C., Bommineni, V. R., and Chowdhury, J. R. (1993). Fate of DNA targeted to the liver by asialoglycoprotein receptor-mediated endocytosis *in vivo*. *J. Biol. Chem.* **268**:11265–11271.

Cooper, D. N., and Youssoufian, H. (1988). The CpG dinucleotide and human genetic disease. *Hum. Genet.* **78**:151–155.

Cotten, M., Wagner, E., Zatloukal, K., Phillips, S., Curiel, D. T., and Birnsteil, M. L. (1992). High-efficiency receptor-mediated delivery of small and large (48 kilobase) gene constructs using the endosome-disruption activity of defective or chemically inactivated adenovirus particles. *Proc. Natl. Acad. Sci. USA* **89**:6094–6098.

Crawfurd, M., Gibbs, D., and Sheppard, D. (1981). Studies on human phenylalanine monooxygenase: Restricted expression. *J. Inherited Metab. Dis.* **4**:191–195.

Cristiano, R. J., Smith, L. C., Kay, M. A., Brinkley, B., and Woo, S. L. C. (1993a). Hepatic gene therapy: efficient gene delivery and expression in primary hepatocytes utilizing a conjugated adenovirus/DNA complex. *Proc. Natl. Acad. Sci. USA* **90**:11548–11552.

Cristiano, R., Smith, L. C., and Woo, S. L. C. (1993b). Hepatic gene therapy: Adenovirus enhancement of receptor-mediated gene delivery and expression in primary hepatocytes. *Proc. Natl. Acad. Sci. USA* **90**:2122–2127.

Curiel, D. T., Agarwal, S., Romer, M. U., Wagner, E., Cotten, M., Birnstiel, M. L., and Boucher, R. C. (1992). Gene transfer to respiratory epithelial cells via the receptor-mediated endocytosis pathway. *Am. J. Respir. Cell. Mol. Biol.* **6**:247–252.

Curiel, D. T., Agarwal, S., Wagner, E., and Cotten, M. (1991). Adenovirus enhancement of transferrin/polylysine-mediated gene delivery. *Proc. Natl. Acad. Sci. USA* **88**:8850–8854.

Dahl, H.-H., and Mercer, J. F. B. (1986). Isolation and sequence of a cDNA clone which contains the complete coding region of rat phenylalanine hydroxylase. Structural homology with tyrosine hydroxylase, glucocorticoid regulation, and use of alternate polyadenylation sites. *J. Biol. Chem.* **261**:4148–4153.

Daiger, S. P., Lidsky, A. S., Chakraborty, R., Koch, R., Güttler, F., and Woo, S. L. C. (1986). Effective use of polymorphic DNA haplotypes at the phenylalanine hydroxylase (PAH) locus in prenatal diagnosis of phenylketonuria. *Lancet* **1**:229–232.

Daiger, S. P., Chakraborty, R., Reed, L., Fekete, G., Schuler, D., Berencsi, G., Nasz, I., Brdicka, R., Kamaryt, J., Pijackova, A., Moore, S., Sullivan, S., and Woo, S. L. C. (1989a). Polymorphic DNA haplotypes at the phenylalanine hydroxylase (PAH) locus in European families with phenylketonuria (PKU). *Am. J. Hum. Genet.* **45**:310–318.

Daiger, S. P., Reed, L., Huang, S.-Z., Zeng, Y.-T., Wang, T., Lo, W. H. Y., Okano, Y., Hase, Y., Fukuda, Y., Oura, T., Tada, K., and Woo, S. L. C. (1989b). Polymorphic DNA haplotypes at the phenylalanine hydroxylase (PAH) locus in Asian families with phenylketonuria (PKU). *Am. J. Hum. Genet.* **45**:319–324.

Dasovich, M., Konecki, D., Lichter-Konecki, U., Eisensmith, R. C., Güttler, F., Naughton, E., Mullins, C., and Woo, S. L. C. (1991). Molecular characterization of a PKU allele prevalent in Southern Europe and Ireland. *Som. Cell. Mol. Genet.* **17**:303–309.

Davidson, B. L., Allen, E. D., Kozarsky, K. F., Wilson, J. M., and Roessler, B. J. (1993). A model system for *in vivo* gene transfer into the central nervous system using an adenoviral vector. *Nature Genet.* **3**:219–223.

Desviat, L. R., Pérez, B., and Ugarte, M. (1993). Phenylketonuria in Spain: RFLP haplotypes and linked mutations. *Hum. Genet.* **92**:254–258.

Dhondt, J. L., Farriaux, J. P., Boudha, A., Largilliere, C., Ringel, J., Roger, M. M., and Leeming, R. J. (1985). Neonatal hyperphenylalaninemia presumably caused by guanosine triphosphate-cyclohydrolase activity. *J. Pediatr.* **106**:954–956.

Dhondt, J. L., Guilbaud, P., Rolland, M., Dorche, C., Andre, S., Forzy, G., and Hayte, J. (1988). Neonatal hyperphenylalaninemia presumably caused by a new variant of biopterin synthetase deficiency. *Eur. J. Pediatr.* **147**:153–157.

Dianzani, I., de Sanctis, L., Ferrero, G. B., Alliaudi, C., Ponzone, A., and Camaschella, C. (1992). Molecular analysis of phenylketonuria in Italy. *Am. J. Hum. Genet.* **51**[Suppl.]: A1374.

Dianzani, I., Devoto, M., Camaschella, C., Saglio, G., Ferrero, G. B., Cerone, R., Romano, C., Romeo, G., Giovannini, M., Riva, E., Angeneydt, F., Trefz, F. K., Okano, Y., and Woo, S. L. C. (1990). Haplotype distribution and molecular defects at the phenylalanine hydroxylase locus in Italy. *Hum. Genet.* **86**:69–72.

Dianzani, I., Forrest, S. M., Camaschella, C., Saglio, G., Ponzone, A., and Cotton, R. G. H. (1991). Screening for mutations in the phenylalanine hydroxylase gene from Italian patients with phenylketonuria by using the chemical cleavage method: A new splice mutation. *Am. J. Hum. Genet.* **48**:631–635.

DiLella, A. G., Huang, W. M., and Woo, S. L. C. (1988). Screening for phenylketonuria mutations by DNA amplification with the polymerase chain reaction. *Lancet* **1**:497–499.

DiLella, A. G., Marvit, J., Brayton, K., and Woo, S. L. C. (1987). An amino-acid substitution involved in phenylketonuria is in linkage disequilibrium with DNA haplotype 2. *Nature* **327**: 333–336.

DiLella, A. G., Kwok, S. C. M., Ledley, F. D., Marvit, J., and Woo, S. L. C. (1986a). Molecular structure and polymorphic map of the human phenylalanine hydroxylase gene. *Biochemistry* **25**: 743–749.

DiLella, A. G., Marvit, J., Lidsky, A. S., Güttler, F., and Woo, S. L. C. (1986b). Tight linkage between a splicing mutation and a specific DNA haplotype in phenylketonuria. *Nature* **322**: 799–803.

DiLella, A. G., Ledley, F. D., Rey, F., Munnich, A., and Woo, S. L. C. (1985). Detection of phenylalanine hydroxylase messenger RNA in liver biopsy samples from patients with phenylketonuria. *Lancet* **1**:160–161.

DiSilvestre, D., Koch, R., and Groffen, J. (1991). Different clinical manifestations in three siblings with identical phenylalanine hydroxylase genes. *Am. J. Hum. Genet.* **48**:1014–1016.

Donlon, J., and Kaufman, S. (1978). Glucagon stimulation of rat hepatic phenylalanine hydroxylase through phosphorylation *in vivo*. *J. Biol. Chem.* **253**:6657–6659.

Dworniczak, B., Aulehla-Scholz, C., and Horst, J. (1989). Phenylketonuria: Detection of a frequent haplotype 4 allele mutation. *Hum. Genet.* **84**:95–96.

Dworniczak, B., Aulehla-Scholz, C., Kalaydjieva, L., Ullrich, K., Bartholomé, K., Grudda, K., and Horst, J. (1991a). Aberrant splicing of phenylalanine hydroxylase mRNA: The major cause for phenylketonuria in parts of Southern Europe. *Genomics* **11**:242–246.

Dworniczak, B., Grudda, K., Stümper, J., Bartholomé, K., Aulehla-Scholz, C., and Horst, J.

(1991b). Phenylalanine hydroxylase gene: Novel missense mutation in exon 7 causing severe phenylketonuria. *Genomics* **9**:193–199.

Dworniczak, B., Kalaydjieva, L., Aulehla-Scholz, C., Ullrich, K., Kremensky, I., Radeva, B., and Horst, J. (1991c). Recurrent nonsense mutation in exon 7 of the phenylalanine hydroxylase gene. *Hum. Genet.* **87**:731–733.

Dworniczak, B., Wedemeyer, N., Eigel, A., and Horst, J. (1991d). PCR detection of the Pvull (Ea) RFLP at the human phenylalanine hydroxylase (PAH) locus. *Nucleic Acids Res.* **19**:1958.

Dworniczak, B., Wedemeyer, N., and Horst, J. (1991e). PCR detection of the BglII RFLP at the human phenylalanine hydroxylase (PAH) locus. *Nucleic Acids Res.* **19**:1958.

Dworniczak, B., Kalaydjieva, L., Pankoke, S., Aulehla-Scholz, C., Allen, G., and Horst, J. (1992). Analysis of exon 7 of the human phenylalanine hydroxylase gene: A mutation hot spot? *Hum. Mutat.* **1**:138–146.

Economou-Petersen, E., Henriksen, K. F., Guldberg, P., and Güttler, F. (1992). Molecular basis for nonphenylketonuria hyperphenylalaninemia. *Genomics* **14**:1–5.

Eigel, A., Dworniczak, B., Kalaydjieva, L., and Horst, J. (1991). A frameshift mutation in exon 2 of the phenylalanine hydroxylase gene linked to RFLP haplotype 1. *Hum. Genet.* **87**:739–741.

Eiken, H. G., Knappskog, P. M., Apold, J., Skjelkvåle, L., and Boman, H. (1992a). A de novo phenylketonuria mutation: ATG (Met) to ATA (Ile) in the start codon of the phenylalanine hydroxylase gene. *Hum. Mutat.* **1**:388–391.

Eiken, H. G., Strangeland, K., Skjelkvåle, Knappskog, P., Boman, H., and Apold, J. (1992b). PKU mutations R408Q and F299C in Norway: Haplotype associations, geographical distributions and phenotype characteristics. *Hum. Genet.* **88**:608–612.

Eiken, H. G., Knappskog, P., Boman, H., Motzfeldt, R., and Apold, J. (1993). Prevalence and geographic distribution of PKU mutations in Norway. *25th Annu. Meeting Eur. Soc. Hum. Gene.* A211.

Eisensmith, R. C., and Woo, S. L. C. (1991). Phenylketonuria and the phenylalanine hydroxylase gene. *Mol. Biol. Med.* **8**:3–18.

Eisensmith, R. C., and Woo, S. L. C. (1992a). Molecular basis of phenylketonuria and related hyperphenylalaninemias: Mutations and polymorphisms in the human phenylalanine hydroxylase gene. *Hum. Mutat.* **1**:13–23.

Eisensmith, R. C., and Woo, S. L. C. (1992b). Undated listing of haplotypes at the human phenylalanine hydroxylase (PAH) locus. *Am. J. Hum. Genet.* **51**:1445–1448.

Eisensmith, R. C., Goltsov, A. A., and Woo, S. L. C. (1994). A simple, rapid, and highly informative PCR-based procedure for prenatal diagnosis and carrier screening of phenylketonuria. *Prenatal Diag.,* **in press.**

Eisensmith, R. C., Okano, Y., Dasovich, M., Wang, T., Güttler, F., Lichter-Konecki, U., Konecki, D. S., Svensson, E., Hagenfeldt, L., Rey, F., Munnich, A., Lyonnet, S., Cockburn, F., Conner, J. M., Pembrey, M. E., Smith, I., Gitzelmann, R., Steinmann, B., Apold, J., Eiken, H. G., Giovannini, M., Riva, E., Longhi, R., Romano, C., Cerone, R., Naughten, E. R., Mullins, C., Cahalane, S., Özalp, I., Fekete, G., Schuler, D., Berencsi, G. Y., Nász, I., Brdicka, R., Kamaryt, J., Pijackova, A., Cabalska, B., Bozkowa, K., Schwartz, E., Kalinin, V. N., Jin, L., Chakraborty, R., and Woo, S. L. C. (1992). Multiple origins for phenylketonuria in Europe. *Am. J. Hum. Genet.* **51**:1355–1365.

Eisensmith, R. C., Wang, T., Dasovich, M., Okano, Y., and Woo, S. L. C. (1991). Molecular basis of phenylketonuria (PKU). *Am. J. Hum. Genet.* **49**[Suppl.]: A455.

European Working Group on CF Genetics (1990). Gradient of distribution in Europe of the major CF mutation and of its associated haplotype. *Hum. Genet.* **85**:436–441.

Fang, B., Eisensmith, R. C., Li, X. H. C., Finegold, M. J., Shedlovsky, A., Dove, W., and Woo, S. L. C. (1994). Gene therapy for phenylketonuria: Phenotypic correction in a genetically deficient mouse model by adenovirus-mediated hepatic gene transfer. *Gene Ther.* **1**:241–254.

Ferry, N., Duplessis, O., Houssin, D., Danos, O., and Heard, J.-M. (1991). Retroviral-mediated gene transfer into hepatocytes *in vivo. Proc. Natl. Acad. Sci. USA* **88**:8377–8381.

Fölling, A. (1934a). Utskillelse av fenylpyrodruesyre i urinen som stoffskifteanomali i forbindelse med imbecillitet. *Nord. Med. Tidskr.* **8**:1054–1059.

Fölling, A. (1934b). Über Ausscheidung von Phenylbrenztraubensäure in den Harn als Stoffwechselanomalie in Verbindung mit Imbezillität. *Ztschr. Physiol. Chem.* **227**:169–176.

Forrest, S. M., Dahl, H.-H., Howells, D. W., Dianzani, I., and Cotton, R. G. H. (1991). Mutation detection in phenylketonuria using chemical cleavage of mismatch: Importance of using probes from both normal and patient samples. *Am. J. Hum. Genet.* **49**:175–183.

Friedman, P. A., and Kaufman, S. (1973). Some characteristics of partially purified human liver phenylalanine hydroxylase. *Biochim. Biophys. Acta* **293**:56–61.

Garrod, A. E. (1908). Inborn errors of metabolism. *Lancet* **2**:1–7, 73–79, 142–148, 214–220.

Gibbs, N. K., and Woolf, L. I. (1959). Tests for phenylketonuria: Results of a one-year programme for its detection in infancy and among mental defectives. *Br. Med. J.* **2**:532–535.

Gilardi, P., Courtney, M., Pairani, A., and Perricaudet, M. (1990). Expression of human alpha-1-antitrypsin using a recombinant adenovirus vector. *FEBS Lett.* **267**:60–62.

Goebel-Schreiner, B., and Schreiner, R. (1993). Identification of a new missense mutation in Japanese phenylketonuric patients. *J. Inher. Metab. Dis.* **16**:950–956.

Gold, R. J. M., Maag, U. R., Neal, J. L., and Scriver, C. R. (1974). The use of biochemical data in screening for mutant alleles and in genetic counseling. *Ann. Hum. Genet.* **37**:315–326.

Goldstein, J. L., Brown, M. S., Anderson, R. G. W., Russell, D. W., and Schneider, W. J. (1985). Receptor-mediated endocytosis: Concepts emerging from the LDL receptor system. *Annu. Rev. Cell Biol.* **1**:1–39.

Goltsov, A. A., Eisensmith, R. C., Konecki, D. S., Lichter-Konecki, U., and Woo, S. L. C. (1992a). Linkage disequilibrium between mutations and a VNTR in the human phenylalanine hydroxylase gene. *Am. J. Hum. Genet.* **51**:627–636.

Goltsov, A. A., Eisensmith, R. C., and Woo, S. L. C. (1992b). Detection of the XmnI RFLP at the human phenylalanine hydroxylase locus by PCR. *Nucleic Acids Res.* **20**:927.

Goltsov, A. A., Eisensmith, R. C., Naughten, E. R., and Woo, S. L. C. (1993). A single polymorphic STR system in the human phenylalanine hydroxylase gene permits rapid prenatal diagnosis and carrier screening for phenylketonuria. *Hum. Mol. Genet.* **2**:577–581.

Gottschalk, S., Cristiano, R. J., Smith, L. C., and Woo, S. L. C. (1994). Folate-mediated DNA delivery into tumor cells: Potosomal disruption results in enhanced gene expression. *Gene Ther.* **1**:185–191.

Gregory, D. M., Sovetts, D., Clow, C. L., and Scriver, C. R. (1986). Plasma free amino acid values in normal children and adolescents. *Metabolism* **35**:967–969.

Grenett, H. E., Ledley, F. D., Reed, L. L., and Woo, S. L. C. (1987). Full-length cDNA for rabbit tryptophan hydroxylase: Functional domains and evolution of aromatic amino acid hydroxylases. *Proc. Natl. Acad. Sci. USA* **4**:5530–5534.

Grossman, M., Raper, S. E., Kozarsky, K., Stein, E. A., Engelhardt, J. F., Muller, D., Lupien, P. J., and Wilson, J. M. (1994). Successful *ex vivo* gene therapy directed to liver in a patient with familial hypercholesterolemia. *Nature Genet.* **6**:335–341.

Guldberg, P., and Güttler, F. (1993). A simple method for identification of point mutations using denaturing gradient gel electrophoresis. *Nucleic Acids Res.* **21**:2261–2262.

Guldberg, P., Henriksen, K. F., and Güttler, F. (1993a). Molecular analysis of phenylketonuria in Denmark: 99% of the mutations detected by denaturing gradient gel electrophoresis. *Genomics* **17**:141–146.

Guldberg, P., Lou, H. C., Henriksen, K. F., Mikkelsen, I., Olsen, B., Holck, B., and Güttler, F. (1993b). A novel missense mutation in the phenylalanine hydroxylase gene of a homozygous Pakistani patient with non-PKU hyperphenylalaninemia. *Hum. Mol. Genet.* **2**:1061–1062.

Guldberg, P., Romano, V., Ceratto, N., Bosco, P., Ciuna, M., Indelicato, A., Mollica, F., Meli, C., Giovannini, M., Riva, E., Biasucci, G., Henriksen, K. F., and Güttler, F. (1993c). Mutational spectrum of phenylalanine hydroxylase deficiency in Sicily: Implications for diagnosis of hyperphenylalaninemia in Southern Europe. *Hum. Mol. Genet.* **2**:1703–1707.

Gupta, S., Aragona, E., Vemuru, R. P., Bhargava, K. K., Burk, R. D., and Chowdhury, J. R. (1991). Permanent engraftment and function of hepatocytes delivered to the liver: Implications for gene therapy and liver repopulation. *Hepatology* **14**:144–149.

Guthrie, R. (1961). Blood screening for phenylketonuria. *JAMA* **178**:863.

Guthrie, R., and Susi, A. (1963). A simple phenylalanine method for detecting phenylketonuria in large populations of newborn infants. *Pediatrics* **32**:338–343.

Güttler, F. (1980). Hyperphenylalaninemia: Diagnosis and classification of the various types of phenylalanine hydroxylase deficiency in childhood. *Acta Paediatr. Scand.* **280** [Suppl.]: 1–80.

Güttler, F., Guldberg, P., and Henriksen, K. F. (1993a). Mutation genotype of mentally retarded patients with phenylketonuria. *Dev. Brain Dys.* **6**:92–96.

Güttler, F., Guldberg, P., Henriksen, K. F., Mikkelsen, I., Olsen, B., and Lou, H. (1993b). Molecular basis for the phenotypical diversity of phenylketonuria and related hyperphenylalaninemias. *J. Inher. Metab. Dis.* **16**:602–604.

Güttler, F., Ledley, F. D., Lidsky, A. S., DiLella, A. G., Sullivan, S. E., and Woo, S. L. C. (1987). Correlation between polymorphic DNA haplotypes at phenylalanine hydroxylase locus and clinical phenotypes of phenylketonuria. *J. Pediatr.* **110**:68–71.

Haan, E. A., Jennings, I. G., Cuello, A. C., Nakata, H., Fujisawa, H., Chow, C. W., Kushinsky, R., Brittingham, J., and Cotton, R. G. H. (1987). Identification of serotonergic neurons in human brain by a monoclonal antibody binding to all three aromatic amino acid hydroxylases. *Brain Res.* **426**:19–27.

Herrmann, F., Wulff, K., Wehnert, M., Siedlitz, G., and Güttler, F. (1988). Haplotype analysis of classical and mild phenotype of phenylketonuria in the German Democratic Republic. *Clin. Genet.* **34**:176–180.

Hertzberg, M., Jahromi, K., Ferguson, V., Dahl, H.-H. M., Mercer, J., Mickleson, K. N. P., and Trent, R. J. (1989). Phenylalanine hydroxylase gene haplotypes in Polynesians: Evolutionary origins and absence of alleles associated with severe phenylketonuria. *Am. J. Hum. Genet.* **44**: 382–387.

Herz, J., and Gerard, R. D. (1993). Adenovirus-mediated transfer of low density lipoprotein receptor gene acutely accelerates cholesterol clearance in normal mice. *Proc. Natl. Acad. Sci. USA* **90**: 2812–2816.

Hoeben, R. C., Valerio, D., van der Eb, A. J., and van Ormondt, H. (1992). Gene therapy for human inherited disorders: Techniques and status. *Crit. Rev. Oncol.* **13**:33–54.

Hofman, K. J., Antonarakis, S. E., Missiou-Tsangaraki, S., Boehm, C. D., and Valle, D. (1989). Phenylketonuria in the Greek population. *Mol. Biol. Med.* **6**:245–250.

Hofman, K. J., Steel, G., Kazazian, H. H., and Valle, D. (1991). Phenylketonuria in U.S. blacks: Molecular analysis of the phenylalanine hydroxylase gene. *Am. J. Hum. Genet.* **48**:791–798.

Hommes, F. A. (1993). The effect of hyperphenylalaninemia on the muscarinic acetylcholine receptor in the HPH-5 mouse brain. *J. Inherited Metab. Dis.* **16**:962–974.

Horst, J., Eigel, A., Auleha-Scholz, C., Kalaydjieva, L., Zygulska, M., Kunert, E., and Dworniczak, B. (1991). Molecular basis of phenylketonuria: Report of an extensive study of various Caucasian populations. *Am. J. Hum. Genet.* **49** [Suppl.]: A2302.

Horst, J., Eigel, A., Kalaydjieva, L., and Dworniczak, B. (1993). Phenylketonuria in Germany—Molecular heterogeneity and diagnostic implications. *Dev. Brain Dys.* **6**:32–38.

Huang, S.-Z., Ren, Z.-R., and Zeng, Y.-T. (1990). Application of a new DNA sequence polymorphism as a genetic marker in prenatal diagnosis of phenylketonuria. *J. Med. Genet.* **27**:65–66.

Ishibashi, S., Brown, M. S., Goldstein, J. L., Gerard, R. D., Hammer, R. E., and Herz, J. (1993).

Hypercholesterolemia in LDL receptor knockout mice and its reversal by adenovirus-mediated gene delivery. *J. Clin. Invest* **92**:883–893.

Iwaki, M., Philips, R. S., and Kaufman, S. (1986). Proteolytic modification of the amino-terminal and carboxyl-terminal regions of rat hepatic phenylalanine hydroxylase. *J. Biol. Chem.* **261**: 2051–2056.

Jaffe, H. A., Daniel, C., Longenecker, M., Metzger, M., Setoguchi, Y., Rosenfeld, M. A., Gant, T. W., Thorgeirsson, S. S., Stratford-Perricaudet, L. D., Perricaudet, M., Pavirani, A., Lecocq, J.-P., and Crystal, R. G. (1992). Adenovirus-mediated *in vivo* gene transfer and expression in normal rat liver. *Nature Genet.* **1**:372–378.

Jaruzelska, J., Abadie, V., Marie, J., Lyonnet, S., Brody, E., Rey, F., Rey, J., and Munnich, A. (1992). Illegitimate transcription and *in vitro* splicing of PAH mRNA in a PKU patient carrying an intron point mutation that causes skipping of exon 11. *Am. J. Hum. Genet.* **51**[Suppl.]: A1385.

Jaruzelska, J., Matuszak, R., Lyonnet, S., Rey, F., Rey, J., Filipowicz, J., Borski, K., and Munnich, A. (1993a). Genetic background of clinical heterogeneity of phenylketonuria in Poland. *J. Med. Genet.* **30**:232–234.

Jaruzelska, J., Melle, D., Matuszak, R., Borski, K., and Munnich, A. (1993b). A new 15 bp deletion in exon 11 of the phenylalanine hydroxylase gene in phenylketonuria. *Hum. Mol. Genet.* **1**:763–764.

Jennings, I., and Cotton, R. (1990). Structural similarities among enzyme pterin binding sites as demonstrated by a monoclonal anti-idiotypic antibody. *J. Biol. Chem.* **265**:1885–1889.

Jennings, I. G., Kemp, B. E., and Cotton, R. G. H. (1991). Localization of cofactor binding sites with monoclonal anti-idiotype antibodies: Phenylalanine hydroxylase. *Proc. Natl. Acad. Sci. USA* **88**:5734–5738.

Jervis, G. A. (1939). The genetics of phenylpyruvic oligophrenia. (A contribution to the study of the influence of heredity on mental defect.) *J. Ment. Sci. (London)* **85**:719–762.

Jervis, G. A. (1947). Studies on phenylpyruvic oligophrenia. The position of the metabolic error. *J. Biol. Chem.* **169**:651–656.

Jervis, G. A. (1953). Phenylpyruvic oligophrenia deficiency of phenylalanine-oxidizing system. *Proc. Soc. Exp. Biol. Med.* **82**:514–515.

Jervis, G. A. (1954). Phenylpyruvic oligophrenia (phenylketonuria). *A. Res. Nerv. Ment. Dis.* **33**: 259–282.

Jervis, G. A., Block, R. J., Bolling, D., and Kanze, L. (1940). Phenylalanine content of blood and spinal fluid in phenylpyruvic oligophrenia. *J. Biol. Chem.* **134**:105–113.

John, S. W. M. (1991). Haplotypes and mutations at the phenylalanine hydroxylase locus in French Canadians. Thesis, McGill University, Montreal, Canada.

John, S. W. M., Rozen, R., Laframboise, R., Laberge, C., and Scriver, C. R. (1988). RFLP haplotypes associated with hyperphenylalaninemia alleles at the phenylalanine hydroxylase (PAH) locus in French-Canadians. *Am. J. Hum. Genet.* **43**[Suppl.]: A216.

John, S. W. M., Rozen, R., Laframboise, R., Laberge, C., and Scriver, C. R. (1989). Novel PKU mutation on haplotype 2 in French-Canadians. *Am. J. Hum. Genet.* **45**:905–909.

John, S. W. M., Rozen, R., Scriver, C. R., Laframboise, R., and Laberge, C. (1990). Recurrent mutation, gene conversion, or recombination at the human phenylalanine hydroxylase locus: Evidence in French-Canadians and a catalog of mutations. *Am. J. Hum. Genet.* **46**:970–974.

John, S. W. M., Scriver, C. R., Laframboise, R., and Rozen, R. (1992). *In vitro* and *in vivo* correlations for the 165T and M1V mutations at the phenylalanine hydroxylase locus. *Hum. Mutat.* **1**:147–153.

Kalaydjieva, L., Dworniczak, B., Aulehla-Scholz, C., Kremensky, I., Bronzova, J., Eigel, A., and Horst J. (1990). Classical phenylketonuria in Bulgaria: RFLP haplotypes and frequency of the major mutations. *J. Med. Genet.* **27**:742–745.

Kalaydjieva, L., Dworniczak, B., Aulehla-Scholz, C., Devoto, M., Romeo, G., Sturhmann, M., and Horst J. (1991a). Phenylketonuria mutation in southern Europeans. *Lancet* **337**:865.

Kalaydjieva, L., Dworniczak, B., Aulehla-Scholz, C., Devoto, M., Romeo, G., Sturhmann, M., Kucinskas, V., Yurgelyavicius, V., and Horst, J. (1991b). Silent mutations in the phenylalanine hydroxylase gene as an aid to the diagnosis of phenylketonuria. *J. Med. Genet.* **28**:686–690.

Kalaydjieva, L., Dworniczak, B., Kucinskas, V., Yurgeliavicius, V., Kunert, E., and Horst, J. (1991c). Geographical distribution gradients of the major PKU mutations and the linked haplotypes. *Hum. Genet.* **86**:411–413.

Kalaydjieva, L., Dworniczak, B., Kremensky, I., Koprivarova, R., Radeva, B., Milusheva, R., Aulehla-Scholz, C., and Horst, J. (1992). Heterogeneity of mutations in Bulgarian phenylketonuria haplotype 1 and 4 alleles. *Clin. Genet.* **41**:123–128.

Kalaydjieva, L., Dworniczak, B., Kremensky, I., Radeva, B., and Horst, J. (1993). Population genetics of phenylketonuria in Bulgaria. *Dev. Brain Dys.* **6**:39–45.

Kaleko, M., Garcia, J. V., and Miller, A. D. (1991). Persistent gene expression after retroviral gene transfer into liver cells *in vivo*. *Hum. Gene Ther.* **2**:27–32.

Kamaryt, J., Mrskos, A., Podhradska, D., Kolcova, V., Cabalska, B., Duczynska, N., and Borzymowska, J. (1978). PKU locus: Genetic linkage with human amylase (Amy) loci and assignment to linkage group I. *Hum. Genet.* **43**:205–210.

Kang, E. S., Kaufman, S., and Gerald, P. S. (1970). Clinical and biochemical observations of patients with atypical phenylketonuria. *Pediatrics* **45**:83–92.

Kaufman, S. (1958a). A new cofactor required for the enzymatic conversion of phenylalanine to tyrosine. *J. Biol. Chem.* **230**:931–939.

Kaufman, S. (1958b). Phenylalanine hydroxylation cofactor in phenylketonuria. *Science* **128**: 1506–1508.

Kaufman, S. (1959). Studies on the mechanism of the enzymatic conversion of phenylalanine to tyrosine. *J. Biol. Chem.* **234**:2677–2682.

Kaufman, S. (1963). The structure of the phenylalanine-hydroxylation cofactor. *Proc. Natl. Acad. Sci. USA* **50**:1085–1093.

Kaufman, S. (1970). A protein that stimulates rat liver phenylalanine hydroxylase. *J. Biol. Chem.* **245**:4751–4759.

Kaufman, S., Berlow, S., Summer, G. K., Milstein, S., Schulman, J. D., Orloff, S., Spielberg, S., and Pueschel, S. (1978). Hyperphenylalaninemia due to a deficiency of biopterin. A variant form of phenylketonuria. *N. Engl. J. Med.* **299**:673–679.

Kaufman, S., Holtzman, N. A., Milstein, S., Butler, I. J., and Krumholtz, A. (1975a). Phenylketonuria due to a deficiency of dihydropteridine reductase. *N. Engl. J. Med.* **293**:785–790.

Kaufman, S., Max, E. E., and Kang, E. S. (1975b). Phenylalanine hydroxylase activity in liver biopsies from hyperphenylalaninemia heterozygotes: Deviation from proportionality with gene dosage. *Pediatr. Res.* **9**:632–634.

Kay, M. A., Baley, P., Rothenberg, S., Leland, F., Fleming, L., Ponder, K. P., Liu, T.-J., Finegold, M., Darlington, G., Pokorny, W., and Woo, S. L. C. (1992a). Expression of human $\alpha_1$-antitrypsin in dogs after autologous transplantation of retroviral transduced hepatocytes. *Proc. Natl. Acad. Sci. USA* **89**:89–93.

Kay, M. A., Li, Q., Liu, T.-J., Leland, F., Toman, C., Finegold, M., and Woo, S. L. C. (1992b). Hepatic gene therapy: Persistent expression of human $\alpha_1$-antitrypsin in mice after direct gene delivery *in vivo*. *Hum. Gene Ther.* **3**:641–647.

Kay, M. A., Rothenberg, S., Landen, C., Bellinger, D., Leland, F., Toman, C., Finegold, M., Thompson, A., Read, M., Brinkhous, K., and Woo, S. L. C. (1993). *In vivo* gene therapy of hemophilia B: Sustained partial correction in factor IX deficient dogs. *Science* **262**:117–119.

Kleiman, S., Bernstein, J., Schwartz, G., Eisensmith, R. C., Woo, S. L. C., and Shiloh, Y. (1991).

A defective splice site at the phenylalanine hydroxylase gene in phenylketonuria and benign hyperphenylalaninemia among Palestinian Arabs. *Hum. Mutat.* **1**:340–343.

Kleiman, S., Schwartz, G., Woo, S. L. C., and Shiloh, Y. (1992). A 22-bp deletion in the phenylalanine hydroxylase gene causing phenylketonuria in an Arab family. *Hum. Mutat.* **1**: 344–346.

Kleiman, S., Li, J., Schwartz, G., Eisensmith, R. C., Woo, S. L. C., and Shiloh, Y. (1993). Inactivation of phenylalanine hydroxylase by a missense mutation, R270S, in a Palestinian kinship with phenylketonuria. *Hum. Mol. Genet.* **2**:605–606.

Koch, R., Azen, C., Friedman, E. G., and Williamson, M. L. (1984). Paired comparisons between early treated PKU children and their matched sibling controls on intelligence and school achievement test results at eight years of age. *J. Inher. Metab. Dis.* **7**:86–90.

Konecki, D. S., Schlotter, M., Trefz, F. K., and Lichter-Konecki, U. (1991). The identification of two mis-sense mutations at the PAH gene in a Turkish patient with phenylketonuria. *Hum. Genet.* **87**:389–393.

Kotake, Y., Masai, Y., and Mori, Y. (1922). Über das Verhalten des Phenylalanins im tierischen Organismus. *Ztschr. Physiol. Chem.* **122**:195.

Kwok, S. C. M., Ledley, F. D., DiLella, A. G., Robson, K. J. H., and Woo, S. L. C. (1985). Nucleotide sequence of a full-length complementary DNA clone and amino acid sequence of human phenylalanine hydroxylase. *Biochemistry* **24**:556–561.

Labrune, P., Melle, D., Rey, F., Berthelon, M., Caillaud, C., Rey, J., Munnich, A., and Lyonnet, S. (1991). Single-strand conformation polymorphism for detection of mutations and base substitutions in phenylketonuria. *Am. J. Hum. Genet.* **48**:1115–1120.

Langenbeck, U., Lukas, H. D., Mench-Holinowski, A., Stenzig, K. P., and Lane, J. D. (1988). Correlative study of mental and biochemical phenotypes in never treated patients with classical phenylketonuria. *Brain Dys.* **1**:103–110.

Lazarus, R. A., Benkovic, S. J., and Kaufman, S. (1983). Phenylalanine hydroxylase stimulator protein is a 4α-carbinolamine dehydratase. *J. Biol. Chem.* **258**:10960–10962.

Le Gal La Salle, G., Robert, J. J., Berrard, S., Ridoux, V., Stratford-Perricaudet, L. D., Perricaudet, M., and Mallet, J. (1993). An adenovirus vector for gene transfer into neurons and glia in the brain. *Science* **259**:988–990.

Leamon, C. P., and Low, P. S. (1991). Delivery of macromolecules into living cells: A method that exploits folate receptor endocytosis. *Proc. Natl. Acad. Sci. USA* **88**:5572–5576.

Leamon, C. P., and Low, P. S. (1993). Cytoxicity of momordin–folate conjugates in cultured human cells. *J. Biol. Chem.* **267**:24966–24971.

Leandro, P., Rivera, I., Ribeiro, V., Tavares de Almeida, I., and Lechner, M. C. (1993). Sequencing analysis of PAH genomic DNA reveals 4 novel mutations affecting exons 7 and 11 in a Portugese PKU population. *Abstr. SSIEM, Manchester.*

Ledley, F. D., DiLella, A. G., Kwok, S. C. M., and Woo, S. L. C. (1985a). Homology between phenylalanine hydroxylase and tyrosine hydroxylase reveals common structural and functional determinants. *Biochemistry* **24**:3389–3394.

Ledley, F. D., Grenett, H. E., DiLella, A. G., Kwok, S. C. M., and Woo, S. L. C. (1985b). Gene transfer and expression of human phenylalanine hydroxylase. *Science* **228**:77–79.

Ledley, F. D., Grenett, H., McGinnis-Shelnutt, M., and Woo, S. L. C. (1986a). Retroviral-mediated gene transfer of human phenylalanine hydroxylase into NIH 3T3 and hepatoma cells. *Proc. Natl. Acad. Sci. USA* **83**:409–413.

Ledley, F. D., Levy, H. L., and Woo, S. L. C. (1986b). Molecular analysis of the inheritance of phenylketonuria and mild hyperphenylalaninemia in families with both disorders. *N. Engl. J. Med.* **314**:1276–1280.

Lenke, R. R., and Levy, H. L. (1980). Maternal phenylketonuria and hyperphenylalaninemia: An international survey of the outcome of untreated and treated pregnancies. *N. Engl. J. Med.* **303:** 1202–1208.

Levrero, M., Barban, V., Manteca, S., Ballay, A., Balsamo, C., Avantaggiati, M. L., Natoli, G., Skellekens, H., Tiollais, P., and Perricaudet, M. (1991). Defective and nondefective adenovirus vectors for expressing foreign genes *in vitro* and *in vivo*. *Gene* **101:**195–202.

Li, J., Eisensmith, R. C., Wang, T., Lo, W. H. Y., Huang, S.-Z., Zeng, Y.-T., Yuan, L.-F., Liu, S.-R., and Woo, S. L. C. (1992). Identification of three novel missense PKU mutations among Chinese. *Genomics* **13:**894–895.

Li, J., Eisensmith, R. C., Wang, T., Lo, W. H. Y., Huang, S.-Z., Zeng, Y.-T., Yuan, L.-F., Liu, S.-R., and Woo, S. L. C. (1994). Phenylketonuria in China: Characterization of three novel mutations in the human phenylalanine hydroxylase gene. *Hum. Mutat.* **3:**312–314.

Li, Q.-T., Kay, M., Finegold, M., Stratford-Perricaudet, L., and Woo, S. L. C. (1993). Assessment of recombinant adenoviral vectors for hepatic gene therapy. *Hum. Gene Ther.* **4:**403–409.

Lichter-Konecki, U., Konecki, D. S., DiLella, A. G., Brayton, K., Marvit, J., Hahn, T. M., Trefz, F. K., and Woo, S. L. C. (1988a). Phenylalanine hydroxylase deficiency caused by a single base substitution in an exon of the human phenylalanine hydroxylase gene. *Biochemistry* **27:**2881–2885.

Lichter-Konecki, U., Schlotter, M., Konecki, D. S., Labeit, S., Woo, S. L. C., and Trefz, F. K. (1988b). Linkage disequilibrium between mutation and RFLP haplotype at the phenylalanine hydroxylase locus in the German population. *Hum. Genet.* **78:**347–352.

Lichter-Konecki, U., Schlotter, M., Yaylak, C., Özgüç, M., Çoskun, T., Özalp, I., Wendel, U., Batzler, U., Trefz, F. K., and Konecki, D. (1989). DNA haplotype analysis at the phenylalanine hydroxylase locus in the Turkish population. *Hum. Genet.* **81:**373–376.

Lidsky, A. S., Güttler, F., and Woo, S. L. C. (1985a). Prenatal diagnosis of classical phenylketonuria by DNA analysis. *Lancet* **1:**549–551.

Lidsky, A. S., Law, M. L., Morse, H. G., Kao, F. T., and Woo, S. L. C. (1985b). Regional mapping of the phenylalanine hydroxylase gene and the phenylketonuria locus in the human genome. *Proc. Natl. Acad. Sci. USA* **82:**6221–6225.

Lidsky, A., Ledley, F. D., DiLella, A. G., Kwok, S. C. M., Daiger, S. P., Robson, K. J. H., and Woo, S. L. C. (1985c). Extensive restriction site polymorphism at the human phenylalanine hydroxylase locus and application in prenatal diagnosis of phenylketonuria. *Am. J. Hum. Genet.* **37:**619–634.

Lidsky, A. S., Robson, K., Thirumalachary, C., Barker, P., Ruddle, F., and Woo, S. L. C. (1984). The PKU locus in man is on chromosome 12. *Am. J. Hum. Genet.* **36:**527–533.

Lin, C.-H., Hsiao, K.-J., Tsai, T.-F., Chao, H.-K., and Su, T.-S. (1992). Identification of a missense phenylketonuria mutation at codon 408 in Chinese. *Hum. Genet.* **89:**593–596.

Liu, S.-R., and Zuo, Q.-H. (1986). Newborn screening for phenylketonuria in eleven districts. *Chi. Med. J.* **99:**113–118.

Liu, T.-J., Kay, M. A., Darlington, G. J., and Woo, S. L. C. (1992). Reconstitution of enzymatic activity in hepatocytes of phenylalanine hydroxylase-deficient mice. *Somat. Cell. Mol. Genet.* **18:** 89–96.

Lyonnet, S., Caillaud, C., Rey, F., Berthelon, M., Frezal, J., Rey, J., and Munnich, A. (1989). Molecular genetics of phenylketonuria in Mediterranean countries: A mutation associated with partial phenylalanine hydroxylase deficiency. *Am. J. Hum. Genet.* **44:**511–517.

Lyonnet, S., Melle, D., de Braekeleer, M., Laframboise, R., Rey, F., John, S. W. M., Berthelon, M., Berthelot, J., Journel, H., Le Marec, B., Parent, P., de Parscau, L., Saudubray, J.-M., Rozen, R., Rey, J., Munnich, A., and Scriver, C. R. (1992). Time and space clusters of the French-Canadian M1V phenylketonuria mutation in France. *Am. J. Hum. Genet.* **51:**191–196.

Markwell, M. K., Portner, A., and Schwartz, A. L. (1985). Alternative route of infection for viruses: Entry by the asialoglycoprotein receptor of sendai virus mutant lacking its attachment protein. *Proc. Natl. Acad. Sci. USA* **82**:978–982.

Marvit, J., DiLella, A. G., Brayton, K., Ledley, F. D., Robson, K. J. H., and Woo S. L. C. (1987). GT to AT transition at a splice donor site causes skipping of the preceding exon in phenylketonuria. *Nucleic Acids Res.* **15**:5613–5628.

Matsumoto, H. (1988). Characteristics of Mongoloid and neighboring populations based on the genetic markers of human immunoglobulins. *Hum. Genet.* **80**:207–218.

Matsuo, K., and Hommes, F. A. (1988). The development of the muscarinic acetylcholine receptor in normal and hyperphenylalaninemic rat cerebrum. *Neurochem. Res.* **13**:867–870.

McBurnie, M. A., Kronmal, R. A., Williamson, M., and Roche, A. F. (1991). Physical growth of children treated for phenylketonuria. *Ann. Hum. Biol.* **18**:357–368.

McDonald, D., Bode, V., Dove, W., and Shedlovsky, A. (1990). Pah$^{hph-5}$: A mouse mutant deficient in phenylalanine hydroxylase. *Proc. Natl. Acad. Sci. USA* **87**:1965–1967.

Melle, D., Verelst, P., Rey, F., Berthelon, M., François, B., Munnich, A., and Lyonnet, S. (1991). Two distinct mutations at a single BamHI site in phenylketonuria. *J. Med. Genet.* **28**:38–40.

Menozzi, P., Piazza, A., and Cavalli-Sforza, L. (1978). Synthetic maps of human gene frequencies in Europeans. *Science* **201**:786–792.

Mercer, J. F. B., Grimes, A., Jennings, I., and Cotton, R. G. H. (1984). Identification of two molecular-mass forms of phenylalanine hydroxylase that segregate independently in rats. Specific association of each form with certain rat strains. *Biochem. J.* **219**:891–898.

Michael, S. I., Huang, C., Romer, M. U., Wagner, E., and Curiel, D. T. (1993). Binding-incompetent adenovirus facilitates molecular conjugate-mediated gene transfer by the receptor-mediated endocytosis pathway. *J. Biol. Chem.* **268**:6866–6869.

Miller, A. D., and Rosman, G. J. (1989). Improved retroviral vectors for gene transfer and expression. *BioTechniques* **7**:980–990.

Mitoma, C. (1956). Studies on partially purified phenylalanine hydroxylase. *Arch. Biochem. Biophys.* **60**:476–484.

Mitoma, C., Auld, R. M., and Udenfriend, S. (1957). On the nature of enzymatic defect in phenylpyruvic oligophrenia. *Proc. Soc. Exp. Biol. Med.* **94**:634–635.

Morales, G., Requena, G. M., Jimenez-Ruiz, A., Lopez, M. C., Ugarte, M., and Alonso, C. (1990). Sequence and expression of the *Drosophila* phenylalanine hydroxylase mRNA. *Gene* **93**: 213–219.

Müllbacher, A., Bellett, A. J. D., and Hla, R. T. (1989). The murine cellular immune response to adenovirus type 5. *Immunol. Cell Biol.* **67**:31–39.

Murthy, L. I., and Berry, H. K. (1975). Phenylalanine hydroxylase activity in liver from humans and subhuman primates: Its probable absence in kidney. *Biochem. Med.* **12**:392–397.

Nagatsu, T., and Ichinose, H. (1991). Comparative studies on the structure of human tyrosine hydroxylase with those of the enzyme of various mammals. *Comp. Biochem. Physiolc.* **98**:203–210.

Neda, H., Wu, C. H., and Wu, G. Y. (1991). Chemical modification of an ecotropic murine leukemia virus results in redirection of its target cell specificity. *J. Biol. Chem.* **266**:14143–14146.

Niederweiser, A., Blau, N., Wang, M., Joller, P., Atares, M., and Cardesa-Garcia, J. (1984). GTP cyclohydrolase I deficiency, a new enzyme defect causing hyperphenylalaninemia with neopterin, biopterin, dopamine and serotonin deficiencies and muscular hypotonia. *Eur. J. Pediatr.* **141**: 208–214.

Niederweiser, A., Leimbacher, W., Curtius, H.-C. H., Ponzone, A., Rey, F., and Leupold, D. (1985). Atypical phenylketonuria with "dihydrobiopterine synthetase" deficiency: Absence of phosphate-eliminating enzyme activity demonstrated in liver. *Eur. J. Pediatr.* **144**:13–16.

Niederweiser, A., Shintaku, H., Hasler, T. H., Curtius, H.-C. H., Lehmann, H., Guardamagna, O., and Schmidt, H. (1986). Prenatal diagnosis of "dihydrobiopterine synthetase" deficiency, a variant form of phenylketonuria. *Eur. J. Pediatr.* **145**:176–178.

Okano, Y., Hase, Y., Lee, D.-H., Takada, G., Shigematsu, Y., Oura, T., and Isshiki, G. (1994a). Molecular and population genetics of phenylketonuria in Orientals: Correlation between phenotype and genotype. *J. Inher. Metab. Dis.* **17**:156–159.

Okano, Y., Hase, Y., Shintaku, H., Takada, G., Shigematsu, Y., Araki, K., Oura, T., and Isshiki, G. (1994b). Molecular characterization of phenylketonuric mutations by analysis of phenylalanine hydroxylase mRNA from lymphoblasts in Japanese. *Hum. Mol. Genet.* **3**:659–660.

Okano, Y., Hase, Y., Lee, D.-H., Furuyama, J.-I., Shintaku, Oura, T., and Isshiki, G. (1992). Frequency and distribution of phenylketonuric mutations in Orientals. *Hum. Mutat.* **1**:216–220.

Okano, Y., Eisensmith, R. C., Dasovich, M., Wang, T., Güttler, F., and Woo, S. L. C. (1991a). A prevalent missense mutation in Northern Europe associated with hyperphenylalaninemia. *Eur. J. Pediatr.* **150**:347–352.

Okano, Y., Eisensmith, R. C., Güttler, F., Lichter-Konecki, U., Konecki, D., Trefz, F. K., Dasovich, M., Wang, T., Henriksen, K., Lou, H., and Woo, S. L. C. (1991b). Molecular basis of phenotypic heterogeneity in phenylketonuria. *N. Engl. J. Med.* **324**:1232–1238.

Okano, Y., Wang, T., Eisensmith, R. C., Longhi, R., Giovannini, M., Cerone, R., Romano, C., and Woo, S. L. C. (1991c). Phenylketonuria missense mutations in the Mediterranean. *Genomics* **9**:96–103.

Okano, Y., Wang, T., Eisensmith, R. C., Güttler, F., and Woo, S. L. C. (1990a). Recurrent mutation in the human phenylalanine hydroxylase gene. *Am. J. Hum. Genet.* **46**:919–924.

Okano, Y., Wang, T., Eisensmith, R. C., Steinmann, B., Gitzelmann, R., and Woo, S. L. C. (1990b). Missense mutations associated with RFLP haplotypes 1 and 4 of the human phenylalanine hydroxylase gene. *Am. J. Hum. Genet.* **46**:18–25.

Okano, Y., Wang, T., Eisensmith, R. C., and Woo, S. L. C. (1989). Molecular genetics of PKU among Caucasians. *Am. J. Hum. Genet.* **45** [Suppl.]: A211.

Onishi, A., Liotta, L. J., and Benkovic, S. J. (1991). Cloning and expression of *Chromobacterium violaceum* phenylalanine hydroxylase in *Escherichia coli* and comparison of amino acid sequence with mammalian aromatic amino acid hydroxylases. *J. Biol. Chem.* **266**:18454–18459.

Orkin, S. H., and Kazazian, H. H. (1984). The mutation and polymorphism of the human β-globin gene and its surrounding DNA. *Annu. Rev. Genet.* **18**:131–171.

Özalp, I., Coskun, T., Ceyhan, M., Tokol, S., Oran, O., Erdem, G., Tekinalp, G., Durmus, Z., and Tarikahya, Y. (1986). Incidence of phenylketonuria and hyperphenylalaninemia in a sample of the newborn population. *J. Inher. Metab. Dis.* **9** [Suppl. 2]: 237–239.

PAH Gene Mutation Analysis Consortium, April 1994 (C. R. Scriver, Ed.).

Paul, T. D., Brandt, I. K., Elsas, L. J., Jackson, C. E., Nance, C. S., and Nance, W. E. (1979a). Linkage analysis using heterozygote detection in phenylketonuria. *Clin. Genet.* **16**:217–232.

Paul, T. D., Greco, J., Jr., Brandt, T. K., Jackson, C. E., and Nance, W. E. (1979b). Is there a heterozygote advantage in the birthweight and number of children born to PKU heterozygotes? *Am. J. Hum. Genet.* **31** [Suppl.]: A104.

Peng, H., Armentano, D., MacKenzie-Graham, L., Shen, R.-F., Darlington, G., Ledley, F. D., and Woo, S. L. C. (1988). Retroviral-mediated gene transfer and expression of human phenylalanine hydroxylase in primary mouse hepatocytes. *Proc. Natl. Acad. Sci. USA* **85**:8146–8150.

Penrose, L. S. (1935). Inheritance of phenylpyruvic amentia (phenylketonuria). *Lancet* **2**:192–194.

Penrose, L. S., and Quastel, J. H. (1937). Metabolic studies in phenylketonuria. *Biochem. J.* **31**: 266–274.

Pérez, B., Desviat, L. R., Die, M., Cornejo, V., Chamoles, N. A., Nicolini, H., and Ugarte, M. (1993). Presence of the Mediterranean PKU mutation IVS10 in Latin America. *Hum. Mol. Genet.* **2**:1289–1290.

Pérez, B., Desviat, L. R., Die, M., and Ugarte, M. (1992). Mutation analysis of phenylketonuria in Spain: Prevalence of two Mediterranean mutations. *Hum. Genet.* **89**:341–342.

Pigeon, D., Ferrara, P., Gros, F., and Thibault, J. (1987). Rat pheochromocytoma tyrosine hydroxylase is phosphorylated on serine 40 by an associated protein kinase. *J. Biol. Chem.* **262**: 6155–6158.

Plank, C., Zatloukal, K., Cotten, M., Mechtler, K., and Wagner, E. (1992). Gene transfer into hepatocytes using asialoglycoprotein receptor mediated endocytosis of DNA complexes with an artificial tetra-antennary galactose ligand. *Bioconjugate Chem.* **3**:533–539.

Ponder, K., Gupta, S., Leland, F., Darlington, G., Finegold, M., DeMayo, J., Ledley, F., Chowdhury, J., and Woo, S. L. C. (1991). Mouse hepatocytes migrate to liver parenchyma and function indefinitely after intrasplenic transplantation. *Proc. Natl. Acad. Sci. USA* **88**:1217–1221.

Quantin, B., Perricaudet, L. D., Tajbakhsh, S., and Mandel, J.-L. (1992). Adenovirus as an expression vector in muscle cells *in vivo*. *Proc. Natl. Acad. Sci. USA* **89**:2581–2584.

Ramus, S. J., and Cotton, R. G. H. (1993). A new phenylketonuria (PKU) mutation detected by illegitimate transcription results in RNA mis-splicing: Founder effect and PKU in Australia. *Am. J. Hum. Genet.* **53**:A1218.

Ramus, S. J., Forrest, S. M., Saleeba, J. A., and Cotton, R. G. H. (1992). CpG hotspot causes second mutation in codon 408 of the phenylalanine hydroxylase gene. *Hum. Genet.* **90**:147–148.

Ramus, S. J., Forrest, S. M., Pitt, D. B., Saleeba, J. A., and Cotton, R. G. H. (1993). Comparison of genotype and intellectual phenotype in untreated PKU patients. *J. Med. Genet.* **30**:401–405.

Riess, O., Michael, A., Speer, A., Meiske, W., Cobet, G., and Coutelle, C. (1988). Linkage disequilibrium between RFLP haplotype 2 and the affected PAH allele in PKU families from the Berlin area of the German Democratic Republic. *Hum. Genet.* **78**:343–346.

Rey, F., Berthelon, M., Caillaud, C., Lyonnet, S., Abadie, V., Blandin-Savoja, F., Feingold, J., Saudubray, J. M., Frézal, J., Munnich, A., and Rey, J. (1988). Clinical and molecular heterogeneity of phenylalanine hydroxylase deficiencies in France. *Am. J. Hum. Genet.* **43**:914–921.

Robson, K. J. H., Beattie, W., James, R. J., Cotton, R. C. H., Morgan, F. J., and Woo, S. L. C. (1984). Sequence comparison of rat liver phenylalanine hydroxylase and its cDNA clones. *Biochemistry* **23**:5671–5673.

Robson, K. J. H., Chandra, T., MacGillivray, R. T. A., and Woo, S. L. C. (1982). Polysome immunoprecipitation of phenylalanine hydroxylase mRNA from rat liver and cloning of its cDNA. *Proc. Natl. Acad. Sci. USA* **79**:4701–4705.

Rosenblatt, D., and Scriver, C. R. (1968). Heterogeneity in genetic control of phenylalanine metabolism in man. *Nature* **218**:677–679.

Rosenfeld, M. A., Siegried, W., Yoshimura, K., Yoneyama, K., Fukayama, M., Stier, L. E., Paako, P., Gilardi, P., Stratford-Perricaudet, L. D., Perricaudet, M., Pavirani, A., Lecocq, J.-P., and Crystal, R. G. (1991). Adenovirus-mediated transfer of a recombinant alpha-1-antitrypsin gene to the lung epithelium *in vivo*. *Science* **252**:431–434.

Rosenfeld, M. A., Yoshimura, K., Trapnell, B. C., Yoneyama, K., Rosenthal, E. R., Dalemans, W., Fukayama, M., Bargon, J., Stier, L. Stratford-Perricaudet, L., Perricaudet, M., Guggino, W. B., Pavirani, A., Lecocq, J.-P., and Crystal, R. G. (1992). *In vivo* transfer of the human CFTR gene to the airway epithelium. *Cell* **68**:143–155.

Sarkar, G., and Sommer, S. (1989). Access to a messenger RNA sequence or its protein product is not limited by tissue or species specificity. *Science* **244**:331–334.

Saugstad, L. F. (1973). Increased "reproductive casualty" in heterozygotes for phenylketonuria. *Clin. Genet.* **4**:105–114.

Saugstad, L. F. (1977). Heterozygote advantage for the phenylketonuria allele. *J. Med. Genet.* **14**: 20–24.

Scriver, C. R., Kaufman, S., and Woo, S. L. C. (1988). Mendelian hyperphenylalaninemia. *Annu. Rev. Genet.* **22:**301–321.

Scriver, C. R., Kaufman, S., and Woo, S. L. C. (1989). The hyperphenylalaninemias. *In* "The Metabolic Basis of Inherited Disease, 7th Edition" (C. R. Scriver, A. Beaudet, W. Sly, and D. Valle, Eds.), pp. 495–546. McGraw–Hill, New York.

Scriver, C. R., John, S. W. M., Rozen, R., Eisensmith, R., and Woo, S. L. C. (1993). Associations between populations, PKU mutations and RFLP haplotypes at the PAH locus: An overview. *Dev. Brain Dys.* **6:**11–25.

Shedlovsky, A., McDonald, J. D., Symula, D., and Dove, W. F. (1993). Mouse models of human phenylketonuria. *Genetics* **134:**1205–1210.

Shiman, R., and Gray, D. W. (1980). Substrate activation of phenylalanine hydroxylase. A kinetic characterization. *J. Biol. Chem.* **255:**4793–4800.

Shirahase, W., Oya, N., and Shimada, M. (1991). A new single base substitution in a Japanese phenylketonuria (PKU) patient. *Brain Dev.* **13:**283–284.

Shirahase, W., and Shimada, M. (1992). Genetic study on Japanese classical phenylketonuria (family analysis by PCR-SSCP analysis). *Acta Pediatr. Jpn.* **96:**939–945.

Smith, I., Beasley, M. G., Wolff, O. H., and Ades, A. E. (1988). Behaviour disturbance in 8-year-old children with early-treated phenylketonuria. *J. Pediatr.* **112:**403–408.

Smith, I., Lobascher, M. E., Stevenson, J. E., Woolf, O. H., Schmidt, H., Grubel-Kaiser, S., and Bickel, H. (1978). Effect of stopping low-phenylalanine diet on intellectual progress of children with phenylketonuria. *Br. Med. J.* **2:**723–726.

Smith, S. C., Kemp, B. E., McAdam, W. J., Mercer, J. E. B., and Cotton, R. G. (1984). Two apparent molecular weight forms of human and monkey phenylalanine hydroxylase are due to phosphorylation. *J. Biol. Chem.* **259:**11284–11289.

Sokal, R. R., Oden, N. L., and Wilson, C. (1991). Genetic evidence for the spread of agriculture in Europe by demic diffusion. *Nature* **351:**143–145.

Speer, A., Dahl, H.-H., Riess, O., Cobet, G., Hanke, R., Cotton, R. G. H., and Coutelle, C. (1986). Typing of families with classical phenylketonuria using three alleles of the *Hind* III linked restriction fragment polymorphism, detectable with a phenylalanine hydroxylase cDNA probe. *Clin. Genet.* **29:**491–495.

Stoll, J., and Goldman, D. (1991). Isolation and structural characterization of the murine tryptophan hydroxylase gene. *J. Neurosci. Res.* **28:**457–465.

Stratford-Perricaudet, L. D., Levrero, M., Chasse, J.-F., Perricaudet, M., and Briand, P. (1990). Evaluation of the transfer and expression in mice of an enzyme-encoding gene using a human adenovirus vector. *Hum. Gene Ther.* **1:**241–256.

Stuhrmann, M., Riess, O., Mönch, E., and Kurdoglu, G. (1989). Haplotype analysis of the phenylalanine hydroxylase gene in Turkish phenylketonuria families. *Clin. Genet.* **36:**117–121.

Sullivan, S. E., Lidsky, A. S., Brayton, K., DiLella, A. G., King, M., Connor, M., Cockburn, F., and Woo, S. L. C. (1985). Phenylalanine hydroxylase deletion mutant from a patient with classical PKU. *Am. J. Hum. Genet.* **37** [Suppl.]: A177.

Sullivan, S. E., Moore, S. D., Connors, M., King, M., Cockburn, F., Steinmann, B., Gitzelmann, R., Daiger, S. P., Chakraborty, R., and Woo, S. L. C. (1989). Haplotype distribution of the human phenylalanine hydroxylase locus in Scotland and Switzerland. *Am. J. Hum. Genet.* **44:** 652–659.

Svensson, E., Andersson, B., and Hagenfeldt, L. (1990). Two mutations within the coding sequence of the phenylalanine hydroxylase gene. *Hum. Genet.* **85:**300–304.

Svensson, E., von Döbeln, U., and Hagenfeldt, L. (1991). Polymorphic DNA haplotypes at the phenylalanine hydroxylase locus and their relation to phenotype in Swedish phenylketonuria families. *Hum. Genet.* **87:**11–17.

Svensson, E., Eisensmith, R. C., Dworniczak, B., von Döbeln, U., Hagenfeldt, L., Horst, J., and Woo, S. L. C. (1992). Two missense mutations causing hyperphenylalaninemia associated with DNA haplotype 12. *Hum. Mutat.* **1**:129–137.

Svensson, E., von Döbeln, U., Eisensmith, R. C., Hagenfeldt, L., and Woo, S. L. C. (1993a). Relation between genotype and phenotype in Swedish phenylketonuria and hyperphenylalaninemia patients. *Eur. J. Ped.* **152**:132–139.

Svensson, E., Wang, Y., Eisensmith, R., Hagenfeldt, L., and Woo, S. L. C. (1993b). Three polymorphisms but no disease-causing mutations in the proximal part of the promoter of the phenylalanine hydroxylase gene. *Eur. J. Hum. Gen.* **1**:306–313.

Takahashi, K., Kure, S., Matsubara, Y., and Narisawa, K. (1992). Novel phenylketonuria mutation detected by analysis of ectopically transcribed phenylalanine hydroxylase mRNA from lymphoblast. *Lancet* **340**:1473.

Takahashi, K., Masamune, A., Kure, S., Matsubara, Y., and Narisawa, K. (1994). Ectopic transcription: An application to the analysis of splicing errors in phenylalanine hydroxylase mRNA. *Acta Paediatr.* (**in press**).

Takarada, Y., Kalanin, J., Yamashita, K., Ohtsuka, N., Kagawa, S., and Matsuoka, A. (1993a). Phenylketonuria mutant alleles in different populations: Missense mutation in exon 7 of the phenylalanine hydroxylase gene. *Clin. Chem.* **39**:2354–2355.

Takarada, Y., Yamashita, K., Ohtsuka, N., Kagawa, S., and Matsuoka, A. (1993b). Novel homozygous mutation of phenylalanine hydroxylase gene in a Chinese patient with phenylketonuria. *Clin. Chem.* **39**:1350.

Takarada, Y., Yamashita, K., Ohtsuka, N., Kagawa, S., and Matsuoka, A. (1993c). Novel mutation in exon 7 of the phenylalanine hydroxylase gene in a Chinese patient with phenylketonuria. *Clin. Chem.* **39**:2357.

Thompson, A. J., Smith, I., Kendall, B. E., Youl, B. D., and Brenton, D. (1991). Magnetic resonance imaging changes in early treated patients with phenylketonuria. *Lancet* **337**:1224.

Treacy, E., Byck, S., Clow, C., and Scriver, C. R. (1993). "Celtic" phenylketonuria chromosomes found? Evidence in two regions of Quebec province. *Eur. J. Hum. Genet.* **22**:220–228.

Trefz, F. K., Bartholomè, K., Bickel, H., Lutz, P., Schmidt, H., and Seyberth, H. W. (1981). *In vitro* residual activities of the phenylalanine hydroxylating system in phenylketonuria and variants. *J. Inher. Metab. Dis.* **4**:101–102.

Trefz, F. K., Burgard, P., König, T., Goebel-Schreiner, B., Lichter-Konecki, U., Konecki, D. S., Schmidt, E., Schmidt, H., and Bickel, H. (1993). Genotype–phenotype correlations in phenylketonuria. *Clin. Chim. Acta* **217**:15–21.

Trefz, F. K., Erlenmaier, T., Hunneman, D. H., Bartholomè, K., and Lutz P. (1979). Sensitive *in vivo* assay of the phenylalanine hydroxylating system with a small intravenous dose of heptadeutero L-phenylalanine using high pressure liquid chromatography and capillary gas chromatography/mass fragmentography. *Clin. Chim. Acta* **99**:211–230.

Trefz, F. K., Lichter-Konecki, U., Konecki, D. S., Schlotter, M., and Bickel, H. (1988). PKU and non-PKU hyperphenylalaninemia: Differentiation, indication for therapy and therapeutic results. *Acta Paediatr. Jpn.* **30**:397–404.

Trefz, F. K., Yoshino, M., Nishiyori, A., Aengeneyndt, F., Schmidt-Mader, B., Lichter-Konecki, U., and Konecki, D. S. (1990). RFLP-patterns in Japanese PKU families: New polymorphisms for the mutant phenylalanine hydroxylase gene. *Hum. Genet.* **85**:121–122.

Tsai, T.-F., Hsiao, K.-J., and Su, T.-S. (1990). Phenylketonuria mutation in Chinese haplotype 44 identical with haplotype 2 mutation in northern-European Caucasians. *Hum. Genet.* **84**:409–411.

Turek, J. J., Leamon, C. P., and Low, P. S. (1993). Endocytosis of folate–protein conjugates: Ultrastructural localization in KB cells. *J. Cell Sci.* **106**:4223–4230.

Tyfield, L. A., Osborn, M. J., and Holton, J. B. (1991). Molecular heterogeneity at the phenylalanine hydroxylase locus in the population of the south-west of England. *J. Med. Genet.* **28:** 244–247.

Tyfield, L. A., Osborn, M. J., King, S. K., Jones, M. M., and Holton, J. B. (1993). Molecular basis of phenylketonuria in an English population. *Dev. Brain Dys.* **6:**60–67.

Udenfriend, S., and Cooper, J. R. (1952). The enzymatic conversion of phenylalanine to tyrosine. *J. Biol. Chem.* **194:**503–511.

Verelst, P., François, B., Cassiman, J. J., and Raus, J. (1993). Heterogeneity of phenylketonuria in Belgium. *Dev. Brain Dys.* **6:**97–108.

Verelst, P., Denis, C., Rossius, M., Allaer, D., François, B. Martial, J., and Dahl, H. (1988). Restriction fragment length polymorphism in the phenylalanine hydroxylase locus in the Belgian population. *Abstr. SSIEM, London.*

Wagner, E., Cotten, M., Mechtler, K., Kirlappos, H., and Birnstiel, M. L. (1991). DNA-binding transferrin conjugates as functional gene-delivery agents: Synthesis by linkage of polylysine or ethidium homodimer to the transferrin carbohydrate moiety. *Bioconjugate Chem.* **2:**226–231.

Wagner, E., Zatloukal, K., Cotten, M., Kirlappos, H., Mechtler, K., Curiel, D. T., and Birnstiel, M. L. (1992). Coupling of adenovirus to transferrin-polylysine/DNA complexes greatly enhances receptor-mediated gene delivery and expression of transfected genes. *Proc. Natl. Acad. Sci. USA* **89:**6099–6103.

Wagner, E., Zenke, M., Cotten, M., Beug, H., and Birnstiel, M. L. (1990). Transferrin–polycation conjugates as carriers for DNA uptake into cells. *Proc. Natl. Acad. Sci. USA* **87:**3410–3414.

Wang, T., Okano, Y., Eisensmith, R. C., Huang, S.-Z., Zeng, Y.-T., Lo, W. H. Y., and Woo, S. L. C. (1989). Molecular genetics of PKU in Orientals: Linkage disequilibrium between a termination mutation and haplotype 4 of the phenylalanine hydroxylase gene. *Am. J. Hum. Genet.* **45:** 675–680.

Wang, T., Okano, Y., Eisensmith, R. C., Fekete, G., Schuler, D., Berencsi, G., Nasz, I., and Woo, S. L. C. (1990). Molecular genetics of PKU in Eastern Europe: A nonsense mutation associated with haplotype 4 of the phenylalanine hydroxylase gene. *Somat. Cell. Mol. Genet.* **16:**85–89.

Wang, T., Okano, Y., Eisensmith, R. C., Harvey, M. L., Lo, W. H. Y., Yuan, L.-F., Huang, S.-Z., Furuyama J. I., Oura, T., Sommer, S. S., and Woo, S. L. C. (1991a). Founder effect of a prevalent PKU mutation in the Oriental population. *Proc. Natl. Acad. Sci. USA* **88:**2146–2150.

Wang, T., Okano, Y., Eisensmith, R. C., Lo, W. H. Y., Huang, S.-Z., Zeng, Y.-T., Liu, S.-R., and Woo, S. L. C. (1991b). Missense mutations prevalent in Orientals with phenylketonuria: Molecular characterization and clinical implications. *Genomics* **10:**449–456.

Wang, T., Okano, Y., Eisensmith, R. C., Lo, W. H. Y., Huang, S.-Z., Zeng, Y.-T., and Woo, S. L. C. (1991c). Identification of a novel PKU mutation in Chinese: Further evidence for multiple origins of PKU in Asia. *Am. J. Hum. Genet.* **48:**628–630.

Wang, Y., Okano, Y., Eisensmith, R. C., Lo, W. H. Y., Huang, S.-Z., Zeng, Y.-T., Yuan, L.-F., Liu, S.-R., and Woo, S. L. C. (1992). Identification of three novel PKU mutations among Chinese: Evidence for recombination or recurrent mutation at the PAH locus. *Genomics* **10:** 449–456.

Wedemeyer, N., Dworniczak, B., and Horst, J. (1991). PCR detection of the MspI (Aa) RFLP at the human phenylalanine hydroxylase (PAH) locus. *Nucleic Acids Res.* **19:**1959.

Weinstein, M., Eisensmith, R. C., Abadie, V., Avigad, S., Lyonnet, S., Schwartz, G., Munnich, A., Woo, S. L. C., and Shiloh, Y. (1993). A missense mutation, S349P, completely inactivates phenylalanine hydroxylase and is involved in different hyperphenylalaninemias in North African Jews. *Hum. Genet.* **90:**645–649.

Wilson, J. M., Grossman, M., Wu, C. H., Chowdhury, N. R., Wu, G. Y., and Chowdhury, J. R. (1992). Hepatocyte-directed gene transfer *in vivo* leads to transient improvement in hypercholesterolemia in low density lipoprotein receptor-deficient rabbits. *J. Biol. Chem.* **267**:963–967.

Woo, S. L. C., Lidsky, A. S., Güttler, F., Chandra, T., and Robson, K. J. H. (1983). Cloned human phenylalanine hydroxylase gene allows prenatal diagnosis and carrier detection of classical phenylketonuria. *Nature* **306**:151–155.

Woolf, L. I. (1979). Late onset phenylalanine intoxication. *J. Inher. Metab. Dis.* **2**:19–20.

Woolf, L. I. (1986). The heterozygote advantage of phenylketonuria. *Am. J. Hum. Genet.* **38**:773–774.

Woolf, L. I., and Vulliamy, D. G. (1951). Phenylketonuria with a study of the effect upon it of glutamic acid. *Arch. Dis. Child.* **26**:487–494.

Woolf, L. I., Griffiths, R., and Moncrieff, A. (1955). Treatment of phenylketonuria with a diet low in phenylalanine. *Br. Med. J.* **1**:57–64.

Woolf, L. I., McBea, M. S., Woolf, F. M., and Calahane, S. F. (1975). Phenylketonuria as a balanced polymorphism: The nature of the heterozygote advantage. *Ann. Hum. Genet.* **38**:461–469.

Wu, G. Y., and Wu, C. H. (1987). Receptor-mediated *in vitro* gene transformation by a soluble DNA carrier system. *J. Biol. Chem.* **262**:4429–4432.

Wu, G. Y., and Wu, C. H. (1988). Receptor-mediated gene delivery and expression *in vivo*. *J. Biol. Chem.* **263**:14621–14624.

Wu, G. Y., Wilson, J. M., Shalaby, F., Grossman, M., Shafritz, D. A., and Wu, C. H. (1991). Receptor-mediated gene delivery *in vivo*. *J. Biol. Chem.* **266**:14338–14342.

Wu, C. H., Wilson, J. M., and Wu, G. Y. (1989). Targeting genes: Delivery and persistent expression of a foreign gene driven by mammalian regulatory elements *in vivo*. *J. Biol. Chem.* **264**:16985–16987.

Yamashita, M., Minato, S., Arai, M., Kishida, Y., Nagatsu, T., and Umezawa, H. (1985). Purification of phenylalanine hydroxylase from human adult and foetal livers with a monoclonal antibody. *Biochem. Biophys. Res. Commun.* **133**:202–207.

Yang, Y., Nunes, F. A., Berencsi, K., Furth, E. E., Gönczöl, E., and Wilson, J. M. (1994). Cellular immunity to viral antigens limits E1-deleted adenovirus for gene therapy. *Proc. Natl. Acad. Sci. USA* **91**:4407–4411.

Yang, Y., Raper, S. E., Cohn, J. A., Engelhardt, J. F., and Wilson, J. M. (1993). An approach for treating the hepatobiliary disease of cystic fibrosis by somatic gene transfer. *Proc. Natl. Acad. Sci. USA* **90**:4601–4605.

Zenke, M., Steinlein, P., Wagner, E., Cotten, M., Beug, H., and Birnstiel, M. L. (1990). Receptor-mediated endocytosis of transferrin–polycation conjugates: An efficient way to introduce DNA into hematopoietic cells. *Proc. Natl. Acad. Sci. USA* **87**:3655–3659.

Zhao, T. M., and Lee, T. D. (1989). Gm and Km allotypes in 74 Chinese populations: A hypothesis for the origin of the Chinese nation. *Hum. Genet.* **83**:101–110.

Zygulska, M., Eigel, A., Aulehla-Scholz, C., Pietrzyk, J. J., and Horst, J. (1991). Molecular analysis of PKU haplotypes in the population of southern Poland. *Hum. Genet.* **86**:292–294.

# 7

# The Proterminal Regions and Telomeres of Human Chromosomes

**Nicola J. Royle**
Department of Genetics
University of Leicester
Leicester, LE1 7RH, United Kingdom

## I. INTRODUCTION

Considerable advances have been made in our understanding of the molecular structure of telomeres and other sequences at the ends of eukaryotic chromosomes. Most of the pioneering work has been carried out in unicellular organisms and there is still much to understand about the normal "turnover" of terminal sequences, including telomeres, in higher eukaryotes. Sequences within proterminal regions of human chromosomes support high levels of recombination and conversion and ectopic unequal exchanges probably cause submicroscopic terminal deletions and translocations.

In this chapter I have considered recent advances in the description of terminal sequences of human chromosomes and tried to relate them to the known and proposed functions of telomeres and proterminal regions. The rapid turnover of terminal sequences of human chromosomes in somatic tissues and during evolution are discussed. In addition, chromosome abnormalities which involve the loss or rearrangement of terminal sequences and which contribute to the generation of some contiguous gene syndromes are considered.

## II. THE FUNCTIONS OF TELOMERES AND PROTERMINAL REGIONS

### A. The functions of telomeres

Telomeres have been recognized as functional units since the 1930s and 1940s. From X-irradiation studies on *Drosophila melanogaster*, Hermann Muller hypothesized that a specialized structure called a telomere was required to stabilize a chromosome end. Barbara McClintock observed that the ends of the broken chromosomes in Zea mays were more reactive than natural chromosome ends and fused readily with other broken chromosomes, so entering a cycle of breakage–fusion–anaphase bridge formation. Muller and McClintock's insights into the functions of telomeres based on cytological observations were remarkable and have been fully endorsed by increased knowledge of the molecular structure of telomeres and the conservation of telomeric sequences in most if not all eukaryote species (Blackburn, 1990).

Current views of the functions supported by telomeres can be broadly separated into three areas.

(i) Telomeres form a protective "cap" for the chromosome and this prevents the rapid loss of sequences from the chromosome terminus by exonuclease degradation. It also prevents inappropriate fusion to other chromosomes.

(ii) Telomeres are involved in counteracting the loss of terminal sequences which occurs during replication. Semiconservative DNA replication is achieved by DNA polymerases, but all known DNA polymerases initiate synthesis from a 3′ hydroxyl group in a 5′-to-3′ direction. Consequently, removal of the RNA primer results in the loss of sequences from the 5′ end of the newly synthesized strand of a linear DNA molecule (Zakian, 1989) and this occurs at each cycle of replication. It was thought that telomeres would have a specialized structure to ensure that the linear DNA of a chromosome could be completely replicated (Blackburn and Szostak, 1984; Zakian, 1989). In most species which have been investigated, the DNA component of a telomere is composed of head-to-tail arrays of short repeat units containing three or four consecutive GC base pairs and with the G-rich strand orientated 5′–3′ toward the terminus (Blackburn, 1990, 1991). In 1989, Greider and Blackburn cloned the RNA component of a specialized reverse transcriptase, called telomerase, from the lower eukaryote Tetrahymena. They showed that telomerase could add repeat units onto the end of Tetrahymena chromosomes de novo, using an RNA template. In a few species of Diptera (Okazaki et al., 1993), most notably D. melanogaster (Biessmann and Mason, 1993), terminal arrays of telomeric repeat units have not been identified at chromosome ends and telomerase activity has not been detected. Instead D. melanogaster seems to have acquired an alternative strategy for capping and healing broken chromosomes which utilizes at least two moderately repetitive sequences called Het-A and TART elements (Biessmann and Mason, 1993; Levis et al., 1993). These elements are specialized retrotransposons only found at the ends of chromosomes and it has been shown that Het-A elements can be added onto the ends of broken chromosomes in *Drosophila* (Biessmann and Mason, 1993).

(iii) Telomeres play a role in the nuclear organization. In some cell types of some species, chromosomes adopt a "Rabl" orientation where the telomeres become arranged near one another and near to the nuclear membrane while the centromeres tend to associate on the opposite side of the nucleus. (Biessmann and Mason, 1993; Gilson *et al.*, 1993a). In addition, during leptotene of the first meiotic division of some species, telomeres are clustered together near the nuclear membrane and therefore the chromosomes show a "bouquet" formation (Blackburn and Szostak, 1984; Rasmussen and Holm, 1980). In *Schizosaccharomyces pombe* it has been shown that clustered telomeres are involved in the premeiotic movement of chromosomes (Chikashige *et al.*, 1994).

   *In situ* hybridization studies have shown that telomere repeats lie close to the nuclear membrane (van Dekken *et al.*, 1989) and in fact they are tightly associated with the nuclear matrix throughout interphase in many different human cell types (de Lange, 1992). It is also thought that telomeres are directly involved in terminal associations sometimes observed between homologous and nonhomologous chromosomes.

## B. The functions of proterminal regions of human chromosomes

A proterminal region is defined here as the distal region of a human chromosome up to the beginning of the terminal array of telomere repeat units. The proximal limit of a proterminal region cannot yet be identified, because the physical characterization of these regions is preliminary for most chromosome arms. However, two important functions are supported by proterminal regions; firstly homologous chromosome pairing is initiated in proterminal regions of chromosomes during meiosis in humans (Wallace and Hulten, 1985). Therefore, proterminal regions are presumably involved in the initial homology searches which precede chromosome synapsis. Cytological observations of synaptonemal complexes from 16- to 22-week human fetuses showed that synapsis of chromosomes begins very near the telomere (Rasmussen and Holm, 1980) in zygotene, proceeds toward the centromere, and is completed by pachytene of the first meiotic division (Wallace and Hulten, 1985). The close proximity between homologous and nonhomologous telomeres, while associated with the nuclear maitrix during interphase, may contribute to the initiation of chromosome pairing near chromosome ends (Kipling and Cooke, 1992).

   Secondly, proterminal regions support high levels of meiotic recombination. Extensive linkage analysis in humans using a large panel of three generation pedigrees (The CEPH panel, from Centre d' Etude de Polymorphisme

Humain, Paris) and the formation of detailed linkage maps using polymorphic markers (Donis-Keller et al., 1987; Weissenbach et al., 1992; NIH/CEPH et al., 1992) has generated indirect evidence that proterminal regions are proficient at recombination. Some of the markers used in the linkage analysis have been physically assigned to specific bands or placed on physical maps, making it possible to compare genetic and physical maps. This has revealed linkage map expansion at the ends of many autosomes in both males and females. For example the terminal 40% of the genetic map of chromosome 21 corresponds to 10% of its cytogenetic length (Tanzi et al., 1988; NIH/CEPH et al., 1992). As a rule male linkage maps are shorter than female maps, except toward the ends where male linkage map expansion is considerably greater than female map expansion as, for example, on chromosome 7q (Helms et al., 1992). Of course the underlying cause of terminal linkage map expansion is an increase in recombination toward the ends of chromosomes. Therefore the sex differences in recombination frequencies along chromosomes and particularly within the proterminal regions at the ends of maps must reflect differences in meiosis between male and female germlines. The need for at least one chiasma per meiotic bivalent and the length of time spent at each stage of meiotic prophase I are two factors which may affect the different distribution and frequency of recombination events in the two germlines.

## 1. Proterminal regions, recombination nodules, and chiasmata

The five stages of meiotic prophase I have been conserved during evolution and therefore it is possible to recognize these stages in highly diverged species (for example, Sacharomyces cerevisiae and Homo sapiens), although the timing of the initiation and completion of some stages may vary between species. The first stage, leptotene, is characterized by the appearance of axial elements which are replicated sister chromatids in association with proteins. These axial elements are precursors to the synaptonemal complexes. During zygotene homologous chromosomes are aligned "at a distance," synapsis between homologs begins with the formation of a synaptonemal complex, and "early" recombination nodules can be seen. Recombination nodules are dense bodies which are often seen in association with synaptonemal complexes. At pachytene the homologues are fully aligned, the synaptonemal complex between them is complete, and "late" recombination nodules are visible. The recombination nodules which persist later in prophase (during mid-late pachytene) are thought to be precursors to chiasmata, as the number of "late" recombination nodules corresponds to the number of chiasmata visible during later stages of meiosis. In addition, recombination nodules show a preferential distribution over the terminal regions of human synaptonemal complexes (Solari, 1980).

Desynapsis by the dissociation of the synaptonemal complex begins in diplotene and chiasmata become visible; the cell progresses into diakinesis when the condensed chromosome bivalents, held together by chiasmata, become attached to spindle fibers. Chiasmata tend to be distributed over the terminal regions of chromosomes and they represent recombination events between the paired homologous chromosomes (Hulten, 1974). At least one chiasma per bivalent, or one per chromosome arm for the larger chromosomes, is required for the correct segregation of the chromosomes during meiosis in humans. In the absence of chiasmata, bivalents dissociate to become univalents which can segregate abnormally and lead to the generation of aneuploid daughter cells. Histograms of the distribution of chiasmata along each bivalent showed that a distal or terminal location for a chiasma was most common during male meiosis (Hulten, 1974; Laurie and Hulten, 1985b). In addition, it has been shown that the pattern of chiasma distribution in any given bivalent is regular and similar between different males (Laurie and Hulten, 1985a).

Preferential location of chiasmata in the distal portion of chromosome arms has also been observed in other species, including mice, and is not thought to occur as a result of terminalization of the chiasmata (Maudlin and Evans, 1980). Terminalization is the process by which chiasmata migrate along the chromosome arm towards the telomere, but this process does not begin before or during diakinesis, and so the location of chiasmata during diakinesis probably coincides with the site at which they were formed (Maudlin and Evans, 1980).

## 2. Homologue recognition, recombination, and proterminal regions

Initiation of homologous chromosome pairing and high levels of recombination, the functions attributed to proterminal regions, may in fact be parts of the same process. Carpenter (1979) proposed that homology searches between replicated chromosomes in zygotene of meiosis could be achieved by an initial joining of two chromosomes and the formation of a stretch of synaptonemal complex. This would be followed by a check for extensive sequence homology by single stranded invasion as a D-loop or by a single stranded nick of one chromosome (donor) into the other (recipient) as in a gene conversion-like event. In support of this hypothesis, "early" recombination nodules which occur in association with axial elements of the synaptonemal complex in zygotene are thought to be sites of exchange events (Carpenter, 1979). If extensive and perfect sequence homology is detected, then homologous chromosomes have been identified and the continued formation of a synaptonemal complex would bring the chromatids into alignment. In humans the "zippering" together of the homologs would start from the site where homology was identified, near the chromosome end, and progress towards the centromere. This has been confirmed by cytological observations.

If no homology is detected, the segment of synaptonemal complex dissociates and the chromosomes separate and try to find homology with another chromosome. It has been suggested that some initial alignments are incorrect (Chandley, 1989) and may be represented by early recombination nodules. In Carpenter's model, potential problems arise when some but not perfect sequence similarity is detected; the differences could represent allelic differences or different members of a sequence family present on nonhomologous chromosomes. This is a potential problem for any model where the search for homology is determined by sequence similarity. Increasing the extent of heteroduplex between the donor and recipient may resolve the question but it has been suggested that the extent of heteroduplex formation and the "acceptable" level of sequence mismatch between the chromosomes may be species-specific parameters (Carpenter, 1987). The potential outcome for heteroduplexes between successfully paired homologs could be restoration without exchange, resolution as a gene-conversion event, or resolution into an exchange event which would appear as a chiasma during diplotene.

Analysis of meiotic prophase I in S. cerevisiae has resulted in the isolation of a number of genes, SPO11, MER1, MER2, RED1, HOP1, MEI4, and ZIP1 (Atcheson et al., 1987; Hawley and Arbel, 1993; Menees et al., 1992; Sym et al., 1993), which are expressed during meiosis. ZIP1 codes for a protein that has been localized to the central region of synaptonemal complexes but it is not part of the unpaired axial elements (sister chromatid–protein complex). Part of the ZIP1 sequence shows homology to other proteins which contain coiled coils and therefore it has been suggested that ZIP1 encodes the transverse filaments of the synaptonemal complex which joins the lateral elements, acting like a zipper to bring the paired homologues into close proximity or synapsis (Sym et al., 1993). In addition, zip1 mutants undergo homologous chromosome pairing but are defective in synaptonemal complex formation. Therefore, as predicted by Carpenter's model, homology searching and homolog recognition precede synapsis and the formation of synaptonemal complexes (Hawley and Arbel, 1993). Recent advances in our understanding of the processes and proteins involved in prophase 1 of meiosis have been made in S. cerevisiae and consequently it may be possible to isolate similar genes from the human genome by virtue of their homology to the yeast proteins.

## III. THE ISOLATION AND STRUCTURE OF HUMAN TELOMERES

The human telomeric repeat sequence TTAGGG was isolated from a human repetitive DNA library and shown by in situ hybridization to be present at the ends of all human chromosomes (Moyzis et al., 1988). Hybridization of a (TTAGGG)n repeat probe to human genomic DNA digested with almost any restriction enzyme reveals a smear of hybridizing products, resulting from the length heterogeneity of telomeres within and between cells. In addition, the smears of DNA hybridizing to the TTAGGG repeat unit are sensitive to digestion by *Bal31* which has exonuclease activity, indicating that the TTAGGG repeats are located terminally. In humans the long terminal array of repeat units (2–15 kb in somatic tissues) is not a homogeneous array of TTAGGG repeats but contains many "variant " repeat units such as TTGGGG (Allshire *et al.*, 1988), TTTAGGG, and TGAGGG (Allshire *et al.*, 1989). The variant repeat units seem to be clustered at the proximal ends of the telomeric array of repeats (Allshire *et al.*, 1989, Guerrini *et al.*, 1993) while the distal regions contain only TTAGGG repeats. The nonrandom distribution of repeat units may arise by the loss of telomeric repeat units from the chromosome terminus and the templated addition of TTAGGG repeats by human telomerase (Morin, 1989), which extends the terminal array of repeats. The proximally located variant repeat units in human telomeres probably arise by mutations of the TTAGGG repeat. Then other processes, such as replication slippage and unequal inter- and intrachromosomal exchange events may serve to distribute the variant repeats within and between arrays of telomeric repeats.

Telomeres of some lower eukaryotes, for example *Tetrahymena*, have short homogeneous arrays of a telomeric repeat but as stated above the short telomeres (approximately 300 bp) of *S. cerevisiae* are composed of heterogeneous arrays of repeat units. It has been possible to sequence virtually the entire array of yeast telomere repeats and this has revealed variation in the consensus telomeric repeat unit between the proximal and distal regions of the telomere. In addition, the proximal 120 bp of the telomeric repeat array in yeast is more protected from degradation, recombination, and elongation than the distal 150 bp (Wang and Zakain, 1990a). The proximal location of variant repeat units in human telomeres may correspond to a domain protected from degradation by greater affinity for telomere binding proteins, but it seems equally likely that variant repeat units arise and persist at the proximal end of the telomere because they are a greater distance from the terminus (Guerrini *et al.*, 1993).

Differences between telomere structures of lower and higher eukaryotes have been discovered. For example, telomeres of yeast and other lower eukaryotes, which have short telomeric repeat arrays, have a nonnucleosomal chromatin conformation (Wright *et al.*, 1992; Budarf and Blackburn, 1986) but the bulk of the long arrays of telomere repeats (20–100 kb) in rats are organized into closely spaced nucleosomes (Makarov *et al.*, 1993). The unique features of telomeric chromatin in rats compared with the bulk of the rest of the genome are first, a closer spacing of the nucleosomes (periodicity of 157 bp); second, a different micrococcal nuclease sensitivity; and finally, the absence of binding to histone H1 (Makarov *et al.*, 1993). However, the conformation of the telomeric DNA at the termini of mammalian chromosomes is unknown, and it is quite possible that different DNA–protein conformations exist which cap chromosome ends, these structures have been reviewed in detail elsewhere (Blackburn, 1991; Biessmann and Mason, 1993).

## Telomere binding proteins

Proteins which specifically bind to telomeres have been characterized in lower eukaryotes. In *Oxytricha nova* a heterodimeric protein composed of an α-subunit (56 kDa) and a β-subunit (41 kDa) binds to the single stranded $3'-(T_4G_4)_2$ extension at the chromosome terminus (Gottschling and Zakian, 1986; Raghuraman *et al.*, 1989; Gray *et al.*, 1991). The α-subunit binds the single-stranded DNA, while the β-subunit stabilizes the DNA–protein complex and in conjunction with the α-subunit protects the telomeric DNA from methylation (Fang *et al.*, 1993). Purification and characterization of a telomere binding protein from another lower eukaryote, *Euplotes crassus*, showed the structure of the telomeric complexes was similar to that of *Oxytricha*, but there were some distinct differences (Price, 1990). A specific $(TTAGGG)_n$ binding protein (PPT) has been partially purified from *Physarum polycephalum*, which has the same telomere repeat sequence as humans. It is not a ribonucleoprotein and therefore it is probably not telomerase, but its role in telomere structure and function has not been fully characterized (Coren *et al.*, 1991).

In *S. cerevisiae* a number of proteins which bind directly or indirectly to yeast telomeres have been identified. RAP1 is a well characterized DNA binding protein which binds to the upstream region of some yeast genes, where it plays a role in the activation or repression of transcription. However, it has also been shown that RAP1 binds to the terminal $TG_{1-3}$ repeats of yeast chromosomes (Buchman *et al.*, 1988; Gilson *et al.*, 1993b) where it plays a role in the maintenance of telomere length. A RAP1 interacting factor, RIF1, has also been identified in yeast (Hardy *et al.*, 1992). The C-terminus of RAP1 does not bind DNA and therefore RIF1 probably interacts with this region of RAP1; together these proteins are probably involved in the formation of the nonnucleosome chromatin structure found in yeast telomeres.

To date the isolation of human telomere binding proteins has been less successful. Two research groups have isolated abundant human nuclear proteins which bind to single stranded $(TTAGGG)_n$ DNA in humans (McKay and Cooke, 1992a,b; Ishikawa *et al.*, 1993). Further characterization has shown that these proteins bind to RNA oligonucleotides of r(UUAGGG) repeats with higher affinity than to DNA. Peptide sequencing has shown that they are identical or closely related to the heterogenous nuclear ribonucleoproteins (hnRNPs), some of which bind the 3′ splice site sequence r(UUAG/G). The hnRNPs are involved in the packaging of heterogenous nuclear RNA into spliceosomes and splicing the hnRNA. However, the affinity of some hnRNP proteins for the telomere repeat and the fact that members of the hnRNP groups D and E are not stably associated with spliceosomes has led to the suggestion that they, like the RAP1 protein in yeast, have a dual role within the cell (Ishikawa *et al.*, 1993).

# IV. THE STRUCTURE OF PROTERMINAL REGIONS

## A. The structure of proterminal regions

### 1. General properties

Fractionation of mammalian genomic DNA on cesium chloride density gradients has revealed strong intermolecular compositional heterogeneity which is not exhibited by genomic DNA from cold-blooded vertebrates. In some species, such as mice, it has been possible to fractionate different families of satellite DNA away from the bulk of the genomic DNA using cesium chloride buoyant density gradients (Thiery *et al.*, 1976). Similar analysis of the human genome has shown that it can be divided into two GC-poor fractions, L1 and L2, which constitute about two thirds of the genome. The remaining one-third can be divided into three GC-rich fractions, H1, H2, and H3. The compositional heterogeneity of the human genome seems to occur between long stretches of DNA (about 300 kb) which have been called isochores (Bernardi, 1989). The GC-content is fairly uniform within an isochore but different between isochores; in addition, it seems from sequence analysis of known genes that genes tend to be embedded within an isochore of similar GC-content (Bernardi, 1989). There may also be a relationship between the GC-content of isochores and G- and R-banding patterns produced by cytogenetic techniques. When chromosomes are subjected to trypsin digestion and stained with the Giemsa dye, a reproducible pattern of bands (G-bands) is observed. The G-positive bands take up the stain and appear dark while G-negative bands are light. Under different denaturing conditions the reverse pattern of bands can be produced (R-bands) where the G-negative bands appear as R-positive bands. The GC-poor isochores (AT-rich DNA) tend to be associated with postively staining G-bands, and the GC-rich isochores with positively staining R-bands, although R-positive bands are more heterogeneous in GC content. However, a simple linear relationship is unlikely because two thirds of the human genome has a low GC-content whereas the human karyotype is split evenly between G-positive and R-positive bands. One possible explanation for the discrepancy is that chromatin is packed more densely in G-positive than in R-positive bands (Bernardi, 1989).

The H3 component has the highest GC content and constitutes only 3–5% of the genome. Detailed compositional analysis of chromosome 21 has shown that DNA markers and cloned genes which have been physically localized by a variety of methods to the terminal R-positive band of chromosome

21 (21q22.3) also hybridize to the GC-rich fractions of genomic DNA (Gardiner *et al.*, 1990). Similar analysis of Xq26-Xqter has shown that the most GC-rich region occurs in the terminal R-positive band Xq28 (Pilia *et al.*, 1993). There is some evidence to suggest that the purified H3 fraction of genomic DNA hybridizes to the terminal bands and hence the proterminal region of most chromosome arms (Saccone *et al.*, 1992). Futhermore, cytogenetic analysis using DAPI, an AT-specific fluorochrome, and chromomycin A3, a GC-specific fluorochrome, in conjunction with a variety of denaturing techniques has shown that the GC-richest bands are the terminal R-positve/G-negative bands of most human chromosome arms (Holmquist, 1992).

The distribution of Alu elements, which are the most abundant short interspersed element (SINE) in the human genome, also seems to reflect the relative GC-content of the human karyotype. There are about 900,000 copies of Alu elements in the genome (Hwu *et al.*, 1986), each element is about 300 bp in length and relatively GC-rich (56% GC-content) and together they constitute about 10% of the genome. *In situ* hybridization to human metaphase chromosomes, using an Alu element as a probe, produces a pattern similar to R-banding; but it is not known whether the density of GC-rich Alu elements is higher in the terminal R-positive bands of human chromosomes than in other R-positive bands. Conversely, L1 elements, which are the most abundant human long interspersed element (LINE), seem to reflect the AT-content of the human karyotype. A consensus L1 element of 6.4 kb has an AT content of about 58% and *in situ* hybridization using an LI element produces a pattern similar to G-banding (Kornberg and Rykowski, 1988). .

## 2. Sequences associated with proterminal regions

Proterminal regions contain genes but the number of genes and their distribution within proterminal regions has not been determined for many chromosomes. Recently, a novel family of SINEs called STIR (subtelomeric interspersed repeat) elements has been identified. Autosomal STIR elements have a consensus sequence of 350 bp; but the consensus sequence of Xp/Yp pseudoautosomal STIR elements shows some differences (Rouyer *et al.*, 1990). Unlike other SINEs, STIR elements are not distributed throughout the genome; *in situ* hybridization analysis has shown they are confined to the distal ends of chromosomes although they do not appear to be very close to telomeres.

The term minisatellite was first coined to describe loci composed of arrays of tandemly repeated units (Jeffreys *et al.*, 1985). A subset of such loci has been identified and isolated from the human genome by cross hybridization to the DNA fingerprinting probes 33.6 and 33.15 (Wong *et al.*, 1987; Nakamura *et al.*, 1987; Armour *et al.*, 1990). Sequence analysis of a few of theses clones has shown that they contain GC-rich repeat units. The length and sequence of the repeat unit varies considerably between loci but all show similarity to an 11- to 16-bp GC-rich core sequence (Jeffreys *et al.*, 1985). There are an estimated 1500 GC-rich minisatellite loci within the human genome and about 12% of them are hypervariable with heterozygosities in excess of 85%. The extensive allelic length variation observed at hypervariable minisatellite loci occurs as a result of variation in repeat unit copy number, and the hypervariable loci (heterozygosities in excess of 97%) also have a high germline mutation rate (Jeffreys *et al.*, 1987).

Mapping GC-rich minisatellites by *in situ* hybridization to human metaphase chromosomes (using single locus minisatellite probes) has shown that they are clustered in the proterminal regions, at the ends of chromosomes (Royle *et al.*, 1988). The generation of complete linkage maps of human chromosomes using polymorphic markers, including minisatellite or VNTR loci, has also shown that about 80% of the variable minisatellite loci are located towards the ends of genetic maps (NIH/CEPH *et al.*, 1992). However, it should be remembered that the distribution of monomorphic minisatellites has not been investigated. Sequence analysis of minisatellites isolated from proterminal regions and the DNA which flanks them has shown that not only are they clustered sometimes within a few base pairs of one another, but they are often associated with dispersed repeat elements such as Alu, L1, and retroviral LTR-like sequences (Armour *et al.*, 1989). To date, three minisatellites have been shown to have expanded by tandem repeat amplification from within different dispersed elements (Mermer *et al.*, 1987; Armour *et al.*, 1989, 1992). The presence of GC-rich minisatellites in proterminal regions of human chromosomes which hybridize to the GC-rich H3 component of the genome is consistent with the general properties of isochores outlined above. The association of proterminal minisatellites with Alu (GC-rich) and L1 (GC-poor) elements might reflect the greater heterogeneity of GC-content found within the R-positive bands (Bernardi, 1989). However, if the general properties attributed to isochores are correct, then GC-rich minisatellites in proterminal regions should be associated with GC-rich dispersed elements more frequently than with GC-poor elements. Currently, too few minisatellites have been analyzed in sufficient detail to determine whether that prediction is true.

Several proteins which bind to GC-rich minisatellites have been identified. Minisatellite binding proteins Msbp-1, -2, -3, and -4 are 40, 75, 115, and 35 kDa, respectively (Collick and Jeffreys, 1990; Wahls *et al.*, 1991; Yamazaki *et al.*, 1992), but some of their binding properties are different. For example, Msbp-1 and Msbp-4 are single stranded binding proteins of similar molecular weights, but Msbp-1 binds to the G-rich strand and Msbp-4 to the C-rich strand; in addition, Msbp-1 requires at least two minisatellite repeats for efficient binding (Collick *et al.*, 1991). The function of these minisatellite binding proteins is as yet unknown.

## 3. Minisatellites: Sequences which could support homologue recognition

The germline mutations which cause allele length changes at hypervariable minisatellite loci have been studied in detail at only a small number of loci located at the distal ends of human chromosomes, and several common features underlying the mutation processes at these loci have been identified. A technique (MVR-PCR) for mapping the distribution of variant repeat unit types (which show minor sequence differences from the consensus repeat sequence) along the length of a minisatellite allele (Jeffreys et al., 1990, 1991) has been developed. This has shown that the level of variation at some minisatellite loci is not solely a feature of the number of repeat units. Minisatellite allele codes based on their internal structures have been generated and compared (Monckton et al., 1993); at each of the three loci studied in detail there is a polarity (Jeffreys et al., 1991; Neil and Jeffreys, 1993; Armour et al., 1993) with more differences found between allele codes at one end of the tandem repeat array. In addition, the internal maps of mutant alleles identified within families or in individual sperm have shown that the majority of mutation events which cause a change in the repeat copy number also involve only one end of the allele (Jeffreys et al., 1991). The mechanisms which generate mutations at minisatellite loci are not fully understood, but many mutant alleles appear to arise as small conversions between sister chromatids aligned out of register or by exchanges between alleles. Most or all of the latter are resolved as conversion events between minisatellite alleles, but do not extend into the flanking DNA and therefore exchange of adjacent flanking markers (Jeffreys et al., 1994) or more remote flanking markers is not always observed (Wolff et al., 1988). The MVR-maps of mutant alleles often show complex "patchwork" junctions (Jeffreys et al., 1994), as might be expected for conversion events (Kourilsky, 1986).

In the current hypothesis, hypervariable minisatellites are reporter sequences of "hotspots" for recombination or conversion events initiated in the DNA flanking the minisatellite (Jeffreys et al., 1993). Enhanced recombination has been observed between minisatellite-containing plasmids when introduced into mammalian cells, although the constructs used in these studies lacked any minisatellite flanking DNA (Wahls et al., 1990). In addition, mutant alleles at three different minisatellite loci usually show small gains in repeat units but the significance of this observation or the factors which control minisatellite allele length are unclear. It is also interesting that mutations at some minisatellite loci (for example, D2S90; Vergnaud et al., 1991) arise more frequently in the male than in the female germline.

It is difficult to reconcile the observed terminal linkage map expansion and the terminal localization of recombination nodules and chiasmata with all the properties of minisatellite loci described above. However, minisatellites located at the proximal ends of proterminal regions are chromosome specific and they clearly have some of the properties predicted for sequences which support homologue recognition prior to exchange events in human meiosis.

## B. Subterminal sequence isolation in eukaryotes

The number of telomeres in relation to genome size is high in the macronuclei of some lower eukaryotes (for example *Trypanosoma brucei*) and therefore it has been possible to isolate telomeres from some lower eukaryotes by ligating the natural end of chromosomes onto a vector. Briefly, high-molecular-weight genomic DNA was blunt end ligated onto a vector, this DNA was digested with a restriction endonuclease which did not cleave the vector, and then ligated to favor DNA circularization. The resulting plasmid library was thus enriched for sequences from chromosome ends (Van der Ploeg *et al.*, 1984). An elegant adaptation of this method was used to isolate telomeres from *Arabidopsis thaliana*, which has 10 telomeres, and a genome of 70 Mbp (Richards and Ausubel, 1988), but the ratio of telomeres to genome size (46 telomeres in 3000 Mbp) in humans has limited the usefulness of this strategy for isolating human chromosome ends.

Human telomeres and subterminal sequences have been isolated by enrichment for chromosome ends and other tandemly repeated sequences prior to cloning in plasmids. The strategy entailed enzymatic digestion of high-molecular-weight human genomic DNA with *Bal*31 to ensure that chromosome termini were blunt ended, restriction endonuclease digestion with multiple enzymes which were unlikely to cleave arrays of telomere repeats, size fractionation to remove the bulk of the genomic DNA, and blunt-ended ligation into a plasmid vector. This resulted in the isolation of two clones (pTH2 and pTH14) containing human sequences adjacent to arrays of $(TTAGGG)_n$ repeats (de Lange *et al.*, 1990).

Alternative and more successful strategies for isolating human chromosome ends in yeast have been developed. A yeast artificial chromosome (YAC) cloning vector containing sequences for the propagation of the vector in *Escherichia coli*, a yeast-selectable marker(s), an autonomously replicating sequence, a centromere, and a single terminal array of *Tetrahymena* telomere repeats was ligated to digested human DNA. The ligated DNA was transformed into a suitable yeast strain and grown under selection for the presences of the yeast marker(s). If the recombinant half-YAC terminated in human telomere repeats, it could be rescued by the addition of yeast telomere repeats onto the human telomere repeats (Cross *et al.*, 1989, Brown, 1989). This strategy has been used to isolate small recombinant YACs containing several kilobases of

DNA from human chromosome ends (Cross *et al.*, 1989, Cheng *et al.*, 1989, Brown *et al.*, 1990) as well as much larger YACs containing several hundred kilobases of DNA terminating in a human telomere (Riethman *et al.*, 1989).

The use of PCR amplification for isolating sequences adjacent to human TTAGGG repeated DNA from fluoresence sorted chromosomes is theoretically a powerful method for isolating specific ends. In one such strategy, a primer was annealed to TTAGGG repeats and extended to produce short segments of single-stranded DNA. The second strand was generated by random priming and extension with DNA polymerase and it gave rise to clones, some of which had a complex internal structure of small blocks of TTAGGG and variant repeat arrays interrupted by nontelomeric repeat DNA (Weber *et al.*, 1990). One of the clones hybridized to the end of chromosome 4q but none of the cloned sequences exhibited *Bal*31 sensitivity in the genomic DNA, and therefore they had not originated from a telomere (Weber *et al.*, 1990). A similar strategy was used to isolate sequences from the mouse genome giving rise to the mouse subtelomeric sequence (ST1) which is located at many mouse chromosome ends (Broccoli *et al.*, 1992). In addition, copies of ST1 have been located between pericentromeric arrays of mouse minor satellite and telomere repeat arrays on the telocentric mouse chromosomes (Kipling *et al.*, 1991, Broccoli *et al.*, 1992).

Alternative PCR strategies for isolating sequences adjacent to long arrays of TTAGGG and variant repeat units have been used. A telomere-anchored PCR strategy was designed to prime sequences from the G-rich strand of the telomere and to amplify sequences between the primer and an *Mbo*1 site proximal to the repeat array. Steps were included to enrich significantly for arrays of tandemly repeated sequences resistant to digestion with *Mbo*1 using size fractionation of *Mbo*1 fragments longer than 5.5 kb. Linkers were ligated onto the fractionated DNA and PCR was carried out between the linker and a primer composed of $(CCCTCA)_4$. The majority of the products which hybridized to a synthetic $(TTAGGG)_n$ probe arose by amplification from the proximal end of telomeres into adjacent DNA rather than from adjacent to interstitial blocks of TTAGGG and variant repeats (Royle *et al.*, 1992). A similar strategy has been used to isolate sequences adjacent to telomeres of *Hordeum vulgare* (Killian and Kleinhof, 1992).

It has also been possible to isolate subterminal sequences from the human genome by simply screening genomic cosmid libraries with telomere repeat probes. Terminal restriction fragments including telomeres are mostly absent from conventional cosmid libraries. Therefore cosmids which cross hybridize to a telomere repeat probe almost always originate from interstitial sites, though these are mostly near chromosome ends (Meyne *et al.*, 1990). By screening a human cosmid library with $(TTGGGG)_n$, which is present at the proximal ends of many human telomeres (Allshire *et al.*, 1989), many cross-hybridizing cosmids (about 20 per genome equivalent) have been isolated (Wells *et al.*,

1990). The interstitial blocks of TTAGGG and variant repeats tend to be short repeat arrays and they have been called telomere related sequences (TRS) in order to distinguish them from the terminal arrays of TTAGGG repeats. The cloned TRS sequences were flanked by 'normal' DNA sequences and sequences from one cosmid (cTT1) hybridized to multiple fragments and *Bal*31-sensitive smears of human genomic DNA. In addition, *in situ* hybridization of this probe to human chromosomes showed that most of the cross hybridizing loci in the genome are located near chromosome ends but there was also hybridization to the long arm of chromosome 2 at 2q13 (Wells *et al.*, 1990).

## C. Subterminal repeat sequences in humans

The terms used to describe sequences next to telomeres have included subtelomeric repeats (de Lange *et al.*, 1990), telomere-associated repeated sequences (Brown *et al.*, 1990), subtelomeric sequences (Weber *et al.*, 1990; Wells *et al.*, 1990), subterminal sequences (Cross *et al.*, 1990), human subtelomeric repeats (HST) (Cheng *et al.*, 1991), and telomere-adjacent sequences (Royle *et al.*, 1992). The majority of clones which have been isolated by the methods described above have been partially or completely sequenced and it is possible to align the sequences along at least part of their length. Diagrammatic representation of the alignment of clones pTH2, pTH14 (de Lange *et al.*, 1990), about 1kb of sequence from clones TelSau2.0, TelBam3.4 (Brown *et al.*, 1990), clones TSK6, TSK37 (Royle *et al.*, 1992), with part of the clone pCTT1a (Wells *et al.*, 1990) is shown in Figure 7.1. The pCTT1a clone was isolated from an interstitial site (see above), but most likely from the distal end of a proterminal region. In addition, sequences adjacent to the arrays of TTAGGG and variant repeats from clones pHutel-2-end, pGB4G7 (Cross *et al.*, 1990), yHT1 (Cheng *et al.*, 1991), and cosmid 56.1.1 (Ijdo *et al.*, 1992b) also show homology to this sequence family. Estimates of copy number of these related subterminal sequences indicate that there are a few hundred copies in the genome and therefore it is a low-copy-number repetitive sequence family.

The size and structure of this subterminal repeat sequence is not fully understood, but it is complex and shows internal repetition as it contains several arrays of tandem repeats. Closest to the TTAGGG repeats, there is a discontinuous array of five 37-bp repeats and proximal to the 37-bp repeats lies a tandem array of 29-bp GC-rich repeats, the copy number of the repeats being variable. Proximal to the 29-bp repeats is another tandem array of 61-bp repeats and this array also shows variation in copy number. The 29- and 61-bp repeat arrays have a similar structure to GC-rich minisatellites or VNTRs located more proximally in the proterminal regions of human chromosomes, but they are not locus-specific minisatellites because they are part of the subterminal repeat, which is found on many different chromosome ends. The subterminal repeat

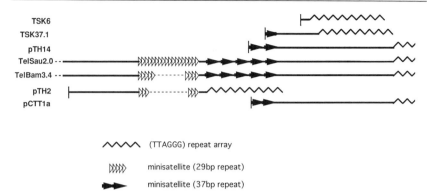

**Figure 7.1.** Diagrammatic representation of the relationship between different isolates of a family of human subterminal repeat sequences. Vertical bars represent the start of the cloned sequence; thick horizontal lines represent DNA which does not contain tandem repeats but is part of the human subterminal repeat and therefore present in a few hundred copies in the genome. Arrays of tandem repeats are represented by types of arrows; the number of 29-bp repeats, shown by the open arrowheads, varies between clones and a dashed line has been introduced to maintain the maximum alignment. The solid arrows represent an array of 37-bp repeat units. The thick zig-zag lines show arrays of TTAGGG and variant repeat units. Clone pCTT1a Wells *et al.*, 1990 was isolated from an interstitial site, and only the sequence flanking one side of the TTAGGG and variant repeat array is included. The relationship between clones TelSau2.0 and TelBam3.4 are shown from positions 1296 and 2313, respectively (Brown *et al.*, 1990); pTH2 and pTH14 (de Lange *et al.*, 1990). From Royle *et al.* (1992).

sequence can be truncated at different positions before the start of the terminal or interstitial array of TTAGGG and variant repeats (see Figure 7.1). Different copies of the subterminal repeat sequence family show less similarity to one another at their proximal ends, though some copies contain Alu elements and truncated portions of L1-like sequences. In addition, some copies of the subterminal repeat sequence family may be transcribed into a 6-kb polyadenylated RNA or different-length nonpolyadenylated RNAs (Cheng *et al.*, 1991).

Several research groups have shown that copies of the subterminal repeat sequence family are distributed over many different chromosomes but are mainly localized to chromosome ends (de Lange *et al.*, 1990; Brown *et al.*, 1990; Cross *et al.*, 1990, and see above discussion; Wells *et al.*, 1990). The combination of chromosome ends which hybridize varies between unrelated individuals. For example, a probe from cosmid 56.1.1 (Ijdo *et al.*, 1992b) consistently hybridized to both homologous chromosome ends of 3q, 9p, 12p, and 20p and at two interstitial sites. The probe also showed hybridization to both, one, or neither homolog at 10 additional chromosome ends among 8 unrelated individuals studied. The chromosome ends that hybridized were consistent within an individual and segregated in a Mendelian manner within families (Ijdo *et al.*, 1992b). The variation in distribution and copy number of the human subterminal repeat sequence between different individuals is similar to the variation in copy number and distribution of the Y' and related but divergent X subterminal repeats of yeast chromosomes. The Y' repeats of yeast have an internally repeated structure; they are present as 0 to 4 copies adjacent to the telomere, separated by short tracts of yeast telomere-like repeat units; the distribution of the Y' repeats differs between laboratory strains of yeast and they are unnecessary for normal telomere function (Louis and Haber, 1990b). The Y' yeast subterminal repeats undergo mitotic and meiotic exchanges between homologous and nonhomologous chromosome ends (Louis and Haber, 1990a). It has been suggested that the variable distribution and copy number of human subterminal repeats between unrelated humans might be explained by unequal exchange events between subterminal repeats and possibly telomeres of homologous and nonhomologous chromosome ends. Such exchange events would tend to homogenize the subterminal repeat family, but the sequence divergence between different copies varies from 2% to about 20% (Cross *et al.*, 1990). The substantial sequence divergence suggests that the subterminal repeat sequence family is ancient—that terminal exchange events between nonhomologous chromosome ends are very infrequent or that they occur more frequently between some subsets of chromosome ends than others.

Some alleles of most human chromosome ends cross-hybridize to the subterminal repeat sequence family (Riethman *et al.*, 1989; Brown *et al.*, 1990; Wells *et al.*, 1990; Cross *et al.*, 1990; Wilkie *et al.*, 1991; Ijdo *et al.*, 1992b), but some chromosome ends do not cross-hybridize. These include the Xp/Yp pseudoautosomal telomere, 7q telomere, and probably telomeres at 2p, 11q, 12q, 13p, 14p, and 18q. These chromosome ends either have no subterminal repeats, highly divergent copies, or sequences from a different subterminal repeat family. YAC and plasmid clones which specifically hybridize to the Xp/Yp pseudoautosomal (Brown *et al.*, 1990; Royle *et al.*, 1992) and 7q (Riethman *et al.*, 1989; Brown *et al.*, 1990) telomeres have been isolated.

## D. Sequence organization of specific chromosome ends

Physical maps of proterminal regions have been produced for only a few chromosome arms and the best characterized are described below.

## 1. Xp/Yp pseudoautosomal region and its telomere

Chromosome synapsis and the formation of a synaptonemal complex is limited between the sex chromosomes, and recombination between them is confined to the two pseudoautosomal regions at the distal ends of the X and Y chromosome arms in the male germline (Rouyer *et al.*, 1986; Freije *et al.*, 1992). In the female germline the X chromosomes undergo synapsis and can recombine along their length. The behavior of the sex chromosomes during gametogenesis also differs between the male and female germlines. In the male germline the majority of the X and Y encoded genes are transcriptionally inactive and the sex chromosomes are condensed during pachytene in a special chromatin conformation called the sex body or XY vesicle (Handel and Hunt, 1992). In contrast, in the female germline the inactive heterochromatic X chromosome is reactivated upon entry into meiotic prophase and transcription from both copies of the X chromosome has been demonstrated. It has been proposed that the condensation of the sex chromosomes during male meiosis helps prevent the initiation of "damaging" recombination between nonhomologous regions of the X and Y chromosomes (McKee and Handel, 1993).

The Xp/Yp pseudoautosomal region supports an obligatory recombination event in the male germline and the formation of a chiasma in this region is believed to be required for normal segregation of the chromosomes as they pass through meiosis. This pseudoautosomal region is about 2.6 Mbp in length, the proximal end is defined by the insertion of an Alu element (Ellis *et al.*, 1989), and the distal end by the telomere (Cooke *et al.*, 1985). Physical and linkage maps have been generated for the whole of the pseudoautosomal region and they confirm the high level of recombination in the male germline, but they also show that there is also a hotspot for recombination in the female germline 20–80 kb from the telomere (Wapenaar *et al.*, 1992; Henke *et al.*, 1993). The Xp/Yp pseudoautosomal region also contains at least six genes (reviewed by Rappold, 1993). It contains STIR elements specific for the pseudoautosomal region (Rouyer *et al.*, 1990) and many hypervariable and unstable GC-rich minisatellites. The most distal variable minisatellite at locus DXYS14 is about 13.4 kb from the start of the telomere (Brown, 1989). The terminal *Sau*3A1 restriction fragment of the Xp/Yp pseudoautosomal region contains four 63-bp repeat units of a monomorphic minisatellite which is specific to the pseudoautosomal region, adjacent to a truncated copy of a recently described family of SINEs (La Mantia *et al.*, 1989), and, of course, the restriction fragment terminates with the telomere (Brown *et al.*, 1990, Royle *et al.*, 1992).

## 2. 4p telomere

A YAC carrying the terminal sequences from chromosome 4p has been rescued in yeast (Bates *et al.*, 1990). The YAC (Y88BT) includes 115 kb of DNA from a *Bss*HII site to the array of telomere repeats. The proximal 52 kb of the YAC sequence are specific to the end of chromosome 4p, but the distal 60–65 kb cross-hybridized to other chromosome ends and in particular to 13pter, 15pter,

21pter, and 22pter. The boundary between chromosome 4-specific and non-specific sequences has been limited to a 1.5-kb region. A pentameric repeat sequence, which may have arisen from the poly(A) tail of an Alu element has been identified in the YAC, Y88BT, from 4p. PCR amplification of DNA from two different somatic cell hybrids showed that the pentameric repeat is present on chromosomes 4 and 21 and it shows variation in repeat copy number (Youngman et al., 1992).

During the search for the dominantly inherited Huntington disease (HD) gene at the distal end of 4p, one affected family was identified with an unusual transmission pattern. Investigation of many marker loci along 4p in this family showed that either a different gene located elsewhere in the genome caused the disease in this family, or more likely, a double crossover or conversion event had occured in the vicinity of the HD gene in the germline of the transmitting parent (Pritchard et al., 1992). There is genetic map expansion at the distal end of 4p, but the increase in recombination is not uniform throughout the 4p proterminal region; rather, it is localized to a few hotspots (Allitto et al., 1991).

## 3. 7q telomere

As stated above, sequences at the distal end of the 7q proterminal region are apparently distinct from those of other chromosome ends and the terminal 240 kb including the telomere has been cloned in a YAC, HTY146, and subcloned into cosmids (Riethman et al., 1989). With the exception of an 11-kb gap which has not been cloned in E. coli, the cosmid contig seems to be a true representation of genomic DNA. Analysis of the YAC and human genomic DNA has shown that this region contains many restriction fragment-length polymorphisms (RFLPs), at least two hypervariable minisatellites (Dietz-Band et al., 1990; Helms et al., 1992), and one unmethylated CpG island which might be adjacent to a gene only 195 kb from the telomere (Riethman et al., 1993). In addition, a variable microsatellite has been identified 10–20 kb from the telomere (Hing et al., 1993). Linkage map expansion and therefore increased meiotic recombination occurs at the distal end of 7q in females and particularly in males, but there is some evidence that the higher recombination rate does not extend to the most terminal 240 kb adjacent to the telomere (Helms et al., 1992).

## 4. 16p telomere

The α-globin gene cluster contains the most distal well characterized genes on 16p in a constitutively open chromatin domain (Vyas et al., 1992); this region also contains at least five hypervariable minisatellites and many base substitu-

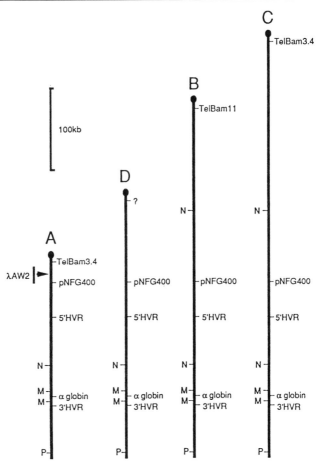

**Figure 7.2.** Long-range restriction map showing the extraordinary length variation between the terminal regions of A, D, B, and C alleles of 16p. Alleles A, C, and B hybridize to different members of the subterminal repeat sequence family (TelBam3.4 or TelBam11) but allele D does not hybridize. The divergence point between allele A and the other alleles is shown by an arrow and the clone (λAW2) containing the divergence point is shown. *Mlu*1 (M), *Nru*1 (N), *Pvu*1 (P). Reproduced with permission from Harris, P. C., and Higgs, D. R. (1993), "Structure of the terminal region of the short arm of chromosome 16." In "Regional physical mapping" (K. E. Davies and S. H. Tilghman, Eds.) Cold Spring Harbor Laboratory Press, New York.

tions which give rise to RFLPs (Higgs *et al.*, 1993). A long range physical map of the distal end of 16p with the restriction enzymes *Mlu*1, *Nru*1, and *Pvu*1, which cut proximal to the α-globin 5′ HVR but not between it and the telomere, has revealed an extraordinary amount of length variation between alleles. Four different length terminal fragments have been identified where the distance between the 5' HVR and the telomere is 170, 245, 350, or 430 kb and these have been designated alleles A, D, B, and C, respectively (Wilkie *et al.*, 1991; see Figure 7.2). Allele A is most common and present on about two thirds of the chromosomes 16 analyzed. The terminal 170-kb fragment (allele A) and 430-kb fragment (allele C) of chromosome 16 hybridized to the copy of the subterminal repeat sequence family represented by TelBam3.4 (Brown *et al.*, 1990), while the 350-kb fragment (allele B) hybridized to the different copy, TelBam11 (Brown *et al.*, 1990). The maps of the four different-length fragments are the same from the ζ2 globin gene to the divergence point located about 145 kb distally. Four genes, which are expressed in a wide variety of tissues, are encoded in the intervening region; one of the genes encodes methyladenine glycosylase (Higgs *et al.*, 1993) and another is only about 20 kb from the divergence point of the 16pter alleles. The divergence point between the most common 16pter alleles A and B comprises a complex segment of about 4 kb which shows partial loss of homology before the point of complete divergence. This segment of DNA includes a CA repeat microsatellite. The lengths of the microsatellite alleles vary with the 16pter alleles, such that the 16pter allele A has a very long array (250–350 repeats) of imperfect CA repeats (Wilkie and Higgs, 1992). Sequences distal to the divergence point of 16pter allele A cross-hybridized to loci at Xqter and Yqter but not to other 16pter alleles, while sequences distal to the divergence point of the 16pter allele B cross-hybridized to loci on 9qter, 10qter, and 18qter. Therefore, the sequences distal to the divergence point of 16pter alleles show more similarity to sequences at the ends of other chromosomes than to one another. The lack of similarity between the 16pter alleles A and B distal to the divergence point and the association of alleles A and B with different copies of the subterminal repeat sequence family (TelBam3.4 and TelBam11, respectively) suggest that the extraordinary terminal length variation has arisen by exchanges between the end of 16pter and other nonhomologous chromosome ends (Higgs *et al.*, 1993).

## 5. 21q telomere

A physical map of the long arm of chromosome 21 extends to the subterminal repeats. The most distal characterized gene, the β-subunit of the neuronal calcium binding protein S100, is between 50 and 200 kb from the start of the subterminal repeats (identified by cross-hybridization to pTH2; de Lange *et al.*, 1990). The subterminal repeats and telomere are probably limited to the distal 50 kb of 21q. There are at least two hot spots for recombination in the terminal 7 Mbp of 21q. One of these is between the CD18 and the collagen genes (COL6A1 and COL6A2) and within 700 kb of the telomere (Burmeister *et al.*, 1991).

In summary, the distal 50–200 kb of proterminal regions of many human chromosome arms are composed of subterminal repeats interspersed with short arrays of TTAGGG and variant repeats. The junction between the proximal chromosome-specific sequences and the distal subterminal repeats may be defined by a relatively short sequence (1–4 kb) of reduced homology. One subterminal repeat sequence family has been characterized but there may be others, as some human chromosome ends (7qter, Xp/Ypter, and others) do not cross-hybridize to it. The length of the subterminal repeat is undefined but it is internally repetitious, containing a variety of GC-rich minisatellites which show length variation. Sequence divergence between the distal ends of different copies of the subterminal repeats can be as high as 20%. The copy number and distribution of the subterminal repeat sequence family varies between unrelated individuals but shows normal Mendelian segregation within families. It has been proposed that a low rate of nonhomologous unequal exchange between chromosome ends could explain the variation in copy number and distribution. One chromosome end (16pter) shows extraordinary terminal length variation and some 16pter alleles are more similar to nonhomologous chromosomes than to other 16pter alleles. Clearly, DNA sequence based homolog recognition during meiosis could not be supported by subterminal repeat sequences which are shared by nonhomologous chromosomes and therefore it is unlikely that they are involved in homolog recognition.

## V. CHROMOSOME ENDS FROM OTHER PRIMATES

Karyotypes of humans, chimpanzees, gorillas, and orangutans are similar and this reflects the close relationship between these species. Most differences between the four species consist of inversions of chromosomal segments and differences in constitutive heterochromatin though other changes including reciprocal translocations, band insertion, and a telomeric fusion have occurred. The reduction to 46 chromosomes in humans compared to 48 chromosomes of the chimpanzee, gorilla, and orangutan has resulted from an ancestral terminal fusion of chimpanzee-like chromosomes 2p and 2q, accompanied by the inactivation of the additional centromere on 2q in man (Yunis and Prakash, 1982). The fusion point on the human chromosome 2 at 2q13 cross-hybridizes to the human subterminal repeat sequence family and the fusion point has been isolated from a genomic cosmid library (Ijdo et al., 1991). The cosmid clone of the fusion point contains a 900-bp sequence composed of two inverted arrays of TTAGGG and variant repeats flanked by copies of the subterminal repeat sequence family (Ijdo et al., 1991). The fusion point is distinct from the rare fragile site (FRA2B), which is also located at 2q13 (Ijdo et al., 1992a).

The human subterminal repeat sequence family can be detected in other apes including chimpanzees and gorillas (Cross et al., 1990; N. Royle

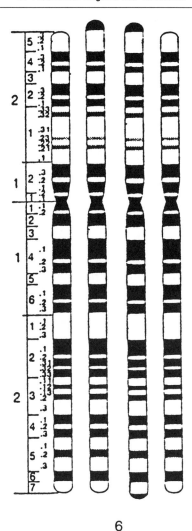

6

**Figure 7.3.** Schematic representation of G-banded chromosome 6 of human, chimpanzee, gorilla, and orangutan arranged from left to right, respectively. The chromosomes are almost identical except for the presence of an additional heterochromatic band at the end of the short arm of chromosome 6 in the chimpanzee and at the end of both arms of chromosome 6 in the gorilla. These bands are absent from the human and orangutan chromosomes. The subterminal satellite isolated from chimpanzee chromosomes is mainly located in terminal bands of chimpanzee and gorilla chromosomes. Reproduced with permission Yunis, J. J., and Prakash, O. (1982). *Science* **215:**1525–1530. (Copyright 1982 by the AAAS).

*et al.*, unpublished results). However, sequences isolated from the human and chimpanzee genomes by the same telomere-anchored PCR strategy has shown that, other than Alu- and L1-like sequences, there was no similarity between the sequences adjacent to arrays of TTAGGG and variant repeats in the two species. Many of the chimpanzee telomere-junction clones contained arrays of an AT-rich 32-bp repeat unit adjacent to the TTAGGG repeat arrays. Southern blot, telomere-anchored PCR and *in situ* hybridization studies all confirmed that arrays of the chimpanzee 32bp repeat are absent from the human genome, but present as a subterminal satellite at 21 chromosome ends and at two interstitial sites of chimpanzee chromosomes. The chimpanzee 32-bp repeat probe hybridizes more strongly to the gorilla but is absent from the orangutan and other monkey genomes (Royle *et al.*, 1994). One karyotypic difference between humans, chimpanzees, and gorillas is the presence of additional heterochromatic G-bands at the ends of some chimpanzee and most gorilla chromosomes (see Figure 7.3) (Yunis and Prakash, 1982; Luke and Verma, 1993). The novel satellite is probably the major component of these additional G-bands.

The similarity between the gorilla, chimpanzee, and human genomes suggests that the ancestor to these three species contained some subterminal satellite which has amplified to different extents and dispersed mainly to chromosome ends in the chimpanzee and gorilla, but it has been lost from the human genome. The physical relationship between the subterminal satellite, the human-like subterminal repeat sequences, and the telomeres in the chimpanzee and gorilla genomes has not yet been determined. The effect of the chimpanzee and gorilla subterminal satellite on chromosome functions, such as homologue recognition and pairing during meiosis, is as yet unknown but subterminal satellites have been identified in other species.

## VI. THE ROLE OF PROTERMINAL REGIONS AND TELOMERES IN SENESCENCE, TUMORIGENESIS, AND GENETICALLY DETERMINED DISEASES

The following section reviews the normal turnover of terminal sequences of human chromosomes in somatic tissues; the destabilization of the genome caused by telomere loss, and the effect of this destabilization in cellular senescence and in tumorigenesis.

### A. The loss of telomere repeats

*Last scene of all,*
*That ends this strange eventful history,*
*Is second childishness and mere oblivion;*
*Sans teeth, sans eyes, sans taste, sans everything.*
*William Shakespeare: "As You Like It," Scene VII, Act II*

The average length of the telomere repeat arrays in human somatic tissues (2–15 kb) is consistently shorter than in fetal tissue, fetal tissue enriched for ova, and in sperm DNA (Allshire et al., 1989; Hastie et al., 1990; Cross et al., 1990; Allsopp et al., 1992). In addition, the length of the telomere repeat array in somatic tissues decreases as the age of the donor increases (Hastie et al., 1990). The rate of reduction has been measured as 33 bp per year in normal blood and colon mucosa (Hastie et al., 1990) or $19.8 \pm 6$ bp in skin (Lindsey et al., 1991), but longer sequences could be lost from individual chromosome ends during a single replication cycle. As the length of the telomere repeat array decreases with increasing age of the donor, the total amount of DNA which hybridizes to the $(TTAGGG)_n$ probe also decreases (Lindsey et al., 1991).

Primary fibroblast cell lines can be established from human skin biopsies but it has been known for many years that they cannot be propagated continuously in culture. After a finite number of generations (cell doublings) the fibroblast cultures show signs of senescence; they grow more slowly, many chromosome rearrangements (including telomeric associations and the formation of dicentrics) are observed, and the senescing cultures die. It has been demonstrated that the average length of the telomere repeat array in primary fibroblast cell lines decreases as the number of cell doublings increases (Harley et al., 1990). A weak negative correlation was found between the age of the donor and the number of cell doublings that a primary cell line would undergo before senescence, such that cell lines from young donors usually undergo more cell doublings. However, there was a more striking correlation between the length of the telomere repeat array when a cell line was established and the number of cell doublings or the replicative capacity of that cell line, such that cell lines with longer telomeres underwent more cell doublings in culture before senescing (Allsopp et al., 1992). Cell lines established from patients with the Hutchinson–Gilford syndrome of premature ageing tended to have shorter telomeres and a reduced replicative capacity in vitro, but the reason for this reduction is unknown. In addition, patients with Down syndrome (DS) showed a higher loss rate of terminal TTAGGG repeats compared with age matched donors. The reason for higher loss rate in DS patients is also unclear, but it has been suggested that the accelerated loss of telomeres is a marker for and could play a role in the premature immunosenesence of DS patients (Vaziri et al., 1993).

These observations are consistent with the hypothesis that telomerase is active in the human male and female germlines, switched off during early embryogenesis, and consequently inactive or possibly very much reduced in adult somatic tissues (Allshire et al., 1989; Hastie et al., 1990; Harley et al., 1990). Therefore, long telomeres are maintained in the germline and extended by the addition of TTAGGG repeats to the termini, whereas the length of the telomere repeat array in somatic tissues reflects the number of cell divisions undergone by that tissue in the absence of telomerase activity. Other factors, such as the expression of telomere binding proteins and their interactions with

telomeric DNA may also play a role in determining telomere length (Saltman
et al., 1993). For example, in yeast a reduction in the level of expression of the
telomere binding protein RAP1 results in a reduction in the telomere length
(Lustig et al., 1990), and overexpression of RAP1 causes an increase in the rate
of cell death and chromosome loss (Conrad et al., 1990). While in Tetrahymena,
alteration of the telomere repeat sequence affected telomere length and chromosome stability, presumably because the interaction between telomeric DNA and
telomere binding proteins was affected (Yu et al., 1990).

There is no obvious difference in the average telomere repeat array
length of the soma and germline in the mouse or between young and old mice.
However, mouse telomeres are considerably longer (50–150 kb) than those of
man and therefore small changes of length might remain undetected (Kipling
and Cooke, 1990; Starling et al., 1990). Telomerase activity has been identified
in protein extracts from several different mouse cell lines, but the mouse enzyme
only synthesized one or two TTAGGG repeats compared to the long tracts of
repeats synthesized by human telomerase under the same conditions (Prowse et
al., 1993). This difference was unexpected considering the length of the mouse
telomere repeat arrays (see discussion above), but it suggests that the control of
telomerase activity is also rather different in these two species. In addition, new
length products are generated at a high rate in the telomere repeat arrays of
inbred strains of mice. The mechanisms which create length changes in telomere repeat arrays are unknown but unequal exchanges between telomeres
could cause length changes in addition to contributing to the maintenance of
the very long telomeres in the mouse (Kipling and Cooke, 1992). Questions
about the regulation of telomere length will not be resolved until more is known
about the role of telomere binding proteins (Runge and Zakain, 1989) and the
control of telomerase expression in a variety of different species.

## 1. Telomeres, telomerase, and cancer

Colorectal carcinomas are clonal in origin (Fearon et al., 1987) and arise from
benign adenomas in the gut; therefore, by the time these carcinomas are diagnosed they have undergone many somatic cell divisions. In most cases the
average lengths of telomere repeat arrays in DNA from colorectal carcinomas are
shorter than in DNA from the normal colonic mucosa from the same individual
(Hastie et al., 1990). In leukemias and solid tumors many terminal fusions are
observed; often 20–30% of the cells show associations between a variety of
chromosome pairs and these associations may be different in each cell (Hastie
and Allshire, 1989). These observations led to the suggestions that the length
reduction of the telomere repeat arrays in tumors reflected the number of cell
doublings in the absence of telomerase activity, or possibly that it simply reflected a growth advantage of the malignant cells with shorter telomeres over

those with longer telomeres (Hastie *et al.*, 1990). However, the HeLa cell line, which was established from a human cervical carcinoma, and other immortalized human cell lines have telomerase activity and the average length of the telomere repeat array in these cell lines is stable (Morin, 1989). As a result it has been proposed that telomerase activity is required for the growth of immortalized cells (Greider, 1994). In addition, as the average telomere length in most tumours is shorter than the average telomere length from the normal somatic tissue of the patient, it has been suggested that telomerase is activated during tumor progression and consequently telomere length is stabilized. This hypothesis is supported by the recent demonstration that telomerase activity is present in the metastatic cells of epithelial ovarian carcinomas with very short telomeres, but not in the normal somatic tissue of the patient (Counter *et al.*, 1994). It also provides exciting prospects for new cancer therapies based on drugs which inhibit telomerase activity and therefore control or slow down the growth of a malignancy.

A study of telomere repeat array length and telomerase activity in mortal cell lines before and after transformation with SV40 or Ad5 viruses (which extend the life span of cell lines in culture), has also been undertaken (Counter *et al.*, 1992). Cell lines expressing viral antigens showed telomere length reduction with increasing cell generations ($\sim$ 65 bp per generation) and a reduction in the quantity of $(TTAGGG)_n$ DNA in the genome. When the telomere repeat arrays were reduced to an average of 1.5 kb, the transformed cells reached a crisis and showed many chromosome abnormalities. If the cell line survived the crisis, both the length of the telomere repeat array and the frequency of dicentric chromosomes stabilized, and telomerase activity could be detected in the immortalized cell line (Counter *et al.*, 1992). Similarily, comparisons of the terminal sequences of untransformed B-lymphocytes that undergo one cell division in culture and EBV-transformed B-lymphocyte cultures, which can be grown continuously in culture, have shown that terminal $(TTAGGG)_n$ repeats were lost from the transformed cells though the rate of loss declined with increasing numbers of cell doublings. The more proximally located $(TGAGGG)_n$ repeats were not lost to the same extent from chromosome ends but some loss of subterminal repeat sequences was detected. Apparently, telomere repeat sequences are completely lost from some chromosome ends in some cells, and the loss of subterminal sequences follows (Guerrini *et al.*, 1993), though other mechanisms could be involved.

The inhibition of DNA synthesis and arrest of senescing cells at the $G_1/S$ boundary of the cell cycle (Goldstein, 1990), the activation of cellular oncogenes, and loss of tumor suppressor genes play major roles in senescence and tumorigenesis. However, the progressive loss of telomere repeats destabilizes the genome, presumably because at some chromosome ends the entire telomere is lost or the array of repeats is so reduced that the telomere is no longer functional. As a result, terminally fused dicentric chromosomes are formed and a cycle of anaphase bridge formation–breakage–fusion is entered. These events may contribute to senescence and the progression of tumorigenesis but the relationships between senescence, tumorigenesis, the loss of telomeres, and reactivation of telomerase are unlikely to be simple.

## 2. Fate of human telomeres reintroduced into mammalian cells

In order to determine whether $(TTAGGG)_n$ could form a functional telomere when introduced into mammalian cells, a linearized plasmid carrying a selectable marker and a cloned human telomere was introduced into a hamster–human somatic cell hybrid containing a human X chromosome. In about 25% of the transformants the telomere constructs were located at the end of a chromosome, where they exhibited Bal31 sensitivity and showed length heterogeneity consistent with a location in a terminal restriction fragment (Farr et al., 1991). Chromosomes with a terminally located telomere construct were stable and therefore a new functional telomere had been formed. Human telomeres can also be rescued in yeast though the human telomere repeats arrays are shortened and yeast telomere repeats are added onto the end (Cross et al., 1989; Brown, 1989). The method of telomere-associated chromosome fragmentation has been exploited to generate a somatic cell hybrid panel with nested terminal deletions of the long arm of the human X chromosome (Farr et al., 1992). The combination of techniques for the targeted integration of a construct by homologous recombination and telomere-associated chromosome fragmentation have been used to truncate selectively the short arm of human chromosome 1 within the interferon-inducible 6-16 gene. Eight clones showed targeted integration of the 6-16/telomere construct into the 6-16 gene, but only one carried a chromosome 1 truncated within the 6-16 gene (Itzhaki et al., 1992). Efficient telomere-associated chromosome fragmentation occurs in established human cell lines and in mouse embryonic stem cells where the newly formed telomeres are extended, probably by the activity of telomerase, and the chromosome is healed. However, telomere-associated chromosome fragmentation is undetectable or very inefficient in human and mouse primary cell lines where telomerase is not active (Barnett et al., 1993).

A variety of mechanisms may be involved in telomere-associated chromosome fragmentation. First, the construct could preferentially integrate and cause fragmentation adjacent to or near existing telomeres and there is some evidence to support this hypothesis (Barnett et al., 1993). Second, the construct integrates randomly into the genome but the presence of a short array of $(TTAGGG)_n$ repeats makes the chromosome prone to breakage at the integration site, and a new telomere is formed. This mechanism seems less likely because stable clones which carry the selectable-telomere construct at intersitial locations have been produced (Farr *et al.*, 1991) and stable interstitial arrays of (TTAGGG)$_n$ repeats already exist in the genome (Wells *et al.*, 1990). Third, telomere constructs preferentially integrate at sites where single- or double-strand breaks have occurred. The chromosome break is healed by the presence of a new telomere and an unstable acentric fragment is formed (Murnane and Yu, 1993). Finally, in established cell lines and somatic cell hybrids, broken chromosomes that have already entered a cycle of anaphase bridge formation–breakage–fusion are rescued by the addition of the telomere construct. Large duplications at the ends of some truncated chromosomes in the X chromosome hybrid panel produced by telomere-associated chromosome fragmentation might be explained by terminal sister chromatid fusion and breakage prior to rescue by the addition of the telomere construct (Farr *et al.*, 1992). In each case the array of (TTAGGG)$_n$ repeats of the new telomere is extended by the addition of telomere repeats and the chromosome is capped and healed.

## B. Chromosome healing *in vivo*

Cytogenetic analysis of human chromosomes can reveal terminal deletions of human chromosomes, but these deletions result in the loss of a relatively large segment of DNA from the chromosome end (minimum 3–5 Mbp of DNA) and the resulting hemizygosity for the many genes encoded within the region is often lethal. However, submicroscopic terminal deletions have recently been identified.

### 1. Healing by telomerase activity

A patient suffering from hemoglobin H disease, a relatively severe form of $\alpha$-thalassemia usually caused by the deletion of three of the four copies of the $\alpha$-globin genes ($-\alpha/--$), was found to carry three intact copies of the $\alpha$-globin genes($-\alpha/\alpha\alpha$). Further investigation showed that the patient's maternally inherited chromosome 16 was truncated 50 kb distal to the $\alpha$-globin gene cluster and all sequences distal to this point including regulatory elements for the globin genes were deleted. The terminally deleted chromosome 16 had been stably inherited from the mother who also carried a normal chromosome 16 and so was not afflicted with $\alpha$-thalassemia. In contrast, the patient also inherited an abnormal paternal chromosome 16 ($-\alpha$) from his father and expressed the disease phenotype. Sequence analysis of the truncated chromosome showed that TTAGGG repeats had been added directly to the break point and therefore the deleted chromosome 16 had been healed by the *de novo* addition of a telomere. No significant similarity (only 4 bp) was identified between sequences at or close to the breakpoint and the (TTAGGG)$_n$ repeats (Wilkie *et al.*, 1990).

In vitro, *Tetrahymena* telomerase does not require the presence of sequences complementary to the telomere repeat at the end of a DNA molecule to enable it to add repeats to the chromosome terminus (Harrington and Greider, 1991). Similarly, protein extracts containing human telomerase activity can add TTAGGG repeats onto synthetic oligonucleotides composed of the sequence from the human chromosome 16p breakpoint (Morin, 1991). These experiments show that telomerase has minimal sequence requirements in order to add telomere repeats onto the end of a DNA molecule. Therefore, it may very well be capable of healing broken chromosomes (see discussion above; Greider, 1991) where there is no homology between the end of the broken chromosome and the telomerase RNA template.

Another patient suffering from mental retardation in addition to α-thalassemia (ATR-16) was investigated for sequence loss from 16pter and was found to have a 1.7 Mb terminal deletion. Again, sequences proximal to the break point showed minimal complementarity (3 bp) to TTAGGG repeats but a new array of TTAGGG repeats was present distal to the breakpoint, and the healed chromosome was somatically stable. The truncated chromosome was paternally inherited and therefore it probably arose as a *de novo* germline deletion healed by telomerase. There are at least 7 genes including the α-globin genes in the 140 kb proximal to the 16p telomere and if this gene density is maintained, up to 80 genes could have been lost in the 1.7-Mb deletion. The mental retardation in the patient could result from hemizygosity at a single gene in the region or the cumulative loss of all the deleted genes (Lamb *et al.*, 1993). In addition, the expression of the genes located adjacent to the newly synthesized telomere may be affected. It is clear that genes adjacent to telomeres in yeast can become transcriptionally inactive or silenced by their proximity to a yeast telomere (see Sandell and Zakain, 1992; Kipling and Cooke, 1992).

The submicroscopic terminal deletions of 16pter are relatively easily diagnosed because the patients usually suffer from α-thalassemia. Submicroscopic terminal deletions at other chromosome ends may remain undiagnosed but cause mental retardation in patients where no other cause has been identified. For example, contiguous gene syndromes such as the Wolf–Hirschorn syndrome at the distal end of 4p and the Miller–Dieker syndrome at 17p could occur by submicroscopic or cryptic terminal deletions (Lamb *et al.*, 1993; Broccoli and Cooke, 1993). It is not yet known how common terminal deletions are but another three 16p terminal deletions have been identified and in each case the deletion was distal to the globin gene cluster (Lamb *et al.*, 1993).

## 2. Why is telomerase inactive in somatic tissues?

The minimal sequence complementarity requirements that telomerase shows prior to adding TTAGGG repeats onto the end of a DNA molecule might be a hindrance in mammalian somatic cells by causing destabilization of the genome. If telomerase competes with normal cellular mechanisms for repairing single- and double-strand breaks, then chromosomes in somatic cells could be truncated by telomerase activity. If this occurred, even at a low rate, acentric fragments would be lost and cell viability affected. The generation of long telomeres by telomerase in the germline and inactivation of telomerase in somatic tissues might avoid the potential somatic instability caused by telomerase. The long telomeres would "buffer" against the loss of coding sequences during the many subsequent mitotic cycles of replication and division after zygote formation. Any destabilization of the genome caused by the chromosome healing activity of telomerase within the germline would mostly lead to the generation of inviable gametes.

## 3. Healing by telomere capture

An alternative way of stabilizing a broken chromosome is by the capture of a telomere from the other homolog or from a nonhomologous chromosome. This process does not require the activity of telomerase.

Terminal deletions are common in many malignancies and can be detected by allele loss from informative marker loci. However, recently it has been shown that some apparently simple deletions found in malignant melanomas are in fact unequal subterminal translocations. The proposed mechanism includes an initial chromosome breakage. The broken chromosome lacks a telomere and is unstable unless it acquires terminal sequences and a telomere from another chromosome end (Meltzer et al., 1993). The mechanism of telomere capture observed in human tumors is unknown but it may be similar to the terminal exchange processes which have been identified in yeast (Wang and Zakain, 1990b; Lundblad and Blackburn, 1993).

Terminal breakpoints, involving a variety of different chromosomes, have been identified in association with other abnormal chromosome structures found in patients. Many of the abnormal chromosomes contained an interstitial block of telomere repeats in addition to two telomeres. In each case it appeared that an acentric fragment from one chromosome was joined onto the telomere of a recipient chromosome with loss of telomere function at the resulting interstitial block of repeats, except in one case where the abnormal chromosome showed somatic instability (Murtif-Park et al., 1992).

## C. Cryptic translocations and contiguous gene syndromes

Reciprocal translocations between the ends of human chromosomes have been described as cryptic because they are undetectable using conventional cytogenetic banding techniques, even on extended prophase chromosomes. As discussed above, there is some evidence that exchanges between subterminal repeat sequences occur between nonhomologous chromosome ends. These events may result in the exchange of subterminal sequences and telomeres but they may have no consequence for the cell. In contrast cryptic translocations occuring between proterminal regions involve the exchange of many genes. As for all reciprocal translocations, the balanced carrier of a cryptic translocation is phenotypically normal and it is only the unbalanced progeny that might be aneuploid, having one or three copies of the genes encoded within the exchanged segment.

A cryptic translocation involving an exchange between 16pter and 1pter has been described (Lamb et al., 1989). In this case the mother carried aa/−a alleles for the α-globin genes and she was a balanced carrier for the translocation t(1p:16p). The son was hemizygous for 16pter and had three copies of 1pter, he carried allele −a at the α-globin locus on his normal paternally inherited chromosome 16 and therefore he suffered hemoglobin H disease and mental retardation. The daughter had a different phenotype; she was hemizygous for 1pter and had three copies of 16pter carrying aa/−a/−a alleles for the α-globin genes and she had normal hematological findings with mental retardation.

Cryptic reciprocal translocations have been observed at other chromosome ends, for example, in patients with the Miller–Dieker syndrome. In one family a cryptic translocation between 17pter and 8qter has been observed (see Figure 7.4) and in another a half-cryptic translocation between 17pter and 3pter has been observed. A half cryptic translocation occurs when additional material can be detected on one chromosome but by cytogenetic analysis alone its origin cannot be determined (Ledbetter, 1992). In addition, abnormal terminal Xp/Yp interchanges are also the underlying cause of XX maleness in a number of patients (Petit et al., 1987).

The frequency of cryptic translocations, as for cryptic terminal deletions, is unknown; but it has been suggested that cryptic translocations should be considered when clinical features of certain contiguous gene syndromes are suspected in a patient (Ledbetter, 1992). Cryptic translocations could be detected using sets of hypervariable marker loci from each chromosome end (Wilkie, 1993). When a family trio (two parents and one child) is fully informative at a hypervariable locus, each of the four allelic chromosome ends can be distinguished. Under these conditions it should be possible to identify a child carrying an unbalanced cryptic translocation (three copies of alleles from one chromosome end and one copy from a different chromosome end), a terminal deletion (only one copy of an allele at one chromosome end), uniparental isodisomy (two copies of only one parental allele); or heterodisomy (one copy of each allele from only one parent). It is also possible, and probably more satisfactory, to detect the location and copy number of each proterminal region by fluorescent in situ hybridization (FISH) of marker loci onto metaphase chromo-

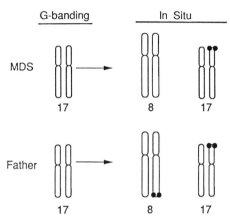

G-banding        In Situ

MDS

17            8    17

Father

17            8    17

**Figure 7.4.** Diagram representing a cryptic translocation in the Miller–Dieker syndrome. Both the patient and her father had normal G-banded chromosome analysis (850 band stage). Fluorescent *in situ* hybridization analysis of the patient showed a deletion of one chromosome 17, while analysis of the father showed a translocation t(8q:17p). Retrospective analysis of the father's G-banded chromosomes did not detect any alteration of 8q or 17p, indicating a completely cryptic exchange between two G-negative terminal bands. Reproduced with permission from by Ledbetter, D. H. (1993). *Am. J. Hum. Genet.* **51**:451-456. (© 1992 by the American Society of Human Genetics. All rights reserved 002-9297/92/5103-0001$02.00.)

somes from the affected individual and parents to look for cryptic translocations. At the present time only a few different fluorescent "tags" are available and so FISH detection of cryptic translocations would be a labor-intensive process (Ledbetter, 1992).

## VII. CONCLUSIONS

In somatic tissues, telomeres of human chromosomes are 5–15 kb heterogeneous arrays of TTAGGG repeats. However, many variant repeat units have been identified and localized to the proximal end of telomeres, while the distal ends are more uniform arrays of TTAGGG repeats. The bulk of telomeric DNA in rats and probably in humans is packaged into nucleosomes which do not bind histone H1, but the structure and the proteins which bind to the DNA at the terminus of human telomeres have not yet been determined. The sequence organization of proterminal regions is less well understood but they contain predominantly chromosome specific sequences at their proximal ends. At least two chromosome ends contain a short stretch of DNA (1–4 kb) which shows a high level of variation and divergence between alleles and distal to it, the terminal 50–200 kb of the proterminal region is composed of subterminal repeat sequences. Copies of the one characterized subterminal repeat sequence family are located at many but not all chromosome ends. At 16pter there is a characterized divergence point and distal to it, some alleles of 16pter show more similarity to the terminal sequences of nonhomologous chromosomes than to other alleles of 16pter.

A rapid turnover of terminal sequences has occurred since the divergence of man from chimpanzees and gorillas (5–7 million years ago). The reorganization has probably resulted in the loss of a subterminal satellite from the human genome. However, the subterminal satellite has been expanded and distributed over many chimpanzee and most gorilla chromosome ends. There is also rapid turnover of telomere repeats in human somatic tissues; this contributes to the destabilization of the genome and has consequences for the ageing individual but not his/her progeny. A high level of homologous meiotic recombination is supported by proterminal regions and there is evidence that terminal exchanges between subterminal repeats sequences of nonhomologous human chromosome ends occurs infrequently. Errors in these processes could result in cryptic terminal translocations and subsequently the generation of unbalanced carriers in the offspring. Such events may be the underlying cause of an unknown number of contiguous gene syndromes. There are also a number of mechanisms by which a broken chromosome can acquire a telomere but the stabilized chromosome is then deleted for many genes. Hemizygosity for a large region of the genome is usually lethal, but cryptic terminal deletions may cause some contiguous gene syndromes. The extent to which cryptic terminal rearrangements of human chromosomes contribute to genetic disease has yet to be determined.

## Acknowledgments

I thank Duncan Baird, Shoajie Di, and Joanne Coleman for their hard work and Meran Owen for his encouragement. I also thank Alec J. Jeffreys for his continuous support and advice, and members of his research group for very helpful discussions. N.J.R. is a UK-HGMP Senior Research Fellow.

## References

Allitto, B. A., MacDonald, M. E., Bucan, M., Richards, J., Romano, D., Whaley, L., Falcone, B., Ianazzi, J., Wexler, N. S., Wasmuth, J. J., Collins, F. S., Lehrach, H., Haines, J. L., and Gusella, J. F. (1991). Increased recombination adjacent to the Huntington disease-linked D4S10 marker. *Genomics* **9:**104–112.
Allshire, R. C., Dempster, M., and Hastie, N. D. (1989). Human telomeres contain at least three types of G-rich repeat distributed non randomly. *Nucleic Acids Res.* **17:**4611–4627.
Allshire, R. C., Gosden, J. R., Cross, S. H., Cranston, G., Rout, D., Sugawara, N., Szostak, J. W.,

Fantes, P. A., and Hastie, N. D. (1988). Telomeric repeat from *T. thermophila* cross hybridizes with human telomeres. *Nature (London)* **332**:656–659.

Allsopp, R. C., Vaziri, H., Patterson, C., Goldstein, S., Younglai, E. V., Futcher, A. B., Greider, C. W., and Harley, C. B. (1992). Telomere length predicts replicative capacity of human fibroblasts. *Proc. Natl. Acad. Sci. USA* **89**:10114–10118.

Armour, J. A., Crosier, M., and Jeffreys, A. J. (1992). Human minisatellite alleles detectable only after PCR amplification. *Genomics* **12**:116–124.

Armour, J. A. L., Harris, P. C., and Jeffreys, A. J. (1993). Allelic diversity at minisatellite MS205 (D16S309): Evidence for polarized variability. *Hum. Mol. Genet.* **2**:1137–1145.

Armour, J. A. L., Povey, S., Jeremiah, S., and Jeffreys, A. J. (1990). Systematic cloning of human minisatellites from ordered array charomid libraries. *Genomics* **8**:501–512.

Armour, J. A. L., Wong, Z., Wilson, V., Royle, N. J., and Jeffreys, A. J. (1989). Sequences flanking the repeat arrays of human minisatellites: Association with tandem and dispersed repeat elements. *Nucleic Acids Res.* **17**:4925–4935.

Atcheson, C., DiDomenico, B., Frackman, S., Easton Esposito, R., and Elder, R. T. (1987). Isolation, DNA sequence, and regulation of a meiosis-specific eukaryotic recombination gene. *Proc. Natl. Acad. Sci. USA* **84**:8035–8039.

Barnett, M. A., Buckle, V. J., Evans, E. P., Porter, A. C. G., Rout, D., Smith, A. G., and Brown, W. R. A. (1993). Telomere directed fragmentation of mammalian chromosomes. *Nucleic Acids Res.* **21**:27–36.

Bates, G. P., MacDonald, M. E., Baxendale, S., Sedlacek, Z., Youngman, S., Romano, D., Whaley, W. L., Allitto, B. A., Poustka, A., Gusella, J. F., and Lehrach, H. (1990). A yeast artificial chromosome telomere clone spanning a possible location of the Huntington Disease Gene. *Am. J. Hum. Genet.* **46**:762–775.

Bernardi, G. (1989). The isochore organization of the human genome. *Annu. Rev. Genet.* **23**:637–661.

Biessmann, H., and Mason, J. M. (1993). Genetics and molecular biology of telomeres. *Adv. Genet.* **30**:185–249.

Blackburn, E. H. (1990). Telomeres: Structure and synthesis. *J. Biol. Chem.* **265**:5919–5921.

Blackburn, E. H. (1991). Structure and function of telomeres. *Nature (London)*. **350**:569–573.

Blackburn, E. H., and Szostak, J. W. (1984). The molecular structure of centromeres and telomeres. *Ann. Rev. Biochem.* **53**:163–194.

Broccoli, D., and Cooke, H. (1993). Ageing, healing, and the metabolism of telomeres. *Am. J. Hum. Genet.* **52**:657–660.

Broccoli, D., Miller, O. J., and Miller, D. A. (1992). Isolation and characterization of a mouse subtelomeric sequence. *Chromosoma.* **101**:442–447.

Brown, W. R. A. (1989). Molecular cloning of human telomeres in yeast. *Nature (London)*. **338**:774–776.

Brown, W. R. A., MacKinon, P. J., Villasante, A., Spurr, N., Buckle, V. J., and Dobson, M. J. (1990). Structure and polymorphism of human telomere-associated DNA. *Cell (Cambridge, MA.)* **63**:119–132.

Buchman, A. R., Kimmerly, W. J., Rine, J., and Kornberg, R. D. (1988). Two DNA binding factors recognize secific sequences at silencers, upstream activating sequences, autonomously replicating sequences, and telomeres in *Saccharomyces cereisiae. Mol. Cell. Biol.* **261**:210–225.

Budarf, M. L., and Blackburn, E. H. (1986). Chromatin structure of the telomeric region and 3′-nontranscribed spacer of *Tetrahymena* ribosomal RNA genes. *J. Biol. Chem.* **261**:363–369.

Burmeister, M., Kim, S., Price, E. R., de Lange, T., Tantravahi, U., Myers, R. M., and Cox, D. R. (1991). A map of the distal region of the long arm of human chromosome 21 constructed by radiation hybrid mapping and pulse-field gel electrophoresis. *Genomics.* **9**:19–30.

Carpenter, A. T. C. (1979). Recombination nodules and synaptonemal complex in recombination-defective females of *Drosophila melanogaster*. *Chromosoma*. **75:**259–292.

Carpenter, A. T. C. (1987). Gene conversion, recombination nodules, and the initiation of meiotic synapsis. *Bioessays* **6:**232–236.

Chandley, A. C. (1989). Asymmetry in chromosome pairing: a major factor in de novo mutation and the production of genetic disease in man. *J. Med. Genet.* **26:**546–552.

Cheng, J.-F., Smith, C. L., and Cantor, C. R. (1989). Isolation and characterization of a human telomere. *Nucleic Acids Res.* **17:**6109–6127.

Cheng, J.-F., Smith, C. L., and Cantor, C. R. (1991). Structural and transcriptional analysis of a human subtelomeric repeat. *Nucleic Acids Res.* **19:**149–153.

Chikashige, Y., Ding, D.-Q., Funabiki, H., Haraguchi, T., Mashiko, S., Yanagida, M., and Hiraoka, Y. (1994). Telomere-led premeiotic chromosome movement in fission yeast. *Science* **264:**270–273.

Collick, A., Dunn, M. G., and Jeffreys, A. J. (1991). Minisatellite binding protein Msbp-1 is a sequence-specific single-stranded DNA binding protein. *Nucleic Acids Res.* **19:**6399–6404.

Collick, A., and Jeffreys, A. J. (1990). Detection of a novel minisatellite-specific DNA-binding protein. *Nucleic Acids Res.* **18:**625–629.

Conrad, M. N., Wright, J. H., Wolf, A. J., and Zakain, V. A. (1990). RAP1 protein interacts with yeast telomeres *in vivo*: Overproduction alters telomere structure and decreases chromosome stability. *Cell (Cambridge, MA)* **63:**739–750.

Cooke, H. J., Brown, W. R. A., and Rappold, G. A. (1985). Hypervariable telomeric sequences from the human sex chromosomes are pseudoautosomal. *Nature (London)* **317:**687–692.

Coren, J. S., Epstein, E. M., and Vogt, V. M. (1991). Characterization of a telomere-binding protein from *Physarum polycephalum*. *Mol. Cell. Biol.* **11:**2282–2290.

Counter, C. M., Avilion, A. A., LeFeuvre, C. E., Stewart, N. G., Greider, C. W., Harley, C. B., and Bacchetti, S. (1992). Telomere shortening associated with chromosome instability is arrested in immortal cells which express telomerase activity. *EMBO J.* **11:**1921–1929.

Counter, C. M., Hirte, H. W., Bacchetti, S., and Harley, C. B. (1994). Telomerase activity in human ovarian carcinoma. *Proc. Natl. Acad. Sci. USA* **91:**2900–2904.

Cross, S., Lindsey, J., Fantes, J., McKay, S., McGill, N., and Cooke, H. (1990). The structure of a subterminal repeated sequence present on many human chromosomes. *Nucleic Acids Res.* **18:** 6649–6657.

Cross, S. H., Allshire, R. C., McKay, S. J., McGill, N. I., and Cooke, H. J. (1989). Cloning of human telomeres by complementation in yeast. *Nature (London)* **338:**771–774.

de Lange, T. (1992). Human telomeres are attached to the nuclear matrix. *EMBO J.* **11:**717–724.

de Lange, T., Shiue, L., Myers, R. M., Cox, D. R., Naylor, S. L., Killery, A. M., and Varmus, H. E. (1990). Structure and variability of human chromosome ends. *Mol. Cell. Biol.* **10:**518–527.

Dietz-Band, J., Riethman, H., Hildebrand, C. E., and Moyzis, R. (1990). Characterization of the polymorphic loci on a telomeric fragment of DNA from the long arm of human chromosome 7. *Genomics* **8:**168–170.

Donis-Keller, H., Green, P., Helms, C., Cartinhour, S., Weiffenbach, B., Stephens, K., Keith, T. P., Bowden, D. W., Smith, D. R., Lander, E. S., Botstein, D., Akots, G., Rediker, K. S., Gravius, T., Brown, V. A., Rising, M. B., Parker, C., Powers, J. A., Watt, D. E., Kauffman, E. R., Bricker, A., Phipps, P., Muller-Kahle, H., Fulton, T. R., Ng, S., Schumm, J. W., Braman, J. C., Knowlton, R. G., Barker, D. F., Crooks, S. M., Lincoln, S. E., Daly, M. J., and Abrahamson, J. (1987). A genetic linkage map of the human genome. *Cell (Cambridge, MA.)* **51:**319–337.

Ellis, N. A., Goodfellow, P. J., Pym, B., Smith, M., Palmer, M., Frischauf, A.-M., and Goodfellow,

P. N. (1989). The pseudoautosomal boundary is defined by an Alu repeat sequence inserted on the Y chromosome. *Nature (London)* **337**:81–84.

Fang, G., Gray, J. T., and Cech, T. R. (1993). Oxytricha telomere-binding protein: separable DNA-binding and dimerization domains of the a-subunit. *Genes Dev.* **7**:870–882.

Farr, C., Fantes, J., Goodfellow, P., and Cooke, H. (1991). Functional reintroduction of human telomeres into mammalian cells. *Proc. Natl. Acad. Sci. USA* **88**:7006–7010.

Farr, C. J., Stevanovic, M., Thompson, E. J., Goodfellow, P. N., and Coofe, H. J. (1992). Telomere-associated chromosome fragmentation: Applications in genome manipulation and analysis. *Nature Genet.* **2**:275–282.

Fearon, E. R., Hamilton, S. R., and Vogelstein, B. (1987). Clonal analysis of human colorectal tumours. *Science* **238**:193–197.

Freije, D., Helms, C., Watson, M. S., and Donis-Keller, H. (1992). Identification of a second pseudoautosomal region near the Xq and Yq telomeres. *Science* **258**:1784–1787.

Gardiner, K., Aissani, B., and Bernardi, G. (1990). A compositional map of human chromosome 21. *EMBO J.* **9**:1853–1858.

Gilson, E., Laroche, T., and Gasser, S. M. (1993a). Telomeres and the functional architecture of the nucleus. *Trends Cell Biol.* **3**:128–134.

Gilson, E., Roberge, M., Giraldo, R., Rhodes, D., and Gasser, S. M. (1993b). Distortion of the DNA double helix by RAP1 at silencers and multiple telomeric binding sites. *J. Mol. Biol.* **231**: 293–310.

Goldstein, S. (1990). Replicative senesence: The human fibroblast comes of age. *Science* **249**: 1129–1133.

Gottschling, D. E., and Zakian, V. A. (1986). Telomere proteins: specific recognition and protection of the natural termini of Oxytricha macronuclear DNA. *Cell (Cambridge, MA.)* **47**:195–205.

Gray, J. T., Celander, D. W., Price, C. M., and Cech, T. R. (1991). Cloning and expression of genes for the Oxytricha telomere-binding protein: Specific subunit interactions in the telomeric complex. *Cell (Cambridge, MA.)* **67**:807–814.

Greider, C. W. (1991). Chromosome first aid. *Cell (Cambridge, MA.)* **67**:645–647.

Greider, C. W. (1994.). Mammalian telomere dynamics: Healing, fragmentation shortening and stabilization. *Curr. Op. Genet. Dev.* **4**:203–211.

Greider, C. W. and Blackburn, E. H. (1989). A telomeric sequence in the RNA of Tetrahymena telomerase required for telomere repeat synthesis. *Nature (London)* **337**:331–337.

Guerrini, A. M., Camponeschi, B., Ascenzioni, F., Piccolella, E., and Donini, F. L. (1993). Subtelomeric as well as telomeric sequences are lost from chromosomes in proliferating B lymphocytes. *Hum. Mol. Genet.* **2**:455–460.

Handel, M. A., and Hunt, P. A. (1992). Sex-chromosome pairing and activity during mammalian meiosis. *BioEssays* **14**:817–822.

Hardy, C. F. J., Sussel, L., and Shore, D. (1992). A RAP1-interacting protein involved in transcriptional silencing and telomere length regulation. *Genes Dev.* **6**:801–814.

Harley, C. B., Futcher, A. B., and Greider, C. W. (1990). Telomeres shorten during ageing of human fibroblasts. *Nature (London)* **345**:458–460.

Harrington, L. A., and Greider, C. W. (1991). Telomerase primer specificity and chromosome healing. *Nature (London)* **353**:451–454.

Hastie, N. D., and Allshire, R. C. (1989). Human telomeres: Fusion and interstitial sites. *Trends Genet.* **5**:326–331.

Hastie, N. D., Dempster, M., Dunlop, M. G., Thompson, A. M., Green, D. K., and Allshire, R. C. (1990). Telomere reduction in human colorectal carcinoma and with ageing. *Nature (London)* **346**:866–868.

Hawley, R. S., and Arbel, T. (1993). Yeast genetics and the fall of the classical view of meiosis. *Cell (Cambridge, MA.)* **72**:301–303.

Helms, C., Mishra, S. K., Reuthman, H., Burgess, A. K., Ramachandra, S., Tierney, C., Dorsey, D., and Donis-Keller, H. (1992). Closure of a genetic linkage map of human chromosme 7q with centromere and telomere polymorphisms. *Genomics* **14**:1041–1054.

Henke, A., Fischer, C., and Rappold, G. A. (1993). Genetic map of the human pseudoautosomal region reveals a high rate of recombination in female meiosis at the Xp telomere. *Genomics* **18**: 478–485.

Higgs, D. R., Wilkie, A. O. M., Vyas, P., Vickers, M. A., Buckle, V. J., and Harris, P. C. (1993). Characterization of the telomeric region of human chromosome 16p. *Chromosomes Today* **11**: 35–47.

Hing, A. V., Helms, C., and Donis-Keller, H. (1993). VNTR and microsatellite polymorphisms within the subtelomeric region of 7q. *Am. J. Hum. Genet.* **53**:509–517.

Holmquist, G. P. (1992). Chromosome bands, their chromatin flavors, and their functional features. *Am. J. Hum. Genet.* **51**:17–37.

Hulten, M. (1974). Chiasma distribution at diakinesis in the normal human male. *Hereditas* **76**: 55–78.

Hwu, H. R., Roberts, J. W., Davidson, E. H., and Britten, R. J. (1986). Insertion and/or deletion of many repeated DNA sequences in human and higher ape evolution. *Proc. Natl. Acad. Sci. USA* **83**:3875–3879.

Ijdo, J. W., Baldini, A., Ward, D. C., Reeders, S. T., and Wells, R. A. (1991). Origin of human chromosome 2: an ancestral telomere-telomere fusion. *Proc. Natl Acad. Sci. USA* **88**:9051–9055.

Ijdo, J. W., Baldini, A., Wells, R. A., Ward, D. C., and Reeders, S. T. (1992a). FRA2B is distinct from inverted repeat arrays at 2q13. *Genomics* **12**:833–835.

Ijdo, J. W., Lindsay, E. A., Wells, R. A., and Baldini, A. (1992b). Multiple variants in subtelomeric regions of normal karyotypes. *Genomics* **14**:1019–1025.

Ishikawa, F., Matunis, M. J., Drefuss, G., and Cech, T. R. (1993). Nuclear proteins that bind the pre-mRNA 3′ splice site sequence r(UUAG/G) and the human telomeric DNA sequence d(TTAGGG)$_n$. *Mol. Cell. Biol.* **13**:4301–4310.

Itzhaki, J. E., Barnett, M. A., MacCarthy, A. B., Buckle, V. J., Brown, W. R. A., and Porter, A. C. G. (1992). Targeted breakage of a human chromosome mediated by cloned human telomeric DNA. *Nature Genet.* **2**:283–287.

Jeffreys, A. J., MacLeod, A., Tamaki, K., Neil, D. L., and Monckton, D. G. (1991). Minisatellite repeat coding as a digital approach to DNA typing. *Nature (London)* **354**:204–209.

Jeffreys, A. J., Monckton, D. G., Tamaki, K., Neil, D. L., Armour, A. J., MacLeod, A., Collick, A., Allen, M., and Jobling, M. (1993). Minisatellite variant repeat mapping: Application to DNA typing and mutation analysis. In "Biochemica et Biophysica Acta DNA fingerprinting: State of the Science," (S. D. J. Pena, R. Chakraborty, J. T. Epplen, and A. J. Jeffreys, Eds.).

Jeffreys, A. J., Neumann, R., and Wilson, V. (1990). Repeat unit sequence variation in minisatellites: A novel source of DNA polymorphism for studying variation and mutation by single molecule analysis. *Cell (Cambridge, MA.)* **60**:473–485.

Jeffreys, A. J., Royle, N. J., Wilson, V., and Wong, Z. (1987). Spontaneous mutation rates to new length alleles at tandem repetitive hypervariable loci in human DNA. *Nature (London)* **332**: 278–281.

Jeffreys, A. J., Tamaki, K., MacLeod, A., Monckton, D. G., Neil, D. L., and Armour, J. A. L. (1994). Complex gene conversion events in germline mutation at human minisatellites. *Nature Genet.* **6**:136–145.

Jeffreys, A. J., Wilson, V., and Thein, S.-W. (1985). Hypervariable "minisatellite" regions in human DNA. *Nature (London)* **314**:67–73.

Killian, A. and Kleinhof, S. (1992). Cloning and mapping telomere associated sequences from *Hordeum vulgare L. Mol. Gen. Genet.* **235**:153–156.

Kipling, D., Ackford, H. E., Taylor, B. A., and Cooke, H. J. (1991). Mouse minor satellite DNA genetically maps to the centromere and is physically linked to the proximal telomere. *Genomics* **11**:235–241.

Kipling, D. and Cooke, H. J. (1990). Hypervariable ultra-long telomeres in mice. *Nature (London)* **347**:400–402.

Kipling, D. and Cooke, H. J. (1992). Beginning or end? Telomere structure, genetics and biology. *Hum. Mol. Genet.* **1**:3–6.

Kornberg, J. R. and Rykowski, M. C. (1988). Human genome organization: Alu, Lines, and the molecular structure of metaphase chromosome bands. *Cell (Cambridge, MA.)* **53**:391–400.

Kourilsky, P. (1986). Molecular mechanisms for gene conversion in higher cells. *Trend. Genet.* **2**: 60–63.

La Mantia, G., Pengue, G., Maglione, D., Pannuti, A., Pascucci, A., and Lania, L. (1989). Identification of new human repetitive sequences: Characterization of the corresponding cDNAs and their expression in embryonal carcinoma cells. *Nucleic Acids Res.* **17**:5913–5922.

Lamb, J., Harris, P. C., Lindenbaum, R. H., and Reeders, S. T. (1989). Detection of breakpoints in submicroscopic chromosomal translocations, illustrating an important mechanism for genetic disease. *Lancet* **2**:819–824.

Lamb, J., Harris, P. C., Wilkie, A. O. M., Wood, W. G., Dauwerse, J. H. G., and Higgs, D. R. (1993). De novo truncation of chromosome 16p and healing with (TTAGGG)$_n$ in the α-thalassemia/mental retardation syndrome (ATR-16). *Am. J. Hum. Genet.* **52**:668–676.

Laurie, D. A., and Hulten, M. A. (1985a). Further studies on bivalent chiasma frequency in human males with normal karyotypes. *Ann. Hum. Genet.* **49**:189–201.

Laurie, D. A., and Hulten, M. A. (1985b). Further studies on chiasma distribution and interference in the human male. *Ann. Hum. Genet.* **49**:203–214.

Ledbetter, D. H. (1992). Minireview: Cryptic translocations and telomere integrity. *Am. J. Hum. Genet.* **51**:451–456.

Levis, R. W., Ganesan, R., Houtchens, K., Tolar, L. A., and Sheen, F.-M. (1993). Transposons in place of telomeric repeats at a *Drosophila telomere. Cell* **75**:1083–1093.

Lindsey, J., McGill, N. I., Lindsey, L. A., Green, D. K., and Cooke, H. J. (1991). *In vivo* loss of telomeric repeats with age in humans. *Mut. Res.* **256**:45–48.

Louis, E. J., and Haber, J. E. (1990a). Mitotic recombination among subtelomeric Y' repeats in Saccharomyces cerevisiae. *Genetics* **124**:547–559.

Louis, E. J., and Haber, J. E. (1990b). The subtelomeric Y' repeat family in Saccharomyces cerevisiae : an experimental system for repeated sequence evolution. *Genetics* **124**:533–545.

Luke, S., and Verma, R. S. (1993). Telomeric repeat (TTAGGG)$_n$ sequences of human chromosomes are conserved in chimpanzee (Pan Troglodytes). *Mol. Gen. Genet.* **237**:460–462.

Lundblad, V., and Blackburn, E. H. (1993). An alternative pathway for yeast telomere maintenance rescues est1$^-$ senescence. *Cell (Cambridge, MA.)* **73**:347–360.

Lustig, A. J., Kurtz, S., and Shore, D. (1990). Involvement of the silencer and UAS binding protein RAP1 in regulation of telomere length. *Science* **250**:549–553.

Makarov, V. L., Lejnine, S., Bedoyan, J., and Langmore, J. P. (1993). Nucleosomal Organization of telomere-specific chromatin in rat. *Cell (Cambridge, MA.)* **73**:775–787.

Maudlin, I., and Evans, E. P. (1980). Chiasma distribution in mouse oocytes during diakinesis. *Chromosoma* **80**:49–56.

McKay, S. J. and Cooke, H. (1992a). hnRNP A2/B1 binds specifically to single stranded vertebrate telomeric repeat (TTAGGG)$_n$. *Nucleic Acids Res.* **20**:6461–6464.

McKay, S. J. and Cooke, H. (1992b). A protein which specifically binds to single stranded TTAGGG$_n$ repeats. *Nucleic Acids Res.* **20**:1387–1391.

McKee, B. D. and Handel, M. A. (1993). Sex chromosomes, recombination, and chromatin conformation. *Chromosoma* **102**:71–80.

Meltzer, P. S., Guan, X.-Y., and Trent, J. M. (1993). Telomere capture stabilizes chromosome breakage. *Nature Genet.* **4**:252–255.

Menees, T. M., Ross-Macdonald, P. B., and Roeder, G. S. (1992). MEI4, a meiosis-specific yeast gene required for chromosome synapsis. *Mol. Cell. Biol.* **12**:1340–1351.

Mermer, B., Colb, M., and Krontiris, T. G. (1987). A family of short, interspersed repeats is associated with tandemly repetitive DNA in the human genome. *Proc. Natl Acad. Sci. USA* **84:** 3320–3324.

Meyne, J., Baker, R. J., Hobart, H. H., Hsu, T. C., Ryder, O. A., Ward, O. G., Wiley, J. E., Wurster-Hill, D. H., Yates, T. L., and Moyzis, R. K. (1990). Distribution of non-telomeric sites of $(TTAGGG)_n$ telomeric sequence in vertebrate chromosomes. *Chromosoma* **99**:3–10.

Monckton, D. G., Tamaki, K., MacLeod, A., Neil, D. L., and Jeffreys, A. J. (1993). Allele-specific MVR-PCR analysis at minisatellite D1S8. *Hum. Mol. Genet.* **2**:513–519.

Morin, G. B. (1989). The human telomere terminal transferase enzyme is a ribonucleoprotein that synthesizes TTAGGG repeats. *Cell (Cambridge, MA.)* **59**:521–529.

Morin, G. B. (1991). Recognition of a chromosome trunction site associated with α-thalassemia by human telomerase. *Nature (London)* **353**:454–456.

Moyzis, R. K., Buckingham, J. M., Cram, L. S., Dani, M., Deaven, L. L., Jones, M. D., Meyne, J., Ratliff, R. L., and Wu, J.-R. (1988). A highly repetitive DNA sequence, $(TTAGGG)_n$, present at the telomeres of human chromosomes. *Proc. Natl Acad. Sci. USA* **85**:6622–6626.

Murnane, J. P., and Yu, L.-C. (1993). Acquisition of telomere repeat sequences by transfected DNA integrated at the site of a chromosome break. *Mol. Cell. Biol.* **13**:977–983.

Murtif-Park, V., Gustashaw, K. M., and Wathen, T. M. (1992). The presence of interstitial telomeric sequences in constitutional chromosome abnormalities. *Am. J. Hum. Genet.* **50**:914–923.

Nakamura, Y., Leppert, M., O'Connell, P., Wolff, R., Holm, T., Culver, M., Martin, C., Fujimoto, E., Hoff, M., Kumlin, E., and White, R. (1987). Variable number of tandem repeat (VNTR) markers for human gene mapping. *Science* **235**:1616–1622.

Neil, D. L., and Jeffreys, A. J. (1993). Digital typing at a second hypervariable locus by minisatellite variant repeat mapping. *Hum. Mol. Genet.* **2**:1129–1135.

NIH/CEPH, Collaborative, Mapping, and Group. (1992). A comprehensive genetic-linkage map of the human genome. *Science* **258**:148–162.

Okazaki, S., Tsuchida, K., Maekawa, H., Ishikawa, H., and Fujiwara, H. (1993). Identification of a pentanucleotide telomeric sequence $(TTAGG)_n$ in the silkworm and in other insects. *Mol. Cell. Biol.* **13**:1424–1432.

Petit, C., de la Chapelle, A., Levilliers, J., Castillo, S., Noel, B., and Weissenbach, J. (1987). An abnormal terminal X–Y interchange accounts for most but not all cases of human XX maleness. *Cell (Cambridge, MA.)* **49**:595–602.

Pilia, G., Little, R. D., Aissani, B., Bernardi, G., and Schlessinger, D. (1993). Isochores and CpG islands in YAC contigs in human Xq26.1-qter. *Genomics* **17**:456–462.

Price, C. M. (1990). Telomere structure in Euplotes crassus: Characterization of DNA-protein interactions and isolation of a telomere-binding protein. *Mol. Cell. Biol.* **10**:3421–3431.

Pritchard, C., Zhu, N., Zuo, J., Bull, L., Pericak-Vance, M. A., Vance, J. M., Roses, A. D., Milatovich, A., Franke, U., Cox, D. R., and Myers, R. M. (1992). Recombination of 4p16 DNA markers in an unusual family with Huntington disease. *Am. J. Hum. Genet.* **50**:1218–1230.

Prowse, K. R., Avilion, A. A., and Greider, C. W. (1993). Identification of a nonprocessive telomerase activity from mouse cells. *Proc. Natl. Acad. Sci. USA* **90**:1493–1497.

Raghuraman, M. K., Dunn, C. J., Hicke, B. J., and Cech, T. R. (1989). Oxytricha telomeric nucleoprotein complexes reconstituted with synthetic DNA. *Nucleic Acids Res.* **17**:4235–4253.

Rappold, G. A. (1993). The pseudoautosomal regions of the human sex chromosomes. *Hum. Genet.* **92**:315–324.

Rasmussen, S. W., and Holm, P. B. (1980). Mechanics of meiosis. *Hereditas* **93**:187–216.

Richards, E. J. and Ausubel, F. M. (1988). Isolation of higher eukaryotic telomere from Arabidopsis thaliana. *Cell (Cambridge, MA.)* **53**:127–136.

Riethman, H. C., Moyzis, R. K., Meyne, J., Burke, D. T., and Olson, M. V. (1989). Cloning human telomeric DNA fragments into Saccharomyces cerevisiae using a yeast-artificial-chromosome vector. *Proc. Natl Acad. Sci. USA* **86**:6240–6244.

Riethman, H. C., Spais, C., Buckingham, J., Grady, D., and Moyzis, R. K. (1993). Physical analysis of the terminal 240kbp of DNA from human chromosme7q. *Genomics* **17**:25–32.

Rouyer, F., de la Chapelle, A., Andersson, M., and Weissenbach, J. (1990). An interspersed repeated sequence specific for human subtelomeric regions. *EMBO J.* **9**:505–514.

Rouyer, F., Simmler, M.-C., Johnsson, C., Vergnaud, G., Cooke, H. J., and Weissenbach, J. (1986). A gradient of sex linkage in the pseudoautosomal region of the human sex chromosomes. *Nature (London)* **319**:291–295.

Royle, N. J., Baird, D. M., and Jeffreys, A. J. (1994). A subterminal satellite, located adjacent to telomeres in chimpanzees is absent from the human genome. *Nature Genet.* **6**:52–56.

Royle, N. J., Clarkson, R. E., Wong, Z., and Jeffreys, A. J. (1988). Clustering of hypervariable minisatellites in proterminal regions of human autosomes. *Genomics* **3**:352–360.

Royle, N. J., Hill, M. C., and Jeffreys, A. J. (1992). Isolation of telomere junction fragments by anchored polymerase chain reaction. *Proc. R. Soc. London. B.* **247**:57–61.

Runge, K. W., and Zakain, V. A. (1989). Introduction of extra telomeric DNA sequences into Saccharomyces cerevisiae results in telomere elongation. *Mol. Cell. Biol.* **9**:1488–1497.

Saccone, S., De Sario, A., Della Valle, G., and Bernardi, G. (1992). The highest gene concentrations in the human genome are in telomeric bands of metaphase chromosomes. *Proc. Natl Acad. Sci. USA* **89**:4913–4917.

Saltman, D., Morgan, R., Cleary, M. l., and de Lange, T. (1993). Telomeric structure in cells with chromosome end associations. *Chromosoma* **102**:121–128.

Sandell, L. L., and Zakain, V. A. (1992). Telomeric position effect in yeast. *Trends Cell Biol.* **2**:10–14.

Solari, A. J. (1980). Synaptonemal complexes and associated structures in microspread human spermatocytes. *Chromosoma* **81**:307–314.

Starling, J. A., Maule, J., Hastie, N. D., and Allshire, R. C. (1990). Extensive telomere repeat arrays in mouse are hypervariable. *Nucleic Acids Res.* **18**:6881–6888.

Sym, M., Engebrecht, J., and Roeder, G. S. (1993). ZIP1 is a synaptonemal complex protein required for meiotic chromosome synapsis. *Cell (Cambridge, MA.)* **72**:365–378.

Tanzi, R. E., Haines, J. L., Watkins, P. C., Stewart, G. D., Wallace, M. R., Hallewells, R., Wong, C., Wexler, N. S., Conneally, P. M., and Gusella, J. F. (1988). Genetic linkage map of human chromosome 21. *Genomics* **3**:129–136.

Thiery, J. P., Macaya, G., and Bernardi, G. (1976). An analysis of eukaryotic genomes by density gradient centrifugation. *J. Mol. Biol.* **108**:219–235.

van Dekken, H., Pinkel, D., Millikin, J., Trask, B., van den Engh, G., and Gray, J. (1989). 3-Dimensional analysis of the organization of human-chromosome domains in human and human hamster hybrid interphase nuclei. *J. Cell Sci.* **94**:299–306.

Van der Ploeg, L. H. T., Liu, A. Y. C., and Borst, P. (1984). Structure of the growing telomeres of trypanosomes. *Cell (Cambridge, MA.)* **36**:459–468.

Vaziri, H., Schachter, F., Uchida, I., Wei, L., Zhu, X., Effros, R., Cohen, D., and Harley, C. B.

(1993). Loss of telomeric DNA during aging of normal and trisomy 21 human lymphocytes. *Am. J. Hum. Genet.* **52**:661–667.

Vergnaud, G., Mariat, D., Apiou, F., Aurias, A., Lathrop, M., and Lauthier, V. (1991). The use of synthetic tandem repeats to isolate new VNTR loci: Cloning of a human hypermutable sequence. *Genomics* **11**:135–144.

Vyas, P., Vickers, M. A., Simmons, D. L., Ayyub, H., Craddock, C. F., and Higgs, D. R. (1992). Cis-acting sequences regulating expression of the human a-globin cluster lie within constitutively open chromatin. *Cell (Cambridge, MA.)* **69**:781–793.

Wahls, W. P., Swenson, G., and Moore, P. D. (1991). Two hypervariable minisatellite DNA binding proteins. *Nucleic Acids Res.* **19**:3269–3274.

Wahls, W. P., Wallace, L. J., and Moore, P. D. (1990). Hypervariable minisatellite DNA is a hotspot for homologous recombination in human cells. *Cell (Cambridge, MA.)* **60**:95–103.

Wallace, B. M. N., and Hulten, M. A. (1985). Meiotic chromosome pairing in the normal human female. *Ann. Hum. Genet.* **49**:215–226.

Wang, S.-S., and Zakain, V. A. (1990a). Sequencing of Saccharomyces telomeres cloned using T4 DNA polymerase reveals two domains. *Mol. Cell. Biol.* **10**:4415–4419.

Wang, S.-S., and Zakain, V. A. (1990b). Telomere-telomere recombination provides an express pathway for telomere acquisition. *Nature (London)* **345**:456–458.

Wapenaar, M. C., Petit, C., Basler, E., Ballabio, A., Henke, A., Rappold, G. A., van Paassen, H. M. B., Blonden, L. A. J., and van Ommem, G. J. B. (1992). Physical mapping of 14 new DNA markers isolated from the human distal Xp region. *Genomics* **13**:167–175.

Weber, B., Collins, C., Robbins, C., Magenis, R. E., Delaney, A. D., Gray, J. W., and Hayden, M. R. (1990). Characterization and organization of DNA sequences adjacent to the human telomere associated repeat (TTAGGG)$_n$. *Nucleic Acids Res.* **18**:3353–3361.

Weissenbach, J., Gyapay, G., Dib, C., Vignal, A., Morisette, J., Millasseau, P., Vaysseix, G., and Lathrop, M. (1992). A second-generation linkage map of the human genome. *Nature (London)* **359**:794–801.

Wells, R. A., Germino, G. G., Krishna, S., Buckle, V. J., and Reeders, S. T. (1990). Telomere-related sequences at interstitial sites in the human genome. *Genomics* **8**:699–704.

Wilkie, A. O., Higgs, D. R., Rack, K. A., Buckle, V. J., Spurr, N. K., Fischel-Ghodsian, N., Ceccherini, I., Brown, W. R. A., and Harris, P. C. (1991). Stable length polymorphism of up to 260kb at the tip of the short arm of human chromosome 16. *Cell (Cambridge, MA.)* **64**:595–606.

Wilkie, A. O. M. (1993). Detection of cryptic chromosomal abnormalities in unexplained mental retardation: A general strategy using hypervariable subtelomeric DNA polymorphisms. *Am. J. Hum. Genet.* **53**:688–701.

Wilkie, A. O. M., and Higgs, D. R. (1992). An unusually large (CA)$_n$ repeat in the region of divergence between subtelomeric alleles of human chromosome 16p. *Genomics* **13**:81–88.

Wilkie, A. O. M., Lamb, J., Harris, P. C., Finney, R. D., and Higgs, D. R. (1990). A truncated chromosome 16 associated with a thalassaemia is stabilized by addition of telomeric repeat (TTAGGG)$_n$. *Nature (London)* **346**:868–871.

Wolff, R. K., Nakamura, Y., and White, R. (1988). Molecular characterization of a spontaneously generated new allele at a VNTR locus: No exchange of flanking DNA sequence. *Genomics* **3**: 347–351.

Wong, Z., Wilson, V., Patel, I., Povey, S., and Jeffreys, A. J. (1987). Characterization of a panel of highly variable minisatellites cloned from human DNA. *Ann. Hum. Genet.* **51**:269–288.

Wright, J. F., Gottschling, D. E., and Zakain, V. A. (1992). Saccharomyces telomeres assume a nonnucleosomal chromatin structure. *Genes Dev.* **6**:197–210.

Yamazaki, H., Nomoto, S., Mishima, Y., and Kominami, R. (1992). A 35-kDa protein binding to a cytosine-rich strand of hypervariable minisatellite DNA. *J. Biol. Chem.* **267**:12311–12316.

Youngman, S., Bates, G. P., Williams, S., McClatchey, A. I., Baxendale, S., Sedlacek, Z.,

Altherr, M., Wasmuth, J. J., MacDonald, M. E., Gusella, J. F., Sheer, D., and Lehrach, H. (1992). The telomeric 60kb of chromosome arm 4p is homologous to telomeric regions on 13p, 15p, 21p, and 22p. *Genomics* **14:**350–356.

Yu, G.-L., Bradley, J. D., Attardi, L. D., and Blackburn, E. H. (1990). *In vivo* alteration of telomere sequences and senescence caused by mutated *Tetrahymena* telomerase RNAs. *Nature (London)* **344:**126–132.

Yunis, J. J. and Prakash, O. (1982). The origin of man: a chromosomal pictorial legacy. *Science* **215:** 1525–1530.

Zakian, V. A. (1989). Structure and function of telomeres. *Annu. Rev. Genet.* **23:**579–604.

# Index

Acid β-glucosidase, Gaucher disease treatment, 29–30, 32, 35
Adenovirus, see Viral vectors
Adverse therapeutic experiences
  clinical identification, 5–8
  ongoing therapy, 8–12
Alglucerase, Gaucher disease treatment, 30–33, 39–41
Aminohydroxypropylidine biphosphonate, bone disease treatment, 28
Apolipoprotein B, 156–164
  gene polymorphisms, 161–164
  post-transcriptional mRNA editing, 159–161
  triglyceride-rich lipoprotein clearance, 156–159
Apolipoprotein E, lipoprotein remnant metabolism, 164–175
Atherosclerosis, 141–191
  apolipoprotein E, 164–175
  atherogenesis, 185–186
  high-density lipoprotein metabolism, 175–183
    cholesterol ester transfer protein, 180–183
    transgenic mouse genetics, 178–180
  historical perspective, 142–147
  lipoprotein (a) metabolism, 183–185
  lipoprotein remnant metabolism, 164–175
  low-density lipoprotein receptor pathway, 147–164
    apolipoprotein B, 156–164
    defect therapy, 153–156
    transgenic mouse genetics, 152–153
  macrophage scavenger receptors, 186–190

Beta cells, function in non-insulin-dependent diabetes mellitus gene, 52–53, 55

Bone disease, treatment with aminohydroxypropylidine biphosphonate, 28
Bone marrow transplantation, Gaucher disease treatment, 34–35

Cancer, telomere role, 298–300
Carrier tests, hemophilia, 110–113, 129–130
Chiasmata, function, 276–277
Cholesterol ester transfer protein, high-density lipoprotein metabolism, 180–183
Chromosomes
  proterminal regions, 273–306
    chromosome ends from primates, 295–296
    function, 275–278
    in genetic disease, 296–305
      cancer, 298–300
      chromosome healing in vivo, 301–303
      contiguous gene syndromes, 303–305
      cryptic translocations, 303–305
      telomere repeat loss, 296–301
    structure, 281–294
      associated sequences, 282–283
      chromosome end sequence, 289–294
      minisatellites, 284–285
      properties, 281–282
      subterminal repeat sequences, 287–289
      subterminal sequence isolation, 285–287
telomeres, 273–306
  binding proteins, 280–281
  in cancer, 298–300
  chromosome ends from other primates, 295–296
  chromosome healing in vivo, 301–303
    telomerase activity, 301–302
    telomerase inactivity in somatic tissues, 302–303
    telomere capture, 303

contiguous gene syndromes, 303–305
cryptic translocation, 303–305
function, 273–275
isolation, 279–281
reintroduction into mammalian cells,
 300–301
repeat loss, 296–301
structure, 279–281
 4p telomere, 290–291
 16p telomere, 291–293
 7q telomere, 291
 21q telomere, 293–294
 Xp/Yp pseudoautosomal region, 289–
 290
Clinical care, long-term, 1–15
 adverse experiences
 clinical identification, 5–8
 ongoing, 8–12
 benefit, 12–13
 gene therapy risks, 3–5, 9, 13–14
 informed consent, 9, 13–14
Cloning
 diabetes-susceptibility gene identification,
 57, 60, 61–62
 phenylalanine hydroxylation cDNA, 204–
 205
Contiguous gene syndromes, 303–305

Deafness, maternally inherited, 78–80
Diabetes Mellitus, *see* Non-insulin-dependent
 diabetes mellitus
Dietary therapy, for phenylketonuria, 202
Differential cloning, diabetes-susceptibility
 gene identification, 61–62
DNA, *see also* Recombinant DNA Advisory
 Committee
 complimentary
 Gaucher disease, 21
 phenylalanine hydroxylation cloning,
 204–205
 polymorphisms, *see specific polymorphism*
 protein complexes in phenylketonuria, 251–
 253
Down syndrome, telomere role, 297–298

Enzyme replacement, Gaucher disease treatment, 29–33

Factor VIII, in hemophilia, 100, 113–116
Factor IX, in hemophilia, 100, 102–104
Fatty acid-binding protein-2 gene, non-insulin-
 dependent diabetes mellitus gene candi-
 date, 87–88
Founder effect, in phenylketonuria, 233–241

Gaucher disease, 17–41
 biology, 20–24
 complementary DNA, 21
 glucocerebrosidase gene characteristics,
 20–21
 mutations, 22–24
 pseudogene characteristics, 20–21
 clinical manifestations, 18–19
 controversies, 39–41
 genotype variability, 36–37
 heterozygotes, 38–39
 in Jewish populations, 19, 25–27
 population genetics, 24–27
 treatment, 27–36
 dosage regulation, 39–41
 enzyme replacement, 29–33
 gene transfer, 35–36
 marrow transplantation, 34–35
 skeletal symptoms, 28
 splenectomy, 28–29
 symptomatic management, 28–29
Gaze palsy, in Gaucher disease, 19
Gene registry
 hemophiliacs, 99–100
 therapy patients, 3, 8, 15
Gene therapy
 adverse experiences
 clinical identification, 5–8
 ongoing, 8–12
 benefits, 12–13
 Gaucher disease, 35–36
 informed consent, 9, 13–14
 low-density lipoprotein receptor pathway defects, 153–156
 ongoing
 adverse experiences, 8–12
 clinical benefit, 12–13
 patient registry, 3, 8
 phenylketonuria, 224, 244–253
 adenoviral vectors, 248–251
 DNA/protein complexes, 251–253
 retroviral vectors, 246–248

regulation, 1–3
risks, 3–5, 9, 13–14
Genome mapping
apolipoprotein B gene, 158–159
diabetes-susceptibility gene identification, 57–61
Gaucher disease, 20–24
hemophilia
factor VIII, 113–116, 124–126
factor IX, 102–103
phenylalanine hydroxylase gene, 204–207
Genotype analysis, for phenylketonuria, 225–228
Glucagon-like peptide-1 receptor gene, non-insulin-dependent diabetes mellitus gene candidate, 81
Glucocerebrosidase, 20–21; see also *Gaucher disease*
Glucokinase gene, non-insulin-dependent diabetes mellitus gene candidate, 86–87
Glucose transporter gene, non-insulin-dependent diabetes mellitus gene candidate, 76–78
Glycogen synthase gene, non-insulin-dependent diabetes mellitus gene candidate, 81–83
Guthrie test, for phenylketonuria, 203

Hemophilia, 99–131
A mutations, 100, 116–130
carrier tests, 129–130
causes
conventional, 118–122
nonconventional, 122–126
detection methods, 116–118
genotype/phenotype correlations, 126–129
prenatal diagnosis, 129–130
B mutations, 100, 104–113
carrier tests, 110–113
detection methods, 104–105
genotype/phenotype correlations, 108–110
prenatal diagnosis, 110–113
types, 105–108
factor VIII, 100, 113–116
factor IX, 100, 102–104
forms, 99–101

genetics, 101–102
treatment, 100–101, 110, 130
Heterozygotes
phenylketonuria selection, 243–244
selective advantage for Gaucher disease, 38–39
High-density lipoproteins, see Lipoprotein metabolism
Horizontal supranuclear gaze palsy, in Gaucher disease, 19
Hutchinson–Gilford syndrome, telomere role, 297–298
Hyperphenylalaninemia
clinical aspects, 203
frequency in human populations, 228–229
genotype/phenotype correlations, 225–228
haplotype/phenotype correlations, 214
mutations
distribution, 217–220
expression analysis, 221–225

Insulin, in non-insulin-dependent diabetes mellitus
gene, 63–64
resistance
leprechaunism, 68–69
Rabson–Mendenhall syndrome, 69–70
type A, 70–74
Insulin receptor gene, 64–75
function, 65–68
leprechaunism, 68–69
mutations, 74–75
Rabson–Mendenhall syndrome, 69–70
structure, 65–68
type A insulin resistance, 70–74
Insulin receptor substrate-1 gene, 75–76
Islet amyloid polypeptide gene, non-insulin-dependent diabetes mellitus gene candidate, 80–81

Leprechaunism, insulin resistance, 68–69
Linkage analysis
chromosome proterminal regions, 275–276
diabetes-susceptibility gene identification, 59–61
Lipoprotein metabolism, 141–191
apolipoprotein B, 156–164

apolipoprotein E, 164–175
atherogenesis, 185–186
high-density lipoprotein metabolism, 175–183
cholesterol ester transfer protein, 180–183
transgenic mouse genetics, 178–180
historical perspective, 142–147
lipoprotein (a) metabolism, 183–185
low-density lipoprotein receptor pathway, 147–164
apolipoprotein B, 156–164
defect therapy, 153–156
transgenic mouse genetics, 152–153, 185
macrophage scavenger receptors, 186–190
remnant metabolism, 164–175
Long-term clinical care, 1–15
adverse experiences
clinical identification, 5–8
ongoing, 8–12
benefit, 12–13
gene therapy risks, 3–5, 9, 13–14
informed consent, 9, 13–14
Low-density lipoproteins, see Lipoprotein metabolism

Macrophage scavenger receptors, low-density lipoprotein uptake, 186–190
Mannose-dependent receptors, Gaucher disease treatment, 30–32
Mapping, see Genome mapping
Maternally inherited diabetes and deafness, 78–80
Maturity-onset diabetes of the young, 83–86
Meiotic prophase I, telomere function, 276–278
Minirepeat polymorphisms, diabetes-susceptibility gene identification, 58
Minisatellite repeat sequences, 282–283
Mouse, see Transgenic mouse models
mRNA, post-transcriptional editing, apolipoprotein B, 159–161
Mutations, see specific disease

Neonatal screening, for phenylketonuria, 203
Niemann–Pick disease, in Jewish populations, 27

Non-insulin-dependent diabetes mellitus, 51–89
candidate genes, 63–88
fatty acid-binding protein-2, 87–88
glucagon-like peptide-1 receptor, 81
glucokinase regulatory protein, 87
glucose transporter genes, 76–78
glycogen synthase, 81–83
insulin gene, 63–64
insulin receptor, 64–75
function, 65–68
leprechaunism, 68–69
mutations, 74–75
Rabson–Mendenhall syndrome, 69–70
structure, 65–68
type A insulin resistance, 70–74
insulin receptor substrate-1, 75–76
islet amyloid polypeptide, 80–81
maternally inherited diabetes and deafness, 78–80
maturity-onset diabetes of the young, 83–86
susceptibility, 55–57
etiology, 54–55
gene susceptibility identification strategies, 55–62
animal models, 62
candidate gene approach, 55–57
differential cloning, 61–62
genome mapping, 57–61
positional cloning, 57, 60
subtraction cloning, 61–62
history of development, 52–54
polymorphism, 88
thrifty gene hypotheses, 88
Norrbottnian disease, 19, 29

Pedigree analysis
diabetes-susceptibility gene identification, 59
hemophilia, 111–112
Phenylketonuria, 199–254
clinical aspects, 203–204
dietary therapy, 202, 244–245
discovery, 199–200
gene therapy, 244–253
adenoviral vectors, 248–251
DNA/protein complexes, 251–253
retroviral vectors, 246–248

molecular genetics, 208–228
  expression analysis, 221–225
  extended haplotypes, 209–210
  genotype/phenotype correlations, 225–228
  haplotype analysis, 210–211
  haplotype/phenotype correlations, 214
  phenylalanine hydroxylation gene mutations, 214–220
    deletion, 214
    distribution, 217–220
    genotype heterogeneity, 214–215
    molecular lesions, 215–217
    mutant chromosomes, 211–214
    normal chromosomes, 211–214
    phenylalanine hydroxylation locus, 208–209
    restriction fragment length polymorphism haplotypes, 208–209
  newborn screening, 203
  phenylalanine hydroxylation, 200–202, 204–208
    cDNA cloning, 204–205
    chromosomal location, 206
    encoding gene, 205–206
    gene structure, 206–207
    pterin-dependent hydroxylase homologs, 207–208
  population genetics, 228–244
    Asian populations, 240–243
    European populations, 233–240
    frequency, 228–229
    heterozygote selection, 243–244
    Middle Eastern populations, 233–240
    mutation/haplotype association, 229–233
Polymerase chain reaction
  hemophilia detection, 104–105, 117–118, 122–123
  phenylalanine hydroxylase gene haplotype detection, 209–210
  telomere subterminal sequence isolation, 286
Polymorphism, DNA, *see specific polymorphism*
Polymorphism hypotheses, 88
Population genetics
  Gaucher disease, 24–27
  phenylketonuria, 228–244
    Asian populations, 240–243
    European populations, 233–240

frequency, 228–229
  heterozygote selection, 243–244
  Middle Eastern populations, 233–240
  mutation/haplotype association, 229–233
Positional cloning, diabetes-susceptibility gene identification, 57, 60
Prenatal diagnosis, hemophilia, 110–113, 129–130
Prophase I, telomere function, 276–278
Proterminal chromosome regions, 273–306
  chromosome ends from other primates, 295–296
  function, 275–278
  genetic disease role, 296–305
    cancer, 298–300
    chromosome healing *in vivo*, 301–303
    contiguous gene syndromes, 303–305
    cryptic translocations, 303–305
    telomere repeat loss, 296–301
  structure, 281–294
    associated sequences, 282–283
    chromosome end sequence organization, 289–294
    minisatellites, 284–285
    properties, 281–282
    subterminal repeat sequences, 287–289
    subterminal sequence isolation, 285–287
Pyruvate kinase gene, in Gaucher disease, 24


Rabson–Mendenhall syndrome, insulin resistance, 69–70
Recombinant DNA Advisory Committee
  ongoing patient therapy, 10–11
  patient registry, 3, 8, 15
Recombination nodules, function, 276–277
Registry, *see* Gene registry
Restriction fragment length polymorphisms
  apolipoprotein B gene, 161–162
  diabetes-susceptibility gene identification, 58, 74
  low-density lipoprotein receptor gene identification, 149
  phenylalanine hydroxylase gene haplotypes, 208–214
Retrovirus, *see* Viral vectors
RNA, post-transcriptional editing, apolipoprotein B, 159–161

Selective advantage
  Gaucher disease, 38–39
  non-insulin-dependent diabetes mellitus, 88
Single-stranded conformational polymorphism
    analysis, diabetes-susceptibility gene iden-
    tification, 55–56
Splenectomy, Gaucher disease treatment, 28–
    29
Sterol regulatory element-1, low-density
    lipoprotein receptor protein transcription,
    150–152
Subterminal sequences
  isolation, 285–287
  repeat sequences, 287–289
Subtraction cloning, diabetes-susceptibility
    gene identification, 61–62
Suicide vectors, risk of use, 4


Tay–Sachs disease, in Jewish populations,
    27
Telomeres, 273–306
  binding proteins, 280–281
  in cancer, 298–300
  chromosome ends from other primates,
    295–296
  chromosome healing *in vivo*, 301–303
    telomerase activity, 301–302
    telomerase inactivity in somatic tissues,
      302–303
    telomere capture, 303
  contiguous gene syndromes, 303–305
  cryptic translocation, 303–305
  function, 273–275

isolation, 279–281
  reintroduction into mammalian cells, 300–
    301
  repeat loss, 296–301
  structure, 279–281
    4p telomere, 290–291
    16p telomere, 291–293
    7q telomere, 291
    21q telomere, 293–294
    Xp/Yp pseudoautosomal region, 289–290
Thrifty gene hypotheses, 88
Transgenic mouse models
  apolipoprotein E analysis, 166–168
  high-density lipoprotein metabolism analy-
    sis, 178–180
  low-density lipoprotein receptor protein
    analysis, 152–153, 185
Tyrosine, conversion from phenylalanine, 200


Variable number tandem repeat polymorphisms
  diabetes-susceptibility gene identification,
    58
  phenylalanine hydroxylase gene haplotype
    identification, 209–211
Very low-density lipoproteins, *see* Lipoprotein
    metabolism
Viral vectors
  gene transfer
    in Gaucher disease, 35–36
    in phenylketonuria
      adenoviral, 248–251
      retroviral, 246–248
  risk of use, 4–5

## DATE DUE

| | AUG 2 5 1995 | | |
|---|---|---|---|
| | DEC 1 4 1998 | | |
| | JAN 0 5 2003 | | |
| | | | |
| | | | |
| | | | |
| | | | |
| | | | |
| | | | |
| | | | |
| | | | |
| | | | |
| | | | |
| | | | |
| | | | |
| | | | |
| | | | Printed in USA |